Thermodynamics for Drug Product Design

Thermodynamics for Drug Product Design

A Guide for Pharmaceutical Scientists

William Craig Stagner
Campbell University
Buies Creek, NC, USA

WILEY

Published by John Wiley & Sons, Inc., Hoboken, New Jersey.
Published simultaneously in Canada.

Library of Congress Cataloging-in-Publication Data applied for:

Hardback: ISBN: 9781119851097

Cover Design: © Amelia Kay Nilles and Ethan Craig Stagner
Cover Image(s): Hot-stage polarized light microscopy photomicrograph by Scott Staton

Set in 9.5/12.5pt STIXTwoText by Straive, Pondicherry, India

In memory of my major professor, J. Keith Guillory, a mentor, confidant, and friend.

In memory of my beloved father and mother, Vernon Stagner and Eunice Stagner Meer; to my loving wife, Nancy, and her fifty-four years of support; our children and grandchildren – Ryan, Beth, Ethan; Tripp, Kelly (Stagner), Amelia Nilles; Justin, Lauren, Henry, Hudson, Hayes; and my brother, Robert.

Contents

SI Defining Constants

SI Defining Constants[a]

Symbol	Defining constant	Exact value
$\Delta\nu_{Cs}$	Hyperfine transition frequency of Cs	9.192631770E9 Hz
c	Speed of light	2.99792458E8 m s^{-1}
h	Planck constant	6.62607015E-34 J s
e	Elementary charge	1.602176634E-19 C
ka_B	Boltzmann constant	1.380649E-23 J K^{-1}
		[b]3.297623E-24 cal K^{-1}
		[b]1.380649E-16 erg K^{-1}
		[b]8.314462E-3 kJ mol^{-1} K^{-1}
N_A	Avogadro constant	$6.02214076 \times 10^{23}$ units mol^{-1}
K_{cd}	Luminous efficacy of 540 THz radiation	683 lm W^{-}
R	Gas constant	8.31446261815324 J K^{-1} mol^{-1}
		1.98720425864083 cal K^{-1} mol^{-1}

[a] Wikipedia. https://en.wikipedia.org/wiki/International_System_of_Units.
[b] Value in different units. https://en.wikipedia.org/wiki/Boltzmann_constant.

Common Derived Quantities

Physical quantity	SI unit	Unit symbol	Base unit terms
Area	—	—	m^2
Volume	—	—	m^3
Concentration	—	—	$mol{\cdot}m^{-3}$
Density	—	—	$kg{\cdot}m^{-3}$
Velocity	—	—	$m{\cdot}s^{-1}$
Energy, heat, work	Joule	J	$N{\cdot}m \equiv m^2{\cdot}kg{\cdot}s^{-2}$
Force	Newton	N	$m{\cdot}kg{\cdot}s^{-2}$
Pressure	Pascal	Pa	$N{\cdot}m^{-2} \equiv m^{-1}{\cdot}kg{\cdot}s^{-2}$

(Continued)

Physical quantity	SI unit	Unit symbol	Base unit terms
Heat capacity	—	C	$J{\cdot}K^{-1} \equiv m^2{\cdot}kg{\cdot}s^{-2}{\cdot}K^{-1}$
Power	Watts	W	$J{\cdot}s^{-1} \equiv m^2{\cdot}kg{\cdot}s^{-3}$
Viscosity	—	Pa·s	$Pa{\cdot}s \equiv m^{-1}{\cdot}kg{\cdot}s$
		Poise (P)	$cm^{-1}{\cdot}g{\cdot}s^{-1}$
		P	$dyne{\cdot}cm^{-2}{\cdot}s$
		P	0.1 Pa·s

Other Frequently Used Constants

Name	Symbol	Numerical expression
Pi	π	[a]3.14159 26535 89793 23846
Euler's number	e	[a]2.71828 18284 59045 23536
Acceleration of gravity	g	[b]9.80665 $m{\cdot}s^{-2}$

[a] https://en.wikipedia.org/wiki/List_of_mathematical_constants.
[b] https://en.wikipedia.org/wiki/Standard_gravity.

Common Unitary Conversion Ratios[a]

1 g/10^{-3} kg
1 L/10^{-3} m^3
1 mL/10^{-3} L
1 µ/10^{-6} m
1 Å/10^{-10} m
1 Å/10^{-8} cm
1 atm/1.01325E5 Pa
1 atm/1.01325 bar
1 atm/760 mmHg
1 bar/750.062 mmHg
1 bar/1E5 Pa
4.184 J/cal
1 $J{\cdot}mol^{-1}$/2.390E-4 kcal·mol
1 $kcal{\cdot}mol^{-1}$/4.184 $J{\cdot}mol^{-1}$
1 erg/10^{-7} J
1 dyn/10^{-5} N
1 centipoise/mPa·s
Ln X/2.303 log X

[a] Unitary conversion ratios equal 1.

SI Unit Prefixes

Prefix	Submultiple	Symbol	Prefix	Multiple	Symbol
Deci	10^{-1}	d	Deca	10^{1}	da
Centi	10^{-2}	c	Hecto	10^{2}	h
Milli	10^{-3}	m	Kilo	10^{3}	k
Micro	10^{-6}	μ	Mega	10^{6}	M
Nano	10^{-9}	n	Giga	10^{9}	G
Pico	10^{-12}	p	Tera	10^{12}	T
Femto	10^{-15}	f	Peta	10^{15}	P
Atto	10^{-18}	a	Exa	10^{18}	E
Zepto	10^{-21}	z	Zetta	10^{21}	Z
Yocto	10^{-24}	y	Yotta	10^{24}	Y
Ronto	10^{-27}	r	Ronna	10^{27}	R
Quecto	10^{-30}	q	Quetta	10^{30}	Q

SI Fundamental/Base Units

Physical quantity	Dimensional symbol	Unit name	Unit symbol
Length	l	Meter	m
Mass	m	Kilogram	kg
Time	t	Second	s
Temperature	T	Kelvin	K
Amount of substance	n	Mole	mol
Luminous intensity	I_V	Candela	cd
Electric current	I	Ampere	A

Selected Derived Physical Quantities

Physical quantity	Formula or name	SI[a] derived unit symbol (s)	Dimensional formula (s)
Area	Length × width	m^2	$[L^2]$
Volume	Length × width × height	m^3	$[L^3]$
Velocity	Distance per second	$m\ s^{-1}$	$[Lt^{-1}]$
Concentration	Amount per volume	$mol\ m^{-3}$	$[nL^{-3}]$
Density	Mass per volume	$kg\ m^{-3}$	$[ML^{-3}]$
Force[b]	Newton (N)	$m\ kg\ s^{-2}$	$[LMt^{-2}]$
Energy, heat, work[c]	Joule (J)	$N\ m = m^2\ kg\ s^{-2}$	$[L^2Mt^{-2}]$
Heat capacity	Joule per kelvin	$N\ m\ K^{-1} = m^2\ kg\ s^{-2}\ K^{-1}$	$[L^2Mt^{-2}K^{-1}]$
Pressure	Pascal (Pa)	$N\ m^{-2} = m^{-1}\ kg\ s^{-2}$	$[L^{-1}Mt^{-2}]$
Surface tension	Newton per length	$N\ m^{-1}$	$[Mt^{-2}]$
Surface energy	Joule per length × length	$J\ m^{-2}$	$[Mt^{-2}]$
Gas constant[d]	Gas constant (R)	$J\ K^{-1}\ mol^{-1} = m^2\ kg\ s^{-2}\ K^{-1}\ mol^{-1}$	$[L^2Mt^{-2}T^{-1}n^{-1}]$
Gas constant	Gas constant (R)	$m^3\ Pa\ K^{-1}\ mol^{-1} = m^2\ kg\ s^{-2}\ K^{-1}\ mol^{-1}$	$[L^2Mt^{-2}T^{-1}n^{-1}]$
Gas constant	Gas constant (R)	$m^2\ kg\ s^{-2}\ K^{-1}\ mol^{-1}$	$[L^2Mt^{-2}T^{-1}n^{-1}]$

[a] SI – Système International.
[b] Force = mass × acceleration.
[c] Work = force × distance.
[d] The numerical value for each gas constant is 8.314 462 618 153 24.

Foreword

The discipline of thermodynamics can strike fear into the most motivated of students and cause the most seasoned professionals to frown. With its countless laws and equations, many named after pioneering scientists from the distant past, it is not surprising that it has gained a formidable reputation in classrooms and laboratories throughout the world. Yet, this seemingly dense and intimidating field of study forms a predictable foundation upon which almost all modern medicines are developed. And, once the layers of complexity have been peeled back, it is an intuitive and elegant discipline that can help practicing scientists and engineers, and students, to understand everything from the stability of drugs to their absorption via the lungs.

In this new textbook, designed especially for pharmaceutical scientists, the author (Professor Emeritus Bill Stagner) has taken a modern, pragmatic, and refreshing approach to teach the discipline of thermodynamics. Based on his long and distinguished career as a teacher and researcher, he has provided a comprehensive yet easy-to-understand map of a seemingly complicated landscape. He has illustrated this map with many practical examples from the clinic, the factory, and the laboratory, and he has provided historical context and commentary along the way to make sure that both the origins and underlying scientific principles are fully understood. To cap it off, a great number of well-thought-out problems sets (along with accompanying solutions) are included to help embed the most important ideas and principles.

In Chapter 1, important foundational scientific concepts (such as uncertainty, propagation of error, and rounding) are explained in easy-to-understand language. This is followed, in Chapter 2, by a description of the basic thermodynamic principles (such as energy conservation, isobaric and isothermal conditions) and their application to pharmaceutical formulation and manufacturing. In Chapter 3, the classic laws of thermodynamics are introduced, along with examples of how they are used in the development and analysis of medicines (such as for assessing purity, phase changes, and product quality). After that, a crucial concept – Gibb's free energy – is introduced, along with examples of its use to predict spontaneous reactions in medicines (such as protein denaturation, liquid-phase separation, and amorphous drug crystallization). At this point in the book (Chapter 5), the various equilibrium states and different phase diagrams are described. Applications include understanding the solubility and solution behavior of common drugs and the use of more complex drug delivery systems to help deliver low-solubility compounds orally in co-solvents and lipids. Chapter 6 delves further into the important topic of drug solubility and key associated phenomena (such as the common ion effect, variation in solubility with pH, and salt disproportionation), all of which are critical to the design of medicines for oral administration. The last two chapters focus on surface chemistry and those surface phenomena that are critical to drug delivery and manufacturing. Many drugs are surface active (amphiphilic), so this is a critical material to include in such a treatise. Surface excess, interfacial tension, wetting, and their importance to the formulation of multiphase systems (such as suspensions and dry-powder inhalers) are described first. Then, in Chapter 8, adsorption phenomena are defined and explained. These are often important for describing the interactions between medicines, atmospheric gases, and packaging materials. To conclude, the practical topics of water vapor sorption by medicines, hygroscopicity, deliquescence, and adsorption of active ingredients on filters during liquid filtration processes are each described in turn.

By authoring this brand-new textbook on thermodynamics, Professor William Stagner has created a highly valuable resource for students and professionals from all segments of the pharmaceutical sciences community. The text is comprehensive, rigorous, and logically organized from start to finish. Bill has been able to eloquently describe developments in this field from its beginnings to the current day and has provided a multitude of pharmaceutical examples and applications. He has also made sure that the reader has access to the most recent research in this field and understands its significance for the development, manufacture, storage, and use of medicines in the 21st century. I am certain that this book will find its way onto the bookshelves of aspiring pharmacists and experienced pharmaceutical scientists in both academia and industry.

Bruno Hancock, PhD, FAAPS
Head of Materials Science at Pfizer Inc. (retired 2024)
Editor, Journal of Pharmaceutical Sciences
Co-Editor, Handbook of Pharmaceutical Excipients
Westerly, RI

Preface

Backstory

As a Pharmaceutics professor teaching Physical Pharmacy and Thermodynamics, I found myself constantly drawing on actual applications and personal anecdotes that accumulated over my 25 plus years of experience as a drug product design scientist and pharmaceutical scientist consultant. COVID-19 created the time and opportunity to share my lectures with a broader audience by writing this textbook.

Purpose of Thermodynamics for Drug Product Design: A Guide for Pharmaceutical Scientists

The purpose of this book is to provide the student and reader with a learning aid to establish a meaningful appreciation and understanding of thermodynamic principles and concepts that underpin the creation of robust drug products. As a means to achieve this purpose, abundant real-life drug product design examples, applications, personal anecdotes, and solved problem sets are provided to highlight the important role that thermodynamic principles and concepts play in drug product creation, and in predicting, preventing, analyzing, and evaluating the root cause of product failures.

Intended Audience

Thermodynamics for Drug Product Design: A Guide for Pharmaceutical Scientists is an easy-to-read and comprehend textbook for higher-level undergraduate and entry-level graduate students. It is also a valuable reference resource for pharmaceutical scientists, chemists, biologists, engineers, regulatory agency personnel, and others engaged in drug dosage form research and development.

How to Use the Book

For instructors and students to optimize comprehension, this textbook is ideally studied sequentially, beginning with Chapter 1 and concluding with Chapter 8. Each chapter addresses new concepts and principles that build upon the content of preceding chapters. This pedagogical approach continually re-exposes students to fundamental thermodynamic principles and concepts, aiming to improve knowledge retention and comprehension. Another pedagogical strategy to improve knowledge retention and comprehension encourages self-review of important thermodynamic concepts presented in an earlier chapter when encountered in a later chapter. To accomplish this, appropriate sections are referenced that direct the reader to earlier and often foundational discussions of the concept.

For professionals involved in drug product design, this approach also makes this book a useful thermodynamics reference. In this case, each chapter can serve as a standalone resource for a specific thermodynamic precept. Here again, the location of the pertinent book section(s) is referenced in case the reader wishes to review the underlying fundamental ideas of a particular thermodynamic topic.

The examples, applications, derivations, and solved problems for each chapter are provided with step-by-step solutions to accommodate any unevenness of an individual's knowledge and skills. Students and professionals may find this detailed approach helpful. If not, the step-by-step derivations and solutions can be skipped.

Book Structure

The book comprises eight chapters:

Chapters included by order are as follows:

- Data Reporting and Analysis
- Underlying Thermodynamic Physical Property Relationships and Definitions
- The Laws of Thermodynamics
- Gibbs Free Energy
- Equilibria
- Drug Solubility Equilibria
- Surface Thermodynamics
- Adsorption Phenomena

As mentioned earlier, each chapter includes numerous new real-life pharmaceutical examples, applications, personal anecdotes, and solved problem sets that contextualize thermodynamic principles and concepts. These different modes of instruction are used to illustrate and illuminate how vital a fundamental understanding of thermodynamics is to drug product creation. As an aside, I found it both interesting and surprising that NaCl, simple table salt, was the cause of catastrophic product failure in 3 of the 7 author anecdotes. This illustrates the importance of understanding the thermodynamic properties of even the "simplest" of compounds.

The chapter flow supports the pedagogical strategy of continually re-exposing students and readers to fundamental thermodynamic principles and concepts, aiming to improve knowledge retention and comprehension. Chapter 1, Data Reporting and Analysis, introduces the idea of dimensional analysis and integrity, along with other important data reporting and analysis topics. Dimensional analysis and these other topic areas are emphasized and used throughout the book. Chapter 2, Underlying Thermodynamic Physical Property Relationships and Definitions, employs a relatively nontraditional approach to introducing thermodynamics to the uninitiated reader. This novel pedagogical approach introduces the naive reader to terms that are commonly associated with the laws of thermodynamics before presenting the laws themselves. Underlying thermodynamic terms such as temperature, energy, heat, work, system, surroundings, boundary, universe, intensive–extensive variables, thermodynamic state variables, reversible/irreversible processes, state function, spontaneous and nonspontaneous processes, and so on, are defined and discussed. Chapter 3 introduces the zeroth, first, second, and third laws of thermodynamics using a more traditional approach, which is enriched by the knowledge, comprehension, and ability to apply the basic thermodynamic definitions and concepts to the laws obtained in Chapter 2. Chapter 4, Gibbs Free Energy, describes the unifying thermodynamic system-oriented theory that describes the interplay between internal energy, enthalpy, and entropy under constant temperature and pressure. Gibbs free energy is a critical thermodynamic principle that underpins the principles and concepts presented in the remaining chapters.

I greatly appreciate feedback and comments that lead to improvements in the book. I am particularly keen interested in adding more real-life drug product design examples and anecdotes that can be used to contextualize thermodynamic principles. I am happy to acknowledge your contribution.

William Craig Stagner
stagnerw@gmail.com

Acknowledgments

The author thanks V. J. Kumar, Lipika Chablani, Bhanu Bejgum, Lipika Chablani, Barath Gowdampally, Dick Winnike, Dale E. Wurster, and Antoine Al-Achi for their critical review of Chapters 1, 2, 3, 4, 5, 6, 7, and 8, respectively.

The author thanks his grandchildren, Amelia Kay Nilles and Ethan Craig Stagner, for the cover layout and design.

Special thanks and gratitude to Scott Staton as illustrator.

1

Data Reporting and Analysis

1.1 Introduction

Science uses experimentation, data collection, and data analysis to become informed about the natural world. The data associated with *physical quantities* are expressed in terms of *units of measure*. The physical quantities and units of measure have related *dimensions*. The methods used for measuring these physical quantities can provide a certain level of accuracy, precision, robustness, linearity, and quantitation limit. Data should be reported with appropriate significant figures (SFs), which reflect the precision and robustness of the analytical method. Often data values need to be *rounded off* to reflect the correct number of SFs.

1.1.1 Fundamental/Base and Derived Physical Quantities

Physical quantities are known as *fundamental* or *derived* (Ade 2021). Fundamental quantities provide a standard of measure of independent physical parameters. Derived physical quantities are a function of fundamental quantities. There are seven fundamental physical quantities: amount of substance, like number of moles or particles, time, length, mass, temperature, electric current, and luminous intensity. The first five quantities are the most frequent fundamental physical quantities discussed in this book. Energy, heat, work, force, chemical potential, pressure, velocity, acceleration, momentum, frequency, area, volume, and density are derived quantities found throughout the text.

Fundamental units are standard units of measure associated with fundamental physical quantities. In pharmaceutical sciences, several unit systems are still used today. The globally recognized scientific standard is the Système International d'Unités (SI) system (SI Reference Point 2025). A second unit system is he IUPAC centimeter-gram-second (cgs) system (IUPAC 1993). The English foot-ounce-second system (Purdue University) is also still used in practice and found in textbooks. A brief discussion of the SI standard units used throughout this book follows.

Since 2019, SI has linked SI units to physical constants. This guarantees their stability and universality.

Mole (mol) is the SI unit defining the amount of a substance. One mole consists of exactly 6.022 140 76E23 *elementary entities*. (The *E* notation has been used by the *SHARE Operating System* since 1958. The *E* means "times ten raised to the power of *n* or $\times 10^n$.") In 2019, Avogadro's number was changed to a constant having units of elementary entities mol^{-1} or mol^{-1}. Avogadro's constant, N_A, is 6.022 140 76E23 mol^{-1}. The elementary entities may be any particle like electrons, atoms, molecules, ions, and others.

The SI unit of time is the *second* (s). One second is the time it takes for a cesium atom to vibrate 9.192 631 770E9 times.

Thermodynamics for Drug Product Design: A Guide for Pharmaceutical Scientists, First Edition. William Craig Stagner.

The meter (m) is the SI unit of length and is defined as the distance traveled by light in a vacuum in 1/2.997 924 58E8 of a second. (The speed of light is 2.997 924 58E8 ms^{-1}.)

The kilogram (kg) is now expressed in terms of the Planck constant (h), 6.626 070 15-34 kg m^2 s^{-1}, with meter and second as defined earlier. The older definition of kg relied on the mass of the kilogram.

The SI unit for temperature is the kelvin (K). In 2019, the definition was changed from fixing the temperature scale using the triple point of water to using the Boltzmann's constant (k) to give an *energy* equivalent. The Boltzmann constant is 1.380 649E-23 J K^{-1} or kg m^2 s^{-2} K^{-1}.

Table 1.1 provides a list of the fundamental physical quantities, SI unit names, SI unit symbols, and the fundamental quantity dimensions.

Table 1.2 lists selected derived quantities, some of which are identified by specific names. As noted earlier, derived physical quantities are a function of or composed of the fundamental quantities, which is evident from their units and dimensions.

Table 1.1 List of Fundamental Physical Quantities

Physical quantity	Name of SI[a] unit	SI unit symbol	Fundamental dimension
Amount of substance	Mole	mol	[n]
Time	Second	s	[t]
Length	Meter	m	[L]
Mass	Kilogram	kg	[M]
Temperature	Kelvin	K	[T]
Electric current	Ampere	A	[I]
Luminous intensity	Candela	cd	[I_v]

[a] SI – Système International.

Table 1.2 Selected Derived Physical Quantities

Physical quantity	Formula or name	SI[a] derived unit symbol(s)	Dimensional formula(s)
Area	Length × width	m^2	[L^2]
Volume	Length × width × height	m^3	[L^3]
Velocity	Distance per second	m s^{-1}	[Lt^{-1}]
Concentration	Amount per volume	mol m^{-3}	[nL^{-3}]
Density	Mass per volume	kg m^{-3}	[ML^{-3}]
Force[b]	Newton (N)	m kg s^{-2}	[LMt^{-2}]
Energy, heat, and work[c]	Joule (J)	N m = m^2 kg s^{-2}	[L^2Mt^{-2}]
Heat capacity	Joule per kelvin	N m K^{-1} = m^2 kg s^{-2} K^{-1}	[L^2Mt^{-2} K^{-1}]
Pressure	Pascal (Pa)	N m^{-2} = m^{-1} kg s^{-2}	[$L^{-1}Mt^{-2}$]
Surface tension	Newton per length	N m^{-1}	[Mt^{-2}]
Surface energy	Joule per length × length	J m^{-2}	[Mt^{-2}]
Gas constant[d]	Gas constant (R)	J K^{-1} mol^{-1} = m^2 kg s^{-2} K^{-1} mol^{-1}	[$L^2Mt^{-2}T^{-1}n^{-1}$]
Gas constant	Gas constant (R)	m^3 Pa K^{-1} mol^{-1} = m^2 kg s^{-2} K^{-1} mol^{-1}	[$L^2Mt^{-2}T^{-1}n^{-1}$]
Gas constant	Gas constant (R)	m^2 kg s^{-2} K^{-1} mol^{-1}	[$L^2Mt^{-2}T^{-1}n^{-1}$]

[a] SI – Système International.
[b] Force = mass × acceleration.
[c] Work = force × distance.
[d] The numerical value for each gas constant is 8.314 462 618 153 24.

1.1.2 Basic Mathematics and Statistics

Mathematics is used to describe or define the nature of our universe. Mathematics can also be viewed as the language of science. Thermodynamics and other areas of science require knowledge of mathematics, especially basic operations of arithmetic, algebra, and geometry; transcendental functions such as logarithms; and calculus to describe and interpret the physical nature of the universe.

Mathematics and statistics are routinely employed to analyze experimental data. Mathematical equations can show fundamental relationships between independent and dependent variables. Many important thermodynamic relationships are described by equations or formulas. Most of the thermodynamic data can be described by linear or exponential functions.

Basic statistics shed light on the central tendency of the data and its precision or variation around the central tendency.

In this chapter, important data reporting considerations such as quantity units, dimensional analysis (DA), significant figures (SFs), and rounding data values are reviewed. Key mathematical functions and statistical parameters that are used in reporting and analyzing experimental data are also reviewed.

1.2 Physical Quantity Dimensions and Dimensional Analysis

The nature of a physical quantity is described by its fundamental dimension or its dimensional formula. In terms of a derived physical quantity, a *dimensional formula is defined by the fundamental dimension(s) raised to an appropriate power necessary to fully describe the quantity of interest.* Tables 1.1 and 1.2 show the fundamental physical quantity dimensions and derived quantity dimensional formulas (Ade 2021a).

1.2.1 Application of Dimensional Analysis

Dimensional analysis (DA) is used throughout the book. DA of a physical quantity shows the association and interrelationships between or among two or more fundamental or derived physical quantities. DA can be used to check for the dimensional consistency of an equation and associated property values. Physical properties of interest can be reduced to their fundamental dimensions of amount, time, length, mass, temperature, electrical current, and luminous intensity by employing DA. DA plays a key role in converting units of one system to another. Dimensions can also be used to determine unknown relationships and equations by utilizing fundamental dimensions for known quantities. Likewise, when deriving a physicochemical equation, dimensional consistency requires that the units and dimensions on the left-hand and right-hand sides be the same.

1.2.2 Limitations of Dimensional Analysis

There are limitations to DA. DA does not apply to transcendental exponential, logarithmic, and trigonometric functions. If one side of an equation has a sum, DA cannot be applied. However, the units on each side of the equation should have the same units. That is, when physical quantities are added or subtracted, they must have the same units, but the units are not added or subtracted. Stating the obvious, DA is not applied to dimensionless numbers.

1.2.3 Dimensionless Numbers

Not all quantities have a dimension. These quantities are known as *dimensionless* quantities. They are also called a *scalar quantity* or *quantity of dimension one*. There are numerous quantities that are dimensionless. Notable dimensionless numbers are pi (π), Euler's number "e," and transcendental values such as the logarithm or trigonometric operations of a quantity. We will talk more about logarithmic functions later in this chapter. Many dimensionless numbers are obtained from ratios of quantities. Pi is a good example, which is the ratio of the circumference of any circle to its diameter.

1.2.4 Dimensional Nomenclature

There are several general rules regarding dimensional quantity formatting and nomenclature. Dimensions are placed in closed brackets as shown in Tables 1.1 and 1.2. Simplified or reduced dimensions are enclosed within closed brackets. For example, density is mass [M] per volume [L^3]. The reported dimension would be [ML^{-3}], not [M] [L^{-3}]. Exponential rules are used in dimensional nomenclature. [L^1] is [L] and [L^0] is 1.

1.2.5 Dimensional Quantity Algebra or Dimensional Algebra

A physical quantity value can be expressed as the *product* of a *numerical value* and a *unit*.

$$\text{Physical quantity} = \text{numerical value} \times \text{unit} \tag{1.1}$$

By convention, for convenience and efficiency, the "times sign" is not used. The commonly accepted formatting without the "times sign" is

$$\text{Temperature} = 298\ \text{K}. \tag{1.2}$$

To employ dimensional algebra, it is important to remember that a physical quantity is the *product* of a numerical value and a unit. From Eqs. (1.1) and (1.2), all three terms can be treated as algebraic quantities amenable to algebraic operations. As an example,

$$\text{Temperature}/\text{K} = 298 \tag{1.3}$$

Dimensional algebra is the operational means by which DA is performed. For purposes of this book, dimensional algebra and DA will be used interchangeably.

As stated earlier, dimensional algebra and analysis find several uses. It is used to convert a physical quantity from one unit system to another unit system. Dimensional algebra is also used to reduce units in an equation to confirm that the left-hand and right-hand sides of an equation are the same. This confirmation also denotes dimensional consistency.

In these cases, one uses *conversion factors* like 1 erg = 1E-7 J or 1 m = 1000 mm. Quantity algebra rearranges these conversion factors into quotients that are equal to 1. These quotients are called *unitary conversion quotients* or *conversion ratios*. The conversion ratio for the above mentioned conversion factor is 1 erg/1E-7 J or 1E-7 J/1 erg. Note, for convenience and ease of cancellation of dimensional terms, conversion ratios can also be displayed as $1\ \text{erg} \cdot 1\text{E}7\ \text{J}^{-1}$ or $1\text{E-}7\ \text{J} \cdot 1\ \text{erg}^{-1}$.

Example 1.1 Conversion Ratios

a) Given the conversion factor 1 L = 1E3 mL; the conversion ratios are

$$1\ \text{L}/1\text{E}3\ \text{mL or } 1\ \text{L}\cdot 1\text{E-}3\ \text{mL}^{-1} \tag{1.4}$$

and

$$1\text{E}3\ \text{mL}/1\ \text{L or } 1\text{E}3\ \text{mL}\cdot 1\ \text{L}^{-1} \tag{1.5}$$

b) Given the conversion factor 1 m = 1E3 mm; the conversion ratios are

$$1\ \text{m}/1\text{E}3\ \text{mm, or } 1\ \text{m}\cdot 1\text{E-}3\ \text{mm}^{-1} \tag{1.6}$$

and

$$1\text{E}3\ \text{mm}/1\ \text{m or } 1\text{E}3\cdot 1\ \text{m}^{-1} \tag{1.7}$$

Example 1.2 Converting a Physical Quantity from One Unit System to Another

Convert 600.0 ergs to joules using conversion ratio dimensional algebra.

$$600.0\ \cancel{\text{ergs}}\left(1\text{E-}7\ \text{J}\cdot 1\ \cancel{\text{erg}^{-1}}\right) = \underline{6.000\text{E-}5\ \text{J}} \tag{1.8}$$

Example 1.3 Using Dimensional Algebra to Confirm Both Sides of an Equation Have the Same Dimensionality

Using the ideal gas law PV = nRT and fundamental units, show that the left-hand and right-hand sides of the equation have the same units and dimensions.

Given from Table 1.2

P = Pa = m^{-1} kg s^{-2}

V = m^3

n = mole

R = m^2 kg s^{-2} K^{-1} mol^{-1}

T = K

J = m^2 kg s^{-2}

1) *Left-hand side* of the ideal gas law equation. Substitute the fundamental units m^{-1} kg s^{-2} for P and m^3 for V. Apply dimensional algebra.

$$\left(\cancel{\text{m}^{-1}}\ \text{kg s}^{-2}\right) \times \text{m}^{\cancel{3}2} = \underline{\text{m}^2\ \text{kg s}^{-2}} \tag{1.9}$$

2) *Right-hand side* of the ideal gas law equation, PV = nRT.

   a) Substitute the fundamental units for n and T. Moles and K are directly substituted for n and T.
   b) Substitute m^2 kg s^{-2} K^{-1} mol^{-1} for R
   c) Perform dimensional algebra.

$$\cancel{\text{mol}} \times \left(\text{m}^2\ \text{kg s}^{-2}\ \cancel{\text{K}^{-1}}\ \cancel{\text{mol}^{-1}}\right) \times \cancel{\text{K}} = \underline{\text{m}^2\ \text{kg s}^{-2}} \tag{1.10}$$

Therefore, the left-hand (Eq. 1.9) and right-hand (Eq. 1.10) sides of the ideal law equation have the same dimensional and unit consistency.

The gas constant, *R*, can be expressed in units of energy per temperature per mole. It is the molar equivalent to the Boltzmann constant. The gas constant value changes depending on the "unit system" employed in making the experimental measurements. As mentioned earlier, the three most used unit systems are the SI, *cgs*, and English systems. Dimensional quantity algebra can be used to convert the gas constant value and its respective units from one unit system to another. Dimensional algebra can also be employed to convert the units from derived units to fundamental units for a particular unit system.

Example 1.4 Use Dimensional Algebra to Convert the Energy Gas Constant, R, from SI Derived Units to Fundamental SI Units

Given

$$\mathrm{R = 8.314\,J\,K^{-1}\,mol^{-1}}$$

$$\mathrm{J = m^2\,kg\,s^{-2}} \text{ and the conversion ratio is } \left(\mathrm{m^2\,kg\,s^{-2}}\right)\cdot \mathrm{J^{-1}}$$

To determine the fundamental SI energy gas constant units, use the conversion ratio for J and perform dimensional algebra.

$$\left(8.314\,\cancel{\mathrm{J}}\,\mathrm{K^{-1}\,mol^{-1}}\right)\cdot\left(\mathrm{m^2\,kg\,s^{-2}}\right)\cdot\cancel{\mathrm{J}^{-1}} = \underline{8.314\,\mathrm{m^2\,kg\,s^{-2}\,K^{-1}\,mol^{-1}}}$$

This is the gas constant in fundamental SI units. It is noted that the "joules" cancel as a result of the quantitative algebra operation.

Example 1.5 Use Dimensional Algebra to Convert *cgs* Energy Gas Constant Units to SI Fundamental Gas Constant Units

Given

$$\mathrm{R = 8.314E7\,erg\,K^{-1}\,mol^{-1}}$$

$$\mathrm{1\,erg = 1E-7\,J;} \text{ conversion ratio is } \mathrm{1\,erg\cdot 1E+7\,J} \text{ or } \mathrm{1E-7\cdot 1\,erg^{-1}}$$

$$\mathrm{J = m^2\,kg\,s^{-2};} \text{ conversion ratio is } \mathrm{J}\cdot\left(\mathrm{m^{-2}\,kg^{-1}\,s^{2}}\right) \text{ or } \left(\mathrm{m^2\,kg\,s^{-2}}\right)\cdot\mathrm{J^{-1}}$$

Employing conversion ratios and dimensional algebra, the energy *cgs* gas constant can be converted to the SI energy gas constant fundamental units.

$$8.314\cancel{\mathrm{E7\,erg}}\,\mathrm{K^{-1}\,mol^{-1}}\times\left(1\cancel{\mathrm{E\text{-}7\,J\cdot erg}^{-1}}\right)\left(\mathrm{kg\,m^2\,s^{-2}}\right)\cdot\cancel{\mathrm{J}^{-1}}\Big) = \underline{8.314\,\mathrm{kg\,m^2\,s^{-2}}}\ \underline{\mathrm{K^{-1}mol^{-1}}},$$

in SI fundamental units.

The reader is referred to Connors and Mecozzi (2010) for additional discussion of dimensional quantitative algebra.

1.3 Data Reporting Using Decimal, Scientific, Normalized Scientific, and E Scientific Numerical Notation

There are several numerical notations that can be used to report and analyze data in base-10 logarithm, including decimal, scientific, normalized (normalized scientific), engineering, and E notations. This book will use decimal, normalized scientific, and E notation systems.

Decimal notation represents base-10 logarithm fractions or real numbers using digits 0 through 9 and plus and minus signs to represent positive and negative numbers, respectively. A decimal point may be used to indicate SFs for whole numbers or represent the fractional portion of a real number.

Scientific notation has been developed to make arithmetic operations easier and improve the convenience of writing large and small numeric values. Scientific notation takes the form:

$$s \times 10^n \tag{1.11}$$

where s is a nonzero real number called the *significand* and n is a positive or negative integer *exponent*. If the significand value is negative, a minus sign precedes s.

Normalized scientific notation employs scientific notation and, by convention, has one nonzero integer before a decimal point. This notation is useful for entering data into tables and to quickly view the magnitude differences of the quantity value.

$$s. \times 10^n \tag{1.12}$$

E scientific notation is an extension of normalized scientific notation, where E means "times ten raised to the power of n, or $\times\ 10^n$." In E scientific notation, Eq. (1.12) would be written as

$$\text{s.En} \tag{1.13}$$

E scientific notation minimizes keystrokes and provides easy readability. It is also the most compact notation for text and table data reporting.

Examples of these four numeric notation methods are summarized in Table 1.3.

Table 1.3 Comparison of Decimal, Scientific, Normalized Scientific, and E Scientific Numerical Notation

Decimal	Scientific	Normalized scientific	E scientific
0.0000000095123	0.95123×10^{-8}	9.5123×10^{-9}	9.5123E-9
5	5×10^0	5×10^0	5.E0
6471.998	6471.998×10^0	6.471998×10^3	6.471998E3
−10 122.3	$-10\,122.3 \times 10^0$	-1.01223×10^4	−1.01223E4
−2000[a]	-2×10^3	-2×10^3	−2.E3

[a] The SFs of this number are ambiguous. The numerical value may have one or as many as four SFs. A decimal point (2000.) indicates the physical quantity has four SFs and is unambiguous. In this example, the value is assumed to have one SF.

1.4 Accuracy

Accuracy of a measurement system is the difference between the true value or target value of a quantity and the measured value or mean of measured values. The FDA has set the accuracy acceptance criteria to be 97.0%–103.0% for a known "spiked" concentration (FDA 2019). Accuracy measurements within this acceptance range are considered to have high accuracy. The further a measured value is from the lower and upper ranges of this acceptance range, the lower the accuracy of the measurement. Most of this section's discussion is based on the United States Food and Drug Administration and USP 43-NF 38.

1.5 Precision as Measurement Variability and Relative Uncertainty

Precision deals with the closeness or difference of measured values independent of the measured accuracy values. Precision is related to the variability of the measured quantities. Precision is also related to the *repeatability* and *reproducibility or robustness* of the analytical method. Repeatability is the variation that occurs under rigorous experimental control, such as using the exact same instrument, with the same operator, in the same lab over a relatively short period of time (days or weeks). Reproducibility represents a more "real-life" scenario, where different instruments, but the same model is used, by different operators, in different labs, over a number of months.

There are two primary types of variability: systematic or determinant variability or error and random or nondeterminate variability or error. Both types of variability can be the result of the operator, analytical instruments, chemicals, environment, and more. Systematic variability results when all the causes of the variability remain constant throughout each measurement. Often poor calibration or instrumental setup can result in systematic variability. On the other hand, random variability can occur unpredictably with each measurement. This random variability or error fluctuates around the mean of the physical quantity.

Precision is commonly reported as *standard deviation (SD)* or *percent relative standard deviation (% RSD)* as an indication of measurement variability. Percent RSD is

$$\text{Percent RSD} = (\text{SD}/\text{mean}) \times 100\% \tag{1.14}$$

where SD is the measurement's standard deviation. For a normal distribution, three SDs account for 99.7% of the measured values. The FDA's precision acceptance criteria for repeatability is 2.0% RSD (FDA 2019).

1.6 Valid Measurements

A reported measurement is considered *valid* when it is both accurate and precise. This concept is illustrated in Figure 1.1. A valid measurement is depicted in the upper left quadrant. The other quadrants show possible scenarios that do not provide a valid physical quantity measurement. The lower left-hand quadrant depicts a set of measurements that are not accurate. This may be due to systematic error since all the data points lie outside the center target value but fall fairly close to each other. The upper right-hand corner shows reasonably high accuracy with physical values scattered around the center. This quadrant exemplifies low precision and relatively high accuracy. Finally, the lower right-hand corner displays poor accuracy and precision. Not one value is in the center target range, and the measurements are spread widely across the measurement space. These data exhibit poor accuracy and large random variability.

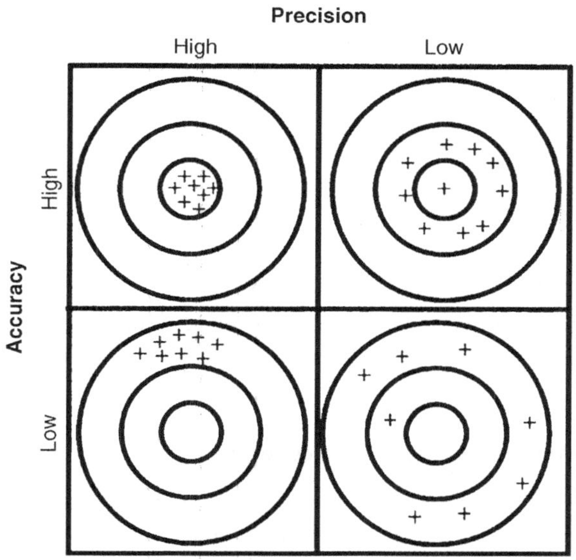

Figure 1.1 Possible accuracy and precision scenarios. The four quadrants represent targets, with the center being the target value or true value. The + sign indicates one of eight individual measured physical quantities.

1.7 Uncertainty

1.7.1 Uncertainty Associated with a Single Measurement

Physical quantity measurement carries a level of *uncertainty or error (σ) or SD*. This level of uncertainty is propagated throughout the calculations that use these measurements. There is uncertainty in the scale of the measuring device, such as a ruler or graduated cylinder. In this case, the uncertainty is equal to the smallest increment divided by 2. The uncertainty for a digital device is the smallest increment or digit position. To acknowledge this uncertainty, the general form of any measurement is written as

$$\text{Measurement or physical quantity value} = x_{best\ estimate\ of\ value} \pm \sigma_{uncertainty\ or\ error}$$

The level of uncertainty determines the number of SFs recorded for measurement estimates. SFs will be discussed in more detail in Section 1.8, Significant Figures (Digits) for Measured Values and Calculations.

There are several rules in reporting uncertainty, $\sigma_{uncertainty\ or\ error}$.

Rule 1- Stating Uncertainties: experimental uncertainties should be expressed to only *one* SF because the uncertainty is an estimate and carries no more significance than one digit.

Example 1.6 Uncertainty Reported to Only One SF

$$\text{Length} = 30.57 \pm 0.034\ \text{cm would be}\ \underline{30.57 \pm 0.03\ \text{cm}} \tag{1.15}$$

The uncertainty of 0.034 cm should be reported to one SF.

Rule 2- Stating *best estimate* measurements or values: the last SF in a reported quantity should have the same place or decimal place value as the uncertainty.

Example 1.7 Reporting the Last SF for a Reported Quantity

$$\text{Velocity} = 2345.67 \pm 200\ \text{cm s}^{-1} \text{ is correctly written } \underline{2300 \pm 200\ \text{cm s}^{-1}} \tag{1.16}$$

In this example, the uncertainty of the reported estimate value lies in the hundred's place. Therefore, the last SF of the reported estimate is at the hundred's place giving 2300 cm s^{-1}. Note that the uncertainty of 200 cm s^{-1} has one SF according to Rule 1.

1.7.2 Uncertainty Associated with Replicate Measurements

As discussed earlier, the FDA acceptance criteria for instrumental methods repeatability precision is 2.0% RSD. The SD and % RSD are statistical calculations that represent the random variability of the replicate sample measurements.

1.7.3 Propagation of Error of Combined Measurements and Calculations

An in-depth discussion of *the propagation of error* is beyond the scope of this book. It is important to be aware that there are propagation of error equations that properly combine uncertainties from different measurements. Propagation of error determination plays a major role in electronic device design. As an example, carbon resistors are widely used in all sorts of electronic circuits. Resistors manufactured to a "specified" resistance will have some variability or error in their "actual" resistance. Resistors connected in series or in parallel will exhibit different propagation of variability or error. In electronic device design, the overall resistance variation/error has a significant impact on the device's performance.

When a calculation requires more than one variable to provide a solution, propagation of errors is required to reflect the appropriate uncertainty. Table 1.4 provides propagation of error SDs for the various arithmetic operations.

In the case of designing resistors in series, the overall SD or error for the series of resistors is the square root of the sum of the squares of the individual resistor's SD or variance. This propagation of error can add significantly to the overall variability of the design circuitry. It is important to be aware that propagation of error occurs for all physical measurements and can have a significant impact on the reported value accuracy.

1.7.4 Author's Anecdote: A Priori Estimation of Dose Variation Using Propagation of Error Equations

It is extremely important that a patient receives the same dose of medication every time the dosage form is administered. My first industrial project was to improve the dose variability of a capsule that contained an active pharmaceutical ingredient (API) in the form of round beads. Inactive beads

Table 1.4 Propagation of Error Equations for Various Arithmetic Operations

Arithmetic operation	Example	Propagated Standard Deviation (SD or σ_x)
Addition or subtraction	$x = a + b - c$	$SD_x = (SD_a{}^2 + SD_b{}^2 + SD_c{}^2)^{0.5}$
Multiplication or division	$x = (ab)/c$	$SD_x = [(SD_a/a)^2 + (SD_b/b)^2 + (SD_c/c)^2]^{0.5}$

a, b, and c are the measured physical quantity values. SD_a, SD_b, and SD_c are the SDs of the physical quantity values (Propagation of Error. Chemistry Libretexts).

were used as a diluent. Dose uniformity is often measured as % RSD, where the SD is divided by the mean, and this quotient is multiplied by 100. Dose uniformity of 3.0% RSD is generally thought to be a relatively stringent measure of acceptable dose uniformity. This allows room for operator-to-operator, day-to-day, and plant-to-plant variation. The United States Pharmacopeia content uniformity limit is 7.8% RSD.

The content uniformity of the product I was assigned typically showed 6.5% RSD, with some batches closer to 5.5% and some batches closer to 7.5% RSD. From time to time batches would fail the USP content uniformity limit.

As I researched what had been done to improve the product content uniformity, I learned the high dose variability had been a chronic problem from the time the product was launched, and the FDA was scrutinizing it very closely. My research revealed that every aspect of the manufacturing process had been investigated, and all improvement attempts had failed. The analytical methods had also undergone extensive method revalidation and gave acceptable results.

I began to think that the propagation of the variabilities due to weight variation, blend inhomogeneity, and assay variability was the reason for the high observed % RSD content/dose uniformity values.

Capsule dose can be generically defined as

$$D = P\,W$$

where D (dose) is the drug amount/capsule, P is the proportion or ratio of drug amount/formulation blend weight and W is the formulation blend weight/capsule. P is dependent on blend homogeneity, which is related to mixing effectiveness, and assay accuracy and precision.

Using the propagation error equation in Table 1.4 for multiplied terms, the dose, SD_D is

$$SD_D = \left[(SD_P/P)^2 + (SD_W/W)^2\right]^{0.5} \tag{1.17}$$

where P and W have been defined earlier, SD_P is the drug formulation proportion SD, and SD_W is the formulation blend weight/capsule SD. SD_P is function blend heterogeneity and assay variability. This is also true for a capsule blend that is a mixture of active beads and placebo beads. The drug proportion standard deviation, SD_P, can be expressed as:

$$SD_P = \left[(SD_{Ph})^2 + (SD_{Pa})^2\right]^{0.5} \tag{1.18}$$

where Ph is the independent measurement of the bead mixture heterogeneity., SD_{Ph} is the bead mixture standard deviation, Pa is the independent measure of the assay content and SD_{Pa} is the assay content standard deviation. Substituting for SD_P from Eq. (1.18) into Eq. (1.17) gives :

$$SD_D = \left\{\left[\left[(SD_{Ph})^2 + (SD_{Pa})^2\right]^{0.5}/P\right]^2 + (SD_W)^2/W\right\}^{0.5} \tag{1.19}$$

Equation (1.19) shows that dose variability is propagated by blend heterogeneity, assay variability, and capsule fill weight variability.

Independent experiments were undertaken to determine SDs estimates. The binomial statistical theory was used to make an independent assessment of the SD_{Ph}. SD_{Pa} was determined from assay validation data. Independent formulation blend weight/capsule SD was determined by weighing beads emptied from individual capsules.

Using Eq. 1.19 and the independent SD measurements an *a priority* SD_D was calculated. It was shown that SD_{Ph} was responsible for 75.3% of the SD_D. 20.1% of the SD_D resulted from SD_W and the remaining 4.6% SD_D resulted from Pa variability (Stagner et al. 1991). From the propagation of error analysis, the blend heterogeneity was determined to be the major cause of the relatively high dose variability. As mentioned earlier, numerous attempts to improve blend uniformity had been undertaken by making changes to blender design, using different types of blenders, and studying the effects of mixing time without satisfactory results.

Interestingly, the binomial distribution model can be applied to this problem. The binomial distribution can be used to determine the probability of a discrete event occurring or not. In this case, if you choose one bead, it will either be a drug bead or a placebo bead. The mean of the binomial distribution is $\mu = Np$, where N is the total number of beads in the capsule and p is the proportion of active beads. The binomial SD of the number of active beads is $SD = (Npq)^{0.5}$, where p is the proportion of active beads, q is the proportion of inactive beads ($1-p$), and N is the total number of beads in the capsule. Employing the equation for the mean and SD shows that one can decrease the drug dose/content % RSD by increasing the total number of beads and increasing the proportion of drug beads (Al-Achi et al. 2023a).

The use of propagation of errors was instrumental in isolating the primary cause of the dose variability to the blend heterogeneity. The binomial distribution theory was used to independently determine the a priori blend heterogeneity. It was also used to resolve the dose content uniformity issue. The binominal distribution showed that blend variability was attributed to two factors, the total number of beads and proportion of active beads. The dose variability % RSD was improved to about 3% RSD from 6.5% RSD by increasing the total number of beads and increasing the proportion of active beads.

1.8 Significant Figures (Digits) for Measured Values and Calculations

As mentioned earlier, *the level of uncertainty determines the number of SFs in a recorded measurement.* The SFs reported for a measurement are equal to the number of digits that are known with a degree of confidence and are considered reliable. The last digit of an SF is an estimate or approximation. As such, 25.14 mg is reported to the 1/100 mg. The digit "4" is an estimate and has the most uncertainty but is still considered reliable. If the exact same amount of material were weighed on a balance capable of measuring the weight to 1/1000 mg, the quantity could be reported to 1/1000 mg, say 25.147 mg. In this case, the "7" would be an estimated measure.

Reporting the last reliable digit depends on the measurement device. For example, if the weight of the material is made on an electronic balance that has digits to the fourth decimal place, the fourth decimal place digit is considered an estimate and recorded as the last SF. This last decimal place is the most uncertain and is still considered reasonably reliable. In the case of reporting the volume using a graduated cylinder that is marked in units of 1 mL, depends on the scientist estimating the last SF to the nearest 0.1 mL.

It is important to note that the category of *exact numbers* is counted, and therefore, exact numbers do not have an uncertainty. For example, a pharmacist counts a specific number of tablets to be dispensed to a patient. Obviously, this example does not account for the possibility of "human error."

1.8.1 Rules for Determining Significant Figures for a Measured or Reported Value

Zeros serve as place holders and set the decimal point, like 0.1, which adds a degree of complexity to determining SFs. The rules to determine SFs are given as follows.

1) All nonzero numbers are significant.
2) A zero within a number is significant.
3) Zeros that do not "set" the decimal point are significant. 5.00 has three SFs.
4) Zeros that "set" the decimal point are not significant. 0.00057 and 0.57 only have two SFs.
5) Zeros that follow digits may or may not be significant. A decimal point notation or E scientific notation must be used to unambiguously indicate significance. For example, 3200 mg has ambiguous SFs compared to 3.200E3 mg (four SFs) or 3200. mg (four SFs).

1.8.2 Determination of Significant Figures for Addition and Subtraction Calculations

The reported value should have the same number of decimal places as the term with the least number of decimal places. To add numbers written in scientific or E scientific notation, the number needs to be written to the same power.

Example 1.8 Addition SFs Using Decimals

$$\begin{array}{l} 135.210\ \text{mg} \\ \underline{+10.2\ \text{mg}} \\ 145.410\ \text{mg} \end{array}$$

10.2 mg contains one decimal place, so the calculated value is reported as <u>145.4 mg or 1.454E2 mg</u> (1.20)

Example 1.9 Subtraction SFs Using E Scientific Notation

$$\begin{array}{l} 1.35210\text{E2 mg} \\ \underline{-0.102\ \text{E2 mg}} \\ 1.25010\text{E2 mg} \end{array}$$

10.2 mg contains one decimal place, so the calculated value is reported as <u>1.250E2 mg,</u> which has one decimal place. (1.21)

Note: Dimensional algebra is not performed for addition and subtraction. The units are not added or subtracted. In the case of Examples 1.8 and 1.9, "mg" units are attached to the calculated physical quantity regardless of the math operation.

1.8.3 Determination of Significant Figures for Multiplication and Division Calculations

The reported value should have the same number of SFs as the term with the least number of SFs.

Example 1.10 Multiplication SFs

$$21.2\ \text{cm} \times 705.06\ \text{cm} = 14,947.272\ \text{cm}^2$$

21.2 cm has the least number of SFs at three SFs. So the calculated value is reported as <u>14,900 cm^2</u> <u>or 1.49E4 cm^2</u>.

(1.22)

Reporting 14,900 leaves some ambiguity regarding the number of SFs this number has. Reporting the answer as "14,900." incorrectly implies that the number has five SFs. The E scientific notation makes it clear that the answer had three SFs.

Example 1.11 Division SFs

705.06 cm/21.2 cm = 33.257

Since 21.2 cm has the fewest SFs at 3, the reported calculation is 33.3. (1.23)

Note: Dimensional algebra is performed for multiplication and division. In the case of multiplication, the units are multiplied as shown in Example 1.10. In the division Example 1.11, the units are divided, resulting in an answer having no reported units.

1.8.4 Logarithm Significant Figures

A logarithm is composed of a *characteristic* and *mantissa*. The characteristic part is an integer, and the mantissa is the decimal portion of the logarithm. The base-10 logarithm ($\log_{10}$) of 339 is 2.530. The "2" is the characteristic, and ".530" is the mantissa. The number of digits in the reported mantissa equals the number of SFs in 339, which is three SFs.

Example 1.12 Logarithm SFs (base-10)

The base-10 logarithm of 0.001237, which has four SFs is

$$\mathrm{Log}_{10}\ 0.001237 = \underline{-2.9076} \tag{1.24}$$

The four digits in mantissa indicate that the original number had four SFs.

Example 1.13 Logarithm SFs (base-e)

The base-e natural logarithm, ln *e*, of 0.001237 is $\underline{-6.6951}$. (1.25)

The number of digits in the mantissa indicates that the logarithm has four SFs.

Note: Logarithms have no associated units because they are a transcendental value.

1.8.5 Antilogarithm Significant Figures

An antilogarithm has *the same number of SFs in the antilogarithm as there are digits in the mantissa of the logarithm value.*

Example 1.14 Antilogarithm SFs (Base-10)

The base-10 antilogarithm ($\mathrm{antilog}_{10}$) of the $\log_{10}$ value −4.470 has three SFs. This is the number of digits −4.470 has in the mantissa.

$$\mathrm{Antilog}_{10} - 4.470 = 10^{-4.470} = \underline{3.39\text{E-}5.} \tag{1.26}$$

The antilogarithm has three SFs, which are highlighted.

As stated earlier, the antilogarithm's SFs equals the number of digits in the mantissa of the log value.

Example 1.15 Antilogarithm SFs (base-e)

The base-e logarithm value −4.470 has three SFs. This is the number of digits it has in its mantissa.

$$\mathrm{Antilog}_{e} - 4.470 = e^{-4.470} = \underline{0.0114.} \tag{1.27}$$

The antilogarithm's three SFs are highlighted. This equals the number of digits in the mantissa of the log value −4.470.

1.9 Rounding Numbers

Rounding should be done on the final calculation to avoid accumulating round-off errors that can occur if intermediate values are rounded. Likewise, in performing the calculations, it is acceptable to carry more SFs than are needed for the SFs in the final calculation.

The digit that is considered for rounding is the last SF. The rounding rules used in this book are as follows: the last significant digit of a number is maintained if the digit is followed by 0, 1, 2, 3, or 4; and the last SF is rounded up one unit if the digit is followed by 5, 6, 7, 8, and 9.

Example 1.16 Rounding 2.4488E4 to Four SFs

Consider the number 2.4488E4. It needs to be rounded to four SFs.
In this example, the underlined 8 is the digit that will be considered for rounding since it is the last SF. Since "8" follows the last SF, the reported value would be rounded up one unit to 2.449E4. (1.28)

Example 1.17 Rounding 2.4484E4 to Four SFs

Consider the number 2.4484E4. It needs to be rounded to four SFs. In this instance, the underlined 8 is the digit that will be rounded. Since "4" follows the last SF, the "8" would be maintained, and the reported value would be 2.448E4. (1.29)

Example 1.18 Rounding 2.4485E4 to Four SFs

Consider the number 2.4485E4. It needs to be rounded to four SFs.
In this case, the underlined 8 is the digit that will be rounded since it is the last SF. Since "5" follows the last SF, the "8" would be rounded up one unit, and the reported value would be 2.449E4. (1.30)

1.10 How Many Digits and Decimals to Use in Research Reports, Tables, and Presentations

Once precision, uncertainty, and SFs have been determined for a measured or calculated value, it would seem straightforward to know how many digits to report. Unfortunately, there are a numerous "rules" that have been suggested for how to best represent data in reports, tables, and presentations (Coghill and Garson 2006; Cole 2015). In general, it is important that the presented number fairly represents the uncertainty of the measurement technique, the data variability, and the significance of the numerical value. As Cole (2015) states, "rules that specify the number of decimal places at the expense of reporting significant figures or vice versa are often found to be unsatisfactory." He suggests that reporting data has a degree of "art" and "common sense." Presenting a large number of digits can overwhelm the reader and the importance of the data lost.

Cole makes the following suggestions:

1) Keep the number of reported digits to no more than 4.
2) The final decimal place should represent the uncertainty in the measurement.
3) The SD and RSD final decimal place should be the same as the reported measured value.
4) The *p value* should be reported to three or four digits.
5) The coefficient of variation should be reported to three or four digits.
6) Scientific or E scientific notation should represent the correct number of SFs and decimal uncertainty while balancing suggestion point number 1.

Table 1.5 follows these suggestions.

Table 1.5 Acyclovir Activation Energy and Arrhenius Frequency Constant at Three pH Values

pH ± SD[a]	Ea^b (Kcal mol^{-1})	A^c (h^{-1})	R^b
1.19 ± 0.02	29.7	40.9	0.939
1.69 ± 0.02	32.9	44.3	0.995
2.19 ± 0.02	31.2	40.6	1.000
Mean	31.3	41.9	
SD[a]	1.6	2.0	

Source: Begum et al. (2018)/with permission of Elsevier.
[a] SD – Standard deviation ($n = 3$).
[b] *Ea* – Arrhenius activation energy.
[c] *A* – Arrhenius frequency constant.

1.11 Exponents

Exponential terms are used extensively in the sciences and especially in thermodynamics. A brief review of exponents is provided with examples to prepare you for the following chapters. The exponent operational rules need to be memorized so they can easily be applied to the reporting and analysis of data.

1.11.1 Exponent Properties and Operations

Exponential properties and operational rules (Anton 1980; Stein 1977):

1. $x^0 = 1$ (1.31)

2. $x^1 = x$ (1.32)

3. $x^m \times x^n = x^{n+n}$ (1.33)

4. $x^{-m} = 1/x^m$ (1.34)

5. $x^m / x^n = x^{m-n}$ (1.35)

6. $(x^m)^2 = x^{2m}$ (1.36)

7. $(x^m)^n = x^{mn}$ (1.37)

8. $(xy)^m = x^m y^m$ (1.38)

In general, x and y can be numbers like 10 or Euler's number, e.

1.11.2 Examples of Scientific Exponent Phenomena

Example 1.19 Richter Scale

The Richter Scale (RS) value is a measure of earthquake power or magnitude. It is a dimensionless number. In this example, let us compare the magnitudes of two earthquakes registered on the RS 8 and 4. The values 8 and 4 represent the exponents of 10. The magnitude of RS values RS_8 and RS_4 are

$$RS_8 = 10^8 = 100,000,000 \tag{1.39}$$

$$RS_4 = 10^4 = 10,000 \tag{1.40}$$

To determine how much more powerful RS_8 is compared to RS_4, we divide RS_8 by RS_4 or $RS_8/RS_4 = 10^8/10^4$. From Eq. (1.35), this quotient equals 10^{8-4}, which equals 10^4 or 10,000. This means that the RS_8 earthquake was 10,000-fold more powerful than the RS_4 earthquake.

Example 1.20 Drug Degradation

It is common for a drug (D) to undergo degradation to form by-products (BP). This can be illustrated by the following degradation equation.

$$D \rightarrow BP \tag{1.41}$$

D is a drug and BP is the degradation by-product. This scheme represents a first-order process that can be described by an exponential function

$$D(t) = D_0 e^{-kt} \tag{1.42}$$

where D is the drug concentration at time t, D_0 is the initial drug concentration, e is the Euler's number, and $k_{(\text{temperature in Kelvin})}$ is the degradation rate constant in units of reciprocal time and a function of Kelvin temperature. The rate constant can be used to predict the drug concentration at any given time. Note the negative sign of the rate constant. This indicates an exponential loss of drug over time. Table 1.6 illustrates this exponential decay, where k at 298.15 K, is 1.05E-2 h^{-1}. The starting concentration is set at 100%.

Table 1.6 Percent *w/v* Drug Degradation Over Time

Time (h)	Drug concentration (% *w/v*)
1.00	98.96[a]
2.00	97.92
4.00	95.89
8.00	91.94
16.00	84.54
32.00	71.15

[a] $D(t) = (100\%\ w/v)e^{-0.0105\ (1/h)(1\ h)} = 98.96\%\ w/v$.

We may be interested in the percent of the drug that remains after 12 hours. This can be calculated by

$$D(t) = (100.\%\ w/v)e^{-0.0105\ (1.00/h)(12.00\ h)} = \underline{88.2\%\ w/v\ \text{drug}} \tag{1.43}$$

1.12 Logarithms

There is an interesting quote by an unknown author regarding the use of logarithms in thermodynamics. It states, "thermodynamics is research on everything for which the negative logarithm is linear with 1/T." As we will see in examples in this chapter and later chapters, this uncanny tongue-in-cheek quip does highlight the importance of becoming fluent in logarithmic properties and operations.

1.12.1 Logarithmic Properties and Operations

When handling logarithmic terms in an equation, the following logarithmic properties and mathematical rules apply (Anton 1980; Stein 1977):

$$1.\ \log 10 = 1 \quad \ln e = 1 \tag{1.44}$$

$$2.\ \log 1 = 0 \quad \ln 1 = 0 \tag{1.45}$$

$$3.\ \log (xy) = \log x + \log y \quad \ln (xy) = \ln x + \ln y \tag{1.46}$$

$$4.\ \log(x/y) = \log x - \log y \quad \ln(x/y) = \ln x - \ln y \tag{1.47}$$

$$5.\ \log x^m = m \log x \quad \ln x^m = m \ln x \tag{1.48}$$

$$6.\ \log x = (\ln x)/2.303 \quad \ln x = 2.303 \log x \tag{1.49}$$

Logarithmic operations can be used to linearize exponential functions. The two most common logarithmic bases used in math and science are the *base-10 logarithm* and the *base-e logarithm*, also known as a *natural logarithm*. The Euler's number *e* has a nonrepeating decimal with a value of 2.71828. The base-10 logarithm operation symbol is "log," which designates it as $\log_{10}$. The base-*e* logarithm operational symbol is "ln," which indicates $\log_e$.

Example 1.21 Derivation of a Conversion Factor Used to Convert a Base-e logarithm to a Base-10 Logarithm and Vice Versa

We can derive the conversion factor, 2.303, by solving Eq. (1.49) for the conversion factor *c*. That is, we will replace 2.302, with *c* in Eq. (1.49) to give Eq. (1.50).

$$\ln x = c \log x \tag{1.50}$$

Solving for *c* gives

$$c = \ln x / \log x \tag{1.51}$$

The conversion factor, *c*, can now be determined by setting *x* to any value. For convenience, let us set $x = 10$. Then,

$$c = \ln 10 / \log 10 = 2.303/1 \text{ or } 2.303 \tag{1.52}$$

Substituting 2.303 for *c* in Eq. (1.50) gives Eq. (1.49) in Rule 6.

1.12.2 Scientific Examples of Base-10 Logarithms (Log_{10} or Log)

Base-10 logarithm or $\log_{10}$ or log conveniently indicates a 10-fold magnitude difference or change at a glance. There are a number of examples in science that use the base-10 logarithm scale such as the RS, microbial growth enumeration, acoustical power, decibel scale, star brightness, and pH scale. An example of the pH scale is shown in the following text.

Example 1.22 pH Is Inversely Proportional to the Molar Hydrogen Ion Concentration.

$$\text{By definition } \mathbf{pH} = -\log\,[\mathbf{H^+}] \tag{1.53}$$

where $[H^+]$ is the molar concentration of the hydrogen ion expressed in moles/liter. The "p" indicates the mathematical operation of "taking" the negative log of a number. That is,

$$p\,(x) = -\log\,(x) \tag{1.54}$$

where x is a number, and $-\log\,(x)$ represents "taking" the negative log of the number, x. From Eq. (1.48) (logarithmic Rule 5), Eq. (1.53) can be rewritten to give

$$\text{pH} = \log\,[H^+]^{-1} \tag{1.55}$$

Using Eq. (1.34), exponential Rule 4, Eq. (1.55) can now be rewritten as

$$\text{pH} = \log 1/[H^+] \tag{1.56}$$

which shows the important property of pH. Equation (1.56) shows that pH is inversely proportional to the molar hydrogen ion concentration. That is, as the $[H^+]$ increases, the pH decreases. This is consistent with the pH scale, which ranges from 0 to 14. A pH of 0 is very acidic, and pH 14 is very alkaline or basic. A pH of 7 is considered a neutral pH, where the $[H^+]$ and [OH] molar concentrations are equal.

Example 1.23 Derivation Showing Neutral pH Is 7

Neutral pH is associated with the autoionization of water shown as follows.

$$H_2O\,(l) \rightleftharpoons H^+\,(aq) + OH^-\,(aq) \tag{1.57}$$

Equation (1.57) shows that water auto dissociates equally into a hydronium ion and a hydroxide ion. Let us assume that the molar concentration of both ions is 1×10^{-7} M. From Eq. (1.53), the pH at this hydronium ion concentration is

$$\text{pH} = -\log\,\left[1 \times 10^{-7}\right] \tag{1.58}$$

The steps to determine the pH at this concentration are shown as follows.

Step 1: Employing Eq. (1.46), logarithmic Rule 3, Eq. (1.58) becomes

$$\text{pH} = -\log 1 + -\log\,\left[10^{-7}\right] \tag{1.59}$$

Step 2: Assessing the first term on the right-hand side of Eq. (1.59)

From Eq. (1.45), logarithm Rule 2, the first term $-\log 1$ becomes

$$-\log 1 = 0 \tag{1.60}$$

Steps 3 to 7: Assessing Second Term of Eq. (1.59).

Step 3: Applying Eq. (1.48), logarithm Rule 5, the second term $-\log\,[10^{-7}]$ results in:

$$\log\,\left[10^{-7}\right]^{-1} \tag{1.61}$$

Step 4: Utilizing Eq. (1.36), exponential Rule 6, Eq. (1.61) gives

$$\log\left[10^{+7}\right] \tag{1.62}$$

Step 5: Using Eq. (1.48), logarithm Rule 5, Eq. (1.62) yields

$$7\,(\log 10) \tag{1.63}$$

Step 6: From Eq. (1.44), logarithmic Rule 1, the second term in Eq. (1.63) converts to

$$\log 10 = 1$$

Step 7: Substituting "1" into Eq. (1.63) for log 10 gives

$$7 \times 1 = 7 \tag{1.64}$$

Step 8: The calculated pH for a hydronium concentration of 1×10^{-7} (1E-7) M combines the values of the first term (Eq. 1.60) and the second term (Eq. 1.64) of Eq. (1.59) to give

$$\text{pH} = 0 + 7 = 7 \tag{1.65}$$

The pH scale ranges from 0 to 14. Zero to 6.99 pH is the acidic range and has an abundance of hydronium ion. pH 7.00 is a neutral pH and has equal molar concentrations of hydronium ion and hydroxide ion. pH 7.01–14.00 is designated the basic range which exhibits a higher portion of hydroxide ion. This derivation confirms that when the hydronium ion and hydroxide ion are in equal concentration, the acid and base neutralize each other giving a neutral pH of 7.

Similar calculations can be undertaken to calculate the pH at any other hydrogen ion molar concentration. Such calculations would confirm that pH is inversely related to hydrogen ion concentration (see Eq. 1.56, pH = log $1/[H^+]$).

1.12.3 Scientific Examples of Base-*e* Logarithms (Log$_e$ or Ln)

There are many natural phenomena that can be described as a function of *e*. Examples are radioactive decay, heat sterilization rates, numerous thermodynamic concepts, and many others. The use of logarithms to linearize drug degradation rate and drug solubility expressions are provided as follows. Linearization of the drug degradation rate equation is an extension to Example 1.20.

Example 1.24 Linearization of the First-Order Drug Degradation Rate Expression
It is common for a drug (*D*) to undergo degradation to form by-products (*BP*). This can be illustrated by the following degradation equation.

$$D \rightarrow BP \tag{1.66}$$

where *D* is the drug and *BP* is the degradation by-product. This scheme represents a first-order process that can described by an exponential function:

$$D(t) = D_0 e^{-kt} \tag{1.67}$$

where *D* is the drug concentration at time *t*, D_0 is the initial drug concentration, *e* is the Euler's number, and *k* is the first-order degradation rate constant. The rate constant can be used to predict how much drug will be left at any given time. Equation (1.67) can be linearized by taking the natural logarithm of both sides.

$$\ln D(t) = \ln\left[D_0 e^{-kt}\right] \tag{1.68}$$

Exponential and logarithmic operations can now be used to derive a linear form of this expression.

Concentrating on the right-hand side of Eq. (1.68) and using Eq. (1.46) (Rule 3, logarithm of a "product term") gives

$$\ln D_0 + \ln e^{-kt} \tag{1.69}$$

Applying Eq. (1.48), Rule 5 to ln e^{-kt} gives

$$-kt \ \ln e \tag{1.70}$$

Employing Eq. (1.44) (Rule 1), $-kt$ ln e becomes

$$-kt\,(1) \text{ where } \ln e = 1 \tag{1.71}$$

The right-hand side of Eq. (1.68), ln $D(t)$ = ln $[D_0e^{-kt}]$, produces

$$\ln D_0 - kt$$

Now, ln $D(t)$ = ln $[D_0e^{-kt}]$ has been converted into the linear form:

$$\ln D(t) = \ln D_0 - kt \tag{1.72}$$

where a plot of drug concentration versus time gives a straight line with a slope of $-k$, the degradation rate constant.

Example 1.25 Linearization of Exponential Ideal Drug Solubility Equation

Equation (1.73) shows that a crystalline drug's solubility is an exponential function of the drug's melting point, the temperature of the solvent, and the molar heat of fusion of the crystalline material.

$$x_2 = e^{\frac{-\Delta H_f}{R}\frac{(T_m - T)}{(T_m T)}} \tag{1.73}$$

Here x_2 is the mole fraction drug solubility, ΔH_f is the molar heat of fusion, R is the gas constant, T_m is the melting point of the drug in kelvin, and T is the temperature of the solvent/solution in kelvin. This solubility equation can be linearized by taking the natural logarithm of both sides and implementing the logarithmic rules and operations provided earlier.

Step1: Take the natural logarithm, ln, of both sides of the equation.

$$\ln x_2 = \ln e^{\frac{-\Delta H_f}{R}\frac{(T_m - T)}{(T_m T)}} \tag{1.74}$$

Step 2: Use Eq. (1.48), logarithmic Rule 5, on the right-hand side of the Eq. (1.74) to give

$$\ln x_2 = -\left(\Delta H_f/R\right)\left[(T_m - T)/(T_m T)\right] \ \ln e \tag{1.75}$$

Step 3: Use Eq. (1.44), logarithmic Rule 1, ln e = 1, to achieve

$$\ln x_2 = -\left(\Delta H_f/R\right)\left[(T_m - T)/(T_m T)\right] \times 1 \tag{1.76}$$

Step 4: From Eq. (1.76), divide $(T_m - T)$ by (TmT) to get

$$\ln x_2 = -\left(\Delta H_f/R\right)\left(1/\mathrm{T} - 1/T_m\right) \tag{1.77}$$

Step 5: The linearized form of $x_2 = e^{\frac{-\Delta H_f}{R}\frac{(T_m - T)}{(T_m T)}}$ (Eq. 1.73) is obtained by multiplying $(1/T - 1/T_m)$ by $-(\Delta H_f/R)$ from Eq. 1.77 to give

$$\ln x_2 = -\left(\Delta H_f/R\right)/\mathrm{T} + \left(\Delta H_f/R\right)/T_m \tag{1.78}$$

This is the linear form of the ideal solubility law, Eq. (1.73). A plot of ln x_2 versus $1/T$ gives a straight line with a slope of $-(\Delta H_f/R)$. The intercept is $(\Delta H_f/R)/T_m$, where $\Delta H_f/T_m$ is ΔS_f, the entropy of fusion.

1.13 Differential and Partial Differential Equations

Differential equations involve *derivatives* of a function that tells how the function changes with respect to changes associated with the function variable or variables. In essence, differential equations are about quantifying change. Differential equations are used to explain how a quantity changes as an independent variable or variables change. For example, pharmaceutical scientists are keen to know how the solubility of a drug changes with temperature, how an aerosol's pressure changes with temperature, or how solubility changes with surface area.

Incremental and *infinitesimal* variable changes are distinguished by the magnitude change of the variables. The slope, m, of a straight line, is defined as the incremental change along the y-axis (Δy), divided by the incremental change along the x-axis (Δx). This slope function is written as

$$m = (y_2 - y_1)/(x_2 - y_1) = \Delta y/\Delta x \tag{1.79}$$

where $\Delta y/\Delta x$ is the slope of the line. Δy and Δx are the incremental changes in y and x, respectively.

As the change in y and x approaches zero, the slope function becomes

$$m = dy/dx \tag{1.80}$$

where dy and dx indicate infinitesimal changes in y and x.

To summarize, a derivative expresses the instantaneous rate change of a function with respect to a change in a function's variable. For example, the expression, $y = f(x)$ describes the relationship between a dependent variable y and an independent variable x. This expression can be stated several ways such as "y is a function of x" or "y depends on x." The derivative of $y = f(x)$ is denoted dy/dx or $f'(x)$. As mentioned in the discussion of Eq. (1.80), the differentials dy and dx represent infinitesimal change in y and x. The differential equation form of a derivative can be integrated.

There are several ways to classify differential equations. Differential equations can be classified by *type* and *by order*. There are two types of differential equations: *ordinary* and *partial* differential equations. These will be discussed after the next section on derivative order.

1.13.1 Differential Order of Ordinary Differential Equations

The derivative order is the number of times a function has been differentiated. Similarly, for a differential equation, its order is determined by the highest order derivative of the equation. The order of a differential equation is defined by the highest order of differentiation present in the equation, not by the total count of derivatives or the number of derivative terms. The generalized differential equation order function is $d^n f(x)/dx^n$ or d $f^{(n)}(x)$, where n is the order of the ordinary differential.

For example, the first derivative of position with respect to time gives velocity. The magnitude of velocity is speed. The second derivative of position with respect to time gives acceleration which describes the how velocity/speed is changing. A second-order derivative is the derivative of the first derivative which represents the rate of change of the rate of change. In the case of speed, the second-order derivative of speed is acceleration given in units of distance per time squared. That is, we are interested in the change of speed with change in time.

1.13.2 First-Order Ordinary Differential Equations

An ordinary differential equation (ODE) is a mathematical equation that involves an unknown function of a single independent variable and one or more of its derivatives with respect to that variable. An example of a first-order ordinary differential equation is the average kinetic energy of an ideal monoatomic gas. In this case, $KE_{ave} = (3/2)kT$, where KE_{ave} is the average kinetic energy of a gas in J, k is the Boltzmann constant in J K^{-1}, and T is the temperature in kelvin. The average kinetic energy is only a function of temperature in kelvin. This relationship between the average kinetic energy function and its single-independent variable is indicated as KE_{ave} (T). It is said that KE_{ave} is a function of T (temperature in Kelvin). The derivative of KE_{ave} can only be taken with respect to the function's single variable, T.

Like addition and subtraction, derivatives follow a set of operating rules. Table 1.7 lists some of the more frequently used first-order ordinary derivative operations and provides derivatives and their differentials. A **derivative** is the rate of change of a function (dy/dx), while a **differential** (dx,dy) represents an infinitesimal change in a variable. The derivative is the ratio of two differentials.

Table 1.7 First-Order Ordinary Derivatives and Their Corresonding Differentials[a]

Derivative	Corresponding Differential	Equation number
$dc/dx = 0$	$dc = 0\ dx = 0$	(1.81)
$dx/dx = 1$	$dx = 1\ dx = dx$	(1.82)
$d(cu)/dx = c\ du/dx$	$d(cu) = c\ (du/dx)\ (dx) = c\ du$	(1.83)
$du^n/dx = nu^{n-1}\ du/dx$	$du^n = (nu^{n-1}\ du/dx)\ dx = nu^{n-1}\ du$	(1.84)
$d\ (ln\ u)/dx = (1/u)\ du/dx$	$d(ln\ u) = [(1/u)\ du/dx]\ dx = (1/u)\ du$	(1.85)
$d(e^u)/dx = e^u\ du/dx$	$d(e^u) = (e^u\ du/dx)dx = e^u\ du$	(1.86)
$d(u + v - w)/dx = du/dx + dv/dx - dw/dx$	$d(u + v) = [(du + dv - dw)/dx]\ dx = du + dv - dw$	(1.87)
$d(uv)/dx = u\ dv/dx + v\ d/dx$	$d(uv) = [(u\ dv + u\ dv)/dx]\ dx = u\ dv + u\ du$	(1.88)
$d(uvw)/dx = uv\ dw/dx + vw\ du/dx + uw\ dv/dx$	$d(uvw) = [(uv\ dw + vw\ du + vw\ du)/dx]\ dx = uv\ dw + vw\ du + vw\ du$	(1.89)
$d(u/v)/dx = (v\ du/dx - u\ dv/dx)/v^2$	$d(u/v) = [(v\ du - u\ dv)/dx)/v^2]\ dx = du/v - u/v^2dv$	(1.90)

Source: From CRC Standard Mathematical Tables.
[a] c and n are constants; u, v, x, and w are variables.

Example 1.26 Temperature Dependence of Monoatomic Gas Kinetic Energy
The average kinetic energy of a monoatomic gas is

$$KE_{ave}(T) = (3/2)kT \tag{1.91}$$

where the average kinetic energy, KE_{ave}, is a function of temperature in Kelvin, indicated by (T). The Boltzmann constant, k, and 3/2 are constants represented by c, in Eq. (1.83). Equation 1.82 and Eq. (1.83) can be used to take the first-order ordinary derivative of Eq. (1.91). The derivative of (1.91) with respect to T is

$$d[KE_{ave}(T)]/\mathrm{dT} = [3/2(k)]\ (dT)/(dT) = [3/2(k)] \tag{1.92}$$

1.13.3 First-Order Partial Derivatives

If a mathematical expression is a function of more than one independent variable it is called a multivariate function. The *partial derivative* delineates the rate of change of the multivariate function with respect to one variable while the other variables are held constant. The partial derivative allows one to understand what effect a single variable has on the change in the function value. In other words, the partial derivative examines the sensitivity of the function to changes in a single variable. The *partial differential* represents a *portion or a part* of the *total function differential value* that results from an infinitesimal change a single variable. Simply put, an individual partial derivative provides only a "partial" view of the function's behavior. The sum of all partial derivatives describes the overall change in the function. The ideal gas law is an example of a multivariate function. The ideal gas law is $PV = nRT$. Solving for P gives $P = (nRT)/V$, where P is the pressure in pascals, n is the number moles of gas, R is the gas constant in J mol^{-1} K^{-1}, T is in K, and V in L. In this case, the number of moles is constant and pressure is a function of temperature and volume. We might be interested in knowing how the pressure changes with a change in the temperature if the volume is held constant. We might also be interested in knowing how the pressure changes with a change in the volume when the temperature is held constant. Finally, we might be interested to know how the pressure changes as both the temperature and volume change. In this case, we gain a partial picture of the total differential by taking the derivative of P with respect to T at constant V. We gain another portion of the picture by taking the derivative of P with respect to V at constant T. To gain a full picture of the effect of temperature and volume changes on pressure, the partial derivatives are added to give the total differential equation.

The ∂ *is* the operator symbol for the partial derivative. This symbol alerts the reader that the function has more than one independent variable. It also denotes that the derivative is being taken with respect to one of the independent variables while the other variables are held constant. Let us say z is a function of w, x, and y. The shorthand version of this statement is $z\,(w,x,y)$. The total differential of z is the sum of the partial derivatives of z, where two of the three variables are held constant. The total differential of z is shown in Eq. (1.93).

$$dz\,(w,x,y). = (\partial z/\partial w)_{x,y} + (\partial z/\partial x)_{w,y} + (\partial z/\partial y)_{w,x} \tag{1.93}$$

where the subscripts indicate which independent variables that are held constant.

Example 1.27 Using the Ideal Gas Law, Determine the Differential Change in the Pressure Caused by Changes in the Temperature and Volume

The differential of P is the sum of the partial derivatives of the pressure with respect to T at constant volume and V at constant temperature.

$$P = (nRT)/V \text{ and } P\,(T,V) = nRT/V \tag{1.94}$$

where nR is a constant.

$$(\partial P/\partial T)_V = nR/V \;\; (\partial T/\partial T) = nR/V \tag{1.95}$$

where $(\partial T/\partial T) = 1$ from Eq. (1.82).

$$(\partial P/\partial V)_T = nRT(\partial V^{-1}/\partial V) = nRT\;\;(-V^{-2}) = -nRT/V^2 \tag{1.96}$$

where $(\partial V^{-1}/\partial V)$ is $-V^{-2}$ from Eq. (1.84).

The sum of the partial derivatives from Eqs. (1.95) and (1.96), gives the *total differential change in pressure* with changes in temperature and pressure.

$$dP = (\partial P/\partial T)_V + (\partial P/\partial V)_T = nR/V - nRT/V^2 \quad (1.97)$$

1.14 Integral Equations

Integrals deal with accumulation and net change, while differentials are associated with infinitesimal change. Integrals combine small changes and can be viewed as an antiderivative. Such thinking is like our discussion of logarithms and antilogarithms. The reverse process of determining a differential is finding the integral, which is called integration.

If the differential is *dx*, then the *indefinite integral* of *dx*, or the antiderivative of *dx* is

$$\int dx = x + c \quad (1.98)$$

where $\int$ *is* the *integration or integral* symbol, *dx* is part of the *integrand* 1 dx, *x* is the integration variable, and *c* is any constant called the integration constant. Indefinite integrals represent a family of functions that differ by a constant because the limits of the accumulation we are interested in are not defined. For *definite integrals,* the accumulation range is specified. In this case, the determined integral value is specific to the specified *accumulation range* or *low* and *high* integration limits.

Several common integration operation rules are shown in Table 1.8. More complex integrals can be found in books covering integral calculus or standard mathematical tables.

Table 1.8 Common Integration Rules for Indefinite Integrals

Differential function to be integrated	Antiderivative or Indefinite integral	Equation number
$\int a\,dx$	$ax + c$	(1.99)
$\int x^n\,dx$; except −1	$(x^{(n+1)})/(n+1) + c$	(1.100)
$\int x^{-1}\,dx$	Ln $\lvert x\rvert + c$	(1.101)
$\int (1/x)\,dx$	Ln $\lvert x\rvert + c$ (same as 1.101)	(1.102)
$\int e^x\,dx$	$e^x + c$	(1.103)
$\int a^x\,dx$	$a^x/ln\,(a) + c$	(1.104)
$\int \ln\,(x)\,dx$	$x \ln\,(x) - x + c$	(1.105)
$\int c\,f(x)\,dx$	$c \int f(x)\,dx + c$	(1.106)
$\int e^{ax}\,dx$	$e^{ax}/a + c$	(1.107)
$\int (f+g)\,dx$	$\int f(x)\,dx + \int g(x)\,dx + c$	(1.108)
$\int (f-g)\,dx$	$\int f(x)\,dx - \int g(x)\,dx + c$	(1.109)

Source: From CRC Standard Mathematical Tables.
Where *a* and *c* are constants. *f* and *g* are functions of *x*.

Example 1.28 Determining the Amount of Reversible Work Obtained during the Isothermal Expansion of an Ideal Gas

Isothermal reversible expansion of an ideal gas is known as $P\Delta V$ work at a constant temperature. $P\Delta V$ work will be discussed in more detail in Chapter 2. The reversible work is given by

$$w_{rev} = -\int_{v1}^{v2} P\,dV \tag{1.110}$$

where P is the ideal gas pressure in pascals and V is the gas volume in liters. Solving the ideal gas law (see Section 1.13.3) for P is $P = nRT/V$. Substituting the expression for P into Eq. (1.110) gives

$$w_{rev} = -\int_{V1}^{V2} \frac{nRT}{V}\,dV \tag{1.111}$$

where n, R, and T are constants, and Eq. (1.111) becomes

$$w_{rev} = -nRT\int_{V1}^{V2} dV/V \tag{1.112}$$

Integrating Eq. (1.112) using Eq. (1.102) from Table 1.8 and evaluating over the limits of V_1 and V_2 give the definite integral, Eq. (1.113).

$$w_{rev} = -nRT(\ln V_2 - \ln V_1) = -nRT \ \ln V_2/V_1 \tag{1.113}$$

From Eq. (1.47), logarithm Rule 4, $\ln V_2 - \ln V_1 = \ln V_2/V_1$. Equation (1.113) tells us that reversible work for an ideal gas is directly proportional to the difference in $\ln V_2 - \ln V_1$. That is, the larger the change in volume, the larger the reversible work.

1.15 Basic Descriptive Statistics

Statistics provides a way to organize and summarize data and provides mathematical techniques to analyze and reach decisions regarding the data. These two aspects of statistics are known as descriptive and inferential statistics.

There are four types of variable data: *nominal, ordinal, interval,* and *ratio*. Nominal data has no order or ranking but can be placed in a specific category like male/female, color, or left-handed/right-handed. Ordinal data exhibits order or hierarchy, but there is no requirement that the distance between each ranking or subcategory is the same. For example, taste can be ranked as "tastes bad" to "tastes great" without intermediate assessment values having equal distance. Ordinal data often uses Hedonic scales of 5–9 rankings or subcategories, which allow for statistical analysis. Interval data has an equal distance between each value but does not have a "true zero." In other words, the data values can be a negative number, as is the case with the Celsius temperature scale. Ratio data displays the equal distance between values and has a true zero, like the Kelvin temperature scale. Nominal and ordinal data variables are called *categorial variables*. Interval and ratio data variables are known as *continuous variables*.

Descriptive statistics summarize the *central tendency* and *dispersion or variation* of values about the central tendency for a set of data. There are several ways to describe the central tendency, which will be discussed further in this section. The *central tendency* is an attempt to determine the *best estimate* of the quantity or function being studied. *Dispersion or data value variation,* on the other hand, provides information about the *variation, uncertainty, and reliability* of the *best estimate*.

Table 1.9 Particulate Counts in Particles per Liter in a Hood on Different Days in Rank Order

Particle Counts per Liter
10
21
33
53
54

1.15.1 Central Tendency of Single Values of Sample Data

There are four main measures of central tendency: arithmetic mean, median, mode, and percentile. The arithmetic mean is the most reported measure of central tendency. The use of the term "*arithmetic mean*" distinguishes it from other means that are also computed, such as the geometric mean and harmonic mean. "*Mean*" will be used for the arithmetic mean for the remainder of this section and book. The mean is sensitive to extreme values in a data set. These extreme values will pull the mean value in their direction. The median, mode, and percentiles are less sensitive to extreme values.

Table 1.9 can be used to illustrate the calculation of mean, median, mode, and percentile of individual data.

Example 1.29 Calculation of Arithmetic Mean

The mean is obtained by adding all the sample values and dividing by the total number of values added.

$$\bar{x} = \frac{\sum x_i}{n} \tag{1.114}$$

where *x*-bar is the mean, x_i is the *i*th observation, and n is the total number of observations.

In this case, the mean is

$$(10 + 21 + 33 + 53 + 54)/5 = 34.2 \text{ particles per liter} \tag{1.115}$$

The number of SFs for this data set is 2, so the answer is rounded to two SFs.

The mean is 34 particles per liter

Example 1.30 Determination of the Median

The median of a data set is the value that divides the data set into two equal parts. The number of values equal to and greater than the median is equal to the number of values equal to or less than the median. If the number of values is *odd,* the median value is the middle number when the values are arranged in rank order. In the case, where there are an even number of values, the mean of the two middle observations is considered the median value. In this example, there is an odd number of values, so the third particle count/liter is the median value.

The median is 33 particles/liter.

Example 1.31 Determination of the Mode or Modes

The mode is the value that occurs most frequently. In Table 1.9, each number is different, so there is no mode. It is also possible for a set of data to have multiple modes. Say that the melting points of 10 lots of a compound from the same supplier reported in rank order are 210.4, 211.0, 211.0, 211.2, 211.7, 211.9, 212.1, 212.1, 212.2, and 212.5. Here, 211.0 and 212.1 occur twice in the data set, which has these two mode values.

Example 1.32 Determination of Percentile
Percentile is another measure of central tendency that is used to understand and interpret data. Percentile is the relative standing of the *k*th percentile that splits the data set into two pieces, one equal to and above the *k*th percentile and one below the *k*th percentile. The following steps are used to determine the *k*th percentile:

1) The values are ordered from smallest to largest.
2) For the *k*th percentile of interest, multiply *k*th percentile (say 20th percentile (0.2) times the number of data points. This number is called the *index number*. In this case, $0.2 \times 5 = 1$. The index value is used to locate the *k*th percentile. The index value notes the number of data points needed to be *counted* to determine the *k*th percentile of interest.
3) From Table 1.9, starting with the smallest number and using the index number of 1, the first number counted is 10, which is the 20th percentile of the five values reported. The number 33 would be the 60th percentile data point.
4) If the index number is not a whole number, the number is rounded up. The 50th percentile index value for Table 1.9 is $0.5 \times 5 = 2.5$. So, 2.5 is rounded up to 3. The third number in Table 1.9 is 33, which is the 50th percentile, where half the values are less than 33 and half the values are greater than 33.

Steps 3 and 4 illustrate the shortcomings of reporting percentile data for small data sets. In the case of Table 1.9, 33 counts per liter is considered the 60th percentile and the 50th percentile.

1.15.2 Dispersion of Single Values of Sample Data

Dispersion is also known as the scatter, variation, and spread of the data set. If all the data points in a set of observations are the same, then there is no dispersion, and the data measurement is highly precise (see Section 1.6 Valid Measurements). Data dispersion is small if the values are close together and greater when the observations are scattered widely. Several often-used statistical measures of dispersion are range, variance, SD, coefficient of variation, and % RSD of the mean or standard error.

Example 1.33 Range
The *range* is the difference between the largest and smallest value in a data set. The range of the data found in Table 1.9 is

$$54 \text{ counts/liter} - 10 \text{ counts/liter} = \underline{34 \text{ counts/liter}} \tag{1.116}$$

The range is the least informative of the dispersion measures. It only uses two data points of the data, which limits usefulness, especially for exceptionally large data sets of hundreds to tens of hundreds of observations.

Example 1.34 Variance
The *variance* measures the dispersion of value around the mean. Computationally, the variance is calculated by subtracting the mean for each observed value. This difference is squared to avoid having negative values. The squared differences are summed or added together and divided by the number of observations minus 1. The computational expression is

$$s^2 = \sum_{i=1}^{n} (x_i - \bar{x})^2/(n-1) \tag{1.117}$$

where s^2 is the sample variance, x_i is the ith observation, and $n-1$ is the number of degrees of freedom. Here, degrees of freedom allude to the idea that if we know the sum of the $n-1$ "observations minus mean differences," the nth difference value is constrained to be a set value. This is because the differences of all the values must equal zero. In this case, there is only 1 value that will give a "zero" sum of the mean differences. The variance for the values given in Table 1.9 leads to

$$s^2 = [(10 - 34.2)^2 + (21 - 34.2)^2 + (33 - 34.2)^2 + (53 - 34.2)^2 + (54 - 34.2)^2]/4 = 1506.8/4 = \underline{3.8\text{E2 counts}^2\ \text{L}^{-2}} \quad (1.118)$$

where the mean of 34.2 counts L^{-1} for the sample data set was determined for Eq. (1.115) before rounding to 34 counts L^{-1}. The variance in this case has been rounded to 2 SFs.

Since the variance does not have the same units as the measured values, it is difficult to understand its physical significance except to say that the larger the variance, the larger the dispersion or variation around the mean. The variance is an important measure of dispersion for inferential statistics.

Example 1.35 Standard Deviation, *s* or SD

As discussed earlier, it is not meaningful to discuss the variance in terms of the original measurement units because the variance is given in terms of squared units. To measure dispersion in the original units, we take the square root of the variance, which is the *standard deviation*, *s* or SD, for the sample.

$$s = \text{SD} = \sqrt{s^2} \quad (1.119)$$

The SD of the data in Table 1.9 is 19.4 counts /liter L^{-1}. The SD is reported as 19. to have two SFs.

The SD is used in conjunction with the mean to describe the dispersion about the mean. For this example, the mean and standard deviation (SD) are noted as 34 ± 19 counts L^{-1}.

The SD is a meaningful measure of data dispersion. It has been shown that most large data sets exhibit a *normal distribution* or *bell-shaped curve* that is symmetrical about its mean. For normal distributions, one SD on both sides of the mean includes approximately 68% of the total area under the bell-shaped curve. Extending the boundaries on both sides of the mean by two SDs covers about 95% of the total area of the curve, and three SDs encompasses 99.7% of the total area.

1.16 Application: Linear Least Squares and Coefficient of Determination

Many natural processes are described by nonlinear functions. Rate processes like chemical degradation, microbial growth, and heat sterilization kill rates are described by nonlinear functions. Equilibrium processes such as solubility and complexation are also described by nonlinear functions. In the past, it was common to derive *linear* forms of these functions by transforming variables. This is done so the data could be plotted using x–y coordinates. Such linear forms and graphic representation of the data are often easier to interpret how specific variables affect the observed values compared to nonlinear data presentation. Today, with ready access to computers, which can easily plot nonlinear data and perform nonlinear regression analysis, the need to derive a linear form of nonlinear equations is not as important.

However, derived linear forms of nonlinear equations will be used throughout this book because the data represented in a linear form is often easier to interpret, and linear least-squares analysis is easy to understand and perform. The well-known linear equation form is

$$y = mx + b \tag{1.120}$$

where y is the dependent variable, x is the independent variable, m is the calculated slope of the line, and b is the calculated y-intercept of the line.

The purpose of linear least squares is to minimize the deviation sum of squares of the individual points from the line. The linear least-squares line is known as the *best-fit line*. The *coefficient of determination, R^2*, is used to measure the *goodness of fit* of the least-squares line and the data points. R^2 determines the strength of the linear association of x and y. R^2 values range from 0 to 1. A value of "0" indicates that the linear model does not explain any of the variation observed between the least-squares line and the data. A "1" denotes that the linear function perfectly explains the data and there is no variation between the least-squares line and the data points.

Figures 1.2 and 1.3 plot the mean of 5 time points (each time point has 144 values) as a zero order and first order dissolution process, respectively. The line in both figures is the least-squares, best-fit line. The coefficient of determination was determined and noted in the graphic.

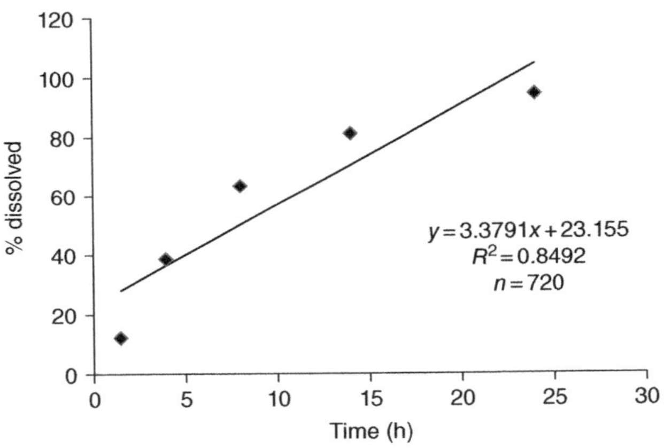

Figure 1.2 Zero-order dissolution model. *Source:* Al-Achi et al. (2023b).

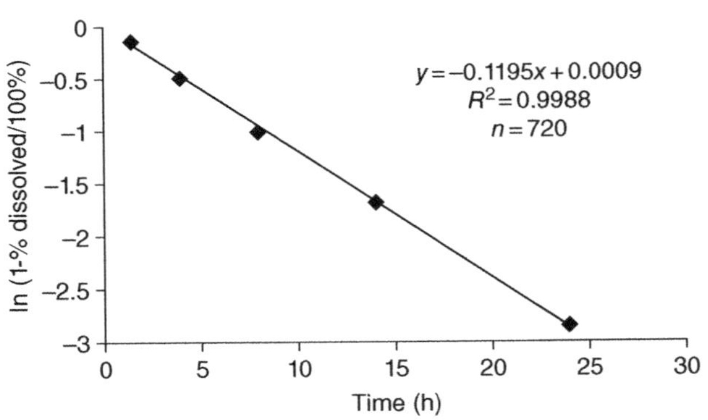

Figure 1.3 First-order dissolution model (fraction of drug remaining to be dissolved). *Source:* Al-Achi et al. (2023b).

Figure 1.2 clearly shows a significant deviation of the data points from the least-squares line. On careful inspection, the zero-order plot shows curvature in the data. The coefficient of determination of 0.8492 is consistent with the data variation seen about the best-fit line. On the other hand, Figure 1.3 exhibits minimal deviation between the data points and the least-squares line. The R^2 of 0.9988 denotes a strong linear association between x and y, indicating that the first-order linear function explains the data and the mechanism of the dissolution process.

References

Ade, S. (2021). *Dimensional Analysis for Dummies: A Simple Approach to Analyzing Dimensional Formulas and Equations Using Physical Quantities.* Made in the USA. Columbia, SC.

Al-Achi, A., Gupta, M.R., and Stagner, W.C. (2023a). *Integrated Pharmaceutics: Applied Preformulation, Product Design, and Regulatory Science*, 2e, 231–232. Hoboken, NJ: John Wiley & Sons.

Al-Achi, A., Gupta, M.R., and Stagner, W.C. (2023b). *Integrated Pharmaceutics: Applied Preformulation, Product Design, and Regulatory Science*, 2e, 467. Hoboken, NJ: John Wiley & Sons.

Anton, W.A. (1980). *Calculus.* New York: Wiley.

Begum, B.C., Johnson, P.R., and Stagner, W.C. (2018). Acyclovir chemical kinetics with the discovery and identification of newly reported degradants and degradation pathways involving formaldehyde as a degradant and reactant intermediate. *Int. J. Pharm.* 535: 172–179.

Coghill, A.M. and Garson, L.R. (ed.) (2006). *The ACS Style Guide: Effective Communications of Scientific Information*, 3e. New York: Oxford University Press.

Cole, T.J. (2015). Too many digits: the presentation of numerical data. *Arch. Dis. Child.* 100: 608–609.

Connors, K.A. and Mecozzi, S. (2010). *Thermodynamics of Pharmaceutical Systems: An Introduction to Theory and Applications*, 2e. Hoboken, NJ: John Wiley & Sons.

FDA (2019). Food and Drug Administration Office Regulatory Affairs Laboratory Manual Volume III Section 4 https://www.fda.gov/media/73535/download (accessed June 2021).

IUPAC (1993). *Mills I, Homann K, Cvitaš T, Kuchitsu K, Quantities, Units, and Symbols in Physical Chemistry*, 2e. Oxford: Blackwell.

IUPAC (2019). https://iupac.org/100/stories/the-new-si/ (accessed May 2021).

NIST (National Institute of Standards and Technology) (2019). SI Redefinition. https://www.nist.gov/si-redefinition (accessed June 2021).

Propagation of Error. Chemistry Libretextx. https://chem.libretexts.org/Bookshelves/Analytical_Chemistry/Supplemental_Modules_(Analytical_Chemistry)/Quantifying_Nature/Significant_Digits/Propagation_of_Error. (assessed Jan 2022).

Purdue University. http://chemed.chem.purdue.edu/genchem/topicreview/bp/ch1/units.html (accessed May 2021).

Selby, M. and Girling, B. (ed.) (1965a). *CRC Standard Mathematical Tables*, 303–304. Clevland: Chemical Rubber Company.

Selby, M. and Girling, B. (ed.) (1965b). *CRC Standard Mathematical Tables*, 308. Clevland: Chemical Rubber Company.

SHARE Operating System. https://en.wikipedia.org/wiki/SHARE_Operating_System (accessed April 2023).

SI Reference Point. Version 1.0 Beta. Bureau International des Poids et Mesures. https://si-digital-framework.org/SI/units?lang=en. 2025.

Stagner, W.C., Cerimele, B.J., and Fricke, G.H. (1991). Content uniformity: Separation and quantitation of sources of dose variation. *Drug Dev. Ind. Pharm.* 17 (2): 233–244.

Stein, K.S. (1977). *Calculus and Analytical Geometry*, 2e. New York: McGraw-Hill.
US Metric Association (2019). https://usma.org/international-system-of-units-si-reference-publications#locale-notification (accessed May 2021).
USP-NF. Rounding: https://www.uspnf.com/sites/default/files/usp_pdf/EN/USPNF/generalNoticesandRequirementsFinal.pdf. Significant Figures: http://ftp.uspbpep.com/v29240/usp29nf24s0_uspgn109.html (accessed April 2023).

Chapter 1 Problem Set

1.1 List the seven fundamental physical quantities.

1.2 What are the fundamental dimensions associated with the fundamental physical quantities?

1.3 What are the formulas for the derived physical quantities: area, volume, velocity, concentration, and density?

1.4 Define dimensional analysis and list three examples where dimensional analysis is useful.

1.5 Explain dimensional quantity algebra.

1.6 Explain conversion ratio. What is another name for conversion ratio?

1.7 Using dimensional analysis, determine the units of the drug complexation equilibrium constant, K_{11}, for the following equation when S_t, S_o, and L_t have units of moles/liter

$$S_t = S_o + (K_{11} S_o L_t)/(1 + K_{11} S_o)$$

S_t is the total molar concentration of the drug after complexation, S_o is the solubility of the drug before a complexing agent (ligand) is added, L_t is the total ligand molar concentration added to the drug liquid, and K_{11} is the 1 : 1 drug: ligand equilibrium constant. Note: to be dimensionally correct each side of the equation needs to have the same units and dimensionality.

1.8 Write the following numbers in E scientific notation

$$-200.12, 0.00347, \text{and } 69201748.$$

1.9 Discuss accuracy and precision and how they relate to a valid measurement.

1.10 How does uncertainty or measurement error relate to how data is reported?

1.11 How many significant figures are in
a) 109.0
b) 0.0190
c) 1090

1.12 Report the correct number of significant figures.

a) 3.40 $*$ 208.0

b) Antilogarithm of 10.800

1.13 Report to the correct number of significant figures.

$$21.1 + 2.037 + 6.13 =$$

1.14 What is the rounded number?

If the final number is reported to three significant figures, and the final number was 89.87, then this value would be rounded off to ________.

If the final number is reported to three significant figures, 89.83 is round off to ______.

1.15 Using exponent rules, calculate the buffer capacity for a buffer having a total buffer concentration of 1 M, $[H_3O^+] = 1.0 \times 10^{-4}$ M, and $K_a = 1.47 \times 10^{-4}$, using the general buffer capacity equation: buffer capacity $= 2.303C\dfrac{K_a[H_3O^+]}{\{K_a + [H_3O^+]\}^2}$

1.16 Convert the nature log form of ln(concentration) = 1.85 – 0.45 × time (h) to its base-10 logarithm form.

1.17 What is the derivative of the zero-order reaction, $[A] = [A_0] - kt$?

1.18 The diffusion coefficient (D) is a function of temperature (T), solution viscosity (η), and particle (r). A is Avogadro's number, and R is the gas constant:

$$D = \frac{RT}{\pi \eta r A}$$

When all variables change simultaneously, what is the equation for the diffusion coefficient? When the temperature is held constant what is the equation for the diffusion coefficient?

1.19 What is the definite integral of zero-order rate expression $\dfrac{dA}{dt} = -k_0A_0$?

1.20 Find the range, mean, median, mode, standard deviation, and % relative standard deviation for this data set.

90, 76, 53, 78, 88, 80, 81, 91, 99, 68, 62, 78, 67, 82, 88, 89, 78, 72, 77, 96, 93, 88, 88

2

Underlying Thermodynamic Physical Property Relationships and Definitions

2.1 Introduction

Matter is governed by the laws of thermodynamics. The science of equilibrium thermodynamics concerns itself with quantitative relationships between heat and other forms of energy such as potential, mechanical, chemical, kinetic, thermal, electrical, and nuclear energy and properties of systems at equilibrium. Equilibrium thermodynamics explains the transformation of heat into mechanical work and the opposite transformation between the two forms of energy.

Thermodynamics answers questions about the spontaneity of a reaction and how far a reaction will go before reaching equilibrium. On the other hand, thermodynamics does not deal with how fast a reaction will be, as that is the domain of kinetics. Thermodynamic concepts are based on the macroscopic properties of material involving large numbers of molecules. It provides the foundation for understanding the effects of macroscopic properties such as temperature, pressure, volume, and concentration on equilibrium. Thermodynamics is not dependent on understanding the molecular or quantum chemistry properties of the material or system.

The zeroth, first, second, and third laws of thermodynamics underpin the concepts that relate to the understanding of Gibbs free energy, equilibrium, equilibrium solubility, physical and chemical stability, ionic equilibria, phase equilibria, and surface and interface phenomena. As such, equilibrium conditions are not dependent on knowing the mechanism that resulted in the equilibrium state. Thermodynamics is concerned with the initial and final states of the system at equilibrium.

2.2 Temperature

Temperature is the degree of hotness or coldness of a body or environment. Temperature is a physical property that determines the direction heat will flow. As a body gets hotter, the particles of the body move faster and have higher kinetic energy. Intuitively, temperature is related to the energy of an object and vice-versa. The kinetic-molecular theory of gases assumes that ideal gas molecules are (1) constantly moving; (2) have negligible volume compared to the space they occupy; (3) have negligible intermolecular interactions; (4) undergo perfectly elastic collisions; and (5) have an average kinetic energy proportional to the ideal gas's absolute temperature. The average kinetic energy of a monatomic ideal gas is directly related to its temperature (Holzer 2010):

$$KE_{ave} = \left(m\, v_{ave}{}^{2}\right)/2 = (3/2)kT \tag{2.1}$$

Thermodynamics for Drug Product Design: A Guide for Pharmaceutical Scientists, First Edition. William Craig Stagner.

where KE_{ave} is the average kinetic energy of the gas in joules (J), m is the mass of the gas in kilograms (kg), v_{ave} is the average molecular speed of the gas in meters (m)/second (s), k is the Boltzmann constant in J/K^{-1}, and T is temperature in kelvin. Using dimensional analysis for Eq. (2.1), the right-hand side of the equation is mv_{ave}^2, which is kg m^2 sec^{-2}, which equals J, and the left-hand side is KE_{ave} in J. The average kinetic energy of a monatomic ideal gas is also related to the average particle momentum, p_{ave}, and temperature by:

$$KE_{ave} = (p_{ave}^2)/2m = (3/2)kT \tag{2.2}$$

where p_{ave} is the average momentum of the particle with m, mass and

$$p_{ave} = mv \tag{2.3}$$

From Eq. (2.1), it is important to note that at zero Kelvin or *absolute zero*, a body or sample of material will be at its lowest possible temperature, where the body has its lowest internal energy, and the kinetic energy cannot be removed. *Note: For thermodynamic discussions throughout this chapter and the remainder of the book, temperature is in degrees Kelvin unless otherwise stated.*

2.3 Energy, Heat, and Work

Heat (q) and *work* (w) are forms of *energy* and have units of J (N m or m^2 kg s^{-2}). As stated earlier, thermodynamics is based on describing the transformation from an initial state to a final state, such as from a solid to its melt. Such a process involves the flow of heat. Heat has been described as the mode of energy transfer and has the units of energy in J. Heat is also described as energy in transit. For example, if a hot object is put in contact with a cooler object, energy will flow from the hotter object to the cooler object. *Work* is defined by:

$$w = FD \tag{2.4}$$

where w is work in J F is the force in Newtons, N, and D is the distance or displacement in meters. *Force* is

$$F = ma \tag{2.5}$$

where m is mass in kg and a is acceleration in m s^{-2}.

The First Law of Thermodynamics (Chapter 3) will discuss in more detail the relationship between *internal energy, heat, and work*. Before we leave this section, it is important to understand that temperature is not a form of heat or energy. Temperature is directly proportional to the average kinetic energy of a gas or material (Eq. 2.1).

2.4 System, Surroundings, Boundary, and Universe

The First Law of Thermodynamics, also known as the conservation of energy law, says the change in the energy of the *universe* can be partitioned into the energy of a *system* and the energy of the system's *surroundings*. The law also states that the overall change in the energy of the universe is zero. This means that energy can neither be created nor destroyed; rather, it can only be transformed or transferred from one form to another. The conservation of energy is described by Eq. (2.6).

$$\Delta E_{Universe} = \Delta E_{System} + \Delta E_{Surroundings} = 0 \tag{2.6}$$

To take some liberty, on a cosmic scale, $\Delta E_{Universe}$ is the change of energy in the universe. For illustrative purposes, a ΔE_{System} could be considered the change in the energy of our sun. In this example, $\Delta E_{Surroundings}$ would be the energy change of everything in the universe except the sun.

It also follows that:

$$\Delta E_{System} = -\Delta E_{Surroundings} \tag{2.7}$$

This thermodynamic concept applies to thermodynamic processes independent of scale. Pharmaceutical product design scientists are often interested in material science and the physical characteristics and behavior of the drugs and formulation excipients. Properties like solubility, hygroscopicity, phase behavior, and compressibility are important properties affecting product design. These properties are also subject to Eqs. (2.6) and (2.7).

A *system* is a part of the physical universe that is a quantity of matter of fixed identity that we are interested in, such as a specified amount of gas, liquid, or solid. It is also the space or region of interest. Systems may be described as *homogeneous*, having a single phase like solutions or as *heterogeneous*, which contains multiple phases like suspensions and emulsions. A system is also described by its composition. Whatever matter or space that is outside the system is the *surroundings*. A *boundary* separates the system from its surroundings. The boundary is usually a surface that separates the system and surroundings. The *universe* is the combination of the system and surroundings. There are three types of systems depending on how the matter and energy interact at the boundary.

2.4.1 Isolated System

Neither matter nor energy can flow into or out of an *isolated* system. This means that the surroundings have no influence on the system, which is the case of the sample in a closed thermos. It is important to emphasize that a "system" is the part of the "universe" that we are interested in studying. The system is defined by the scientist. An example of an isolated pharmaceutical system is a sterile product. The goal of a sterile product is to keep the product inside its container closure system sterile and to keep environmental microbes out.

2.4.2 Closed System

Energy can flow into or out of a *closed* system, but matter cannot cross the boundary between the system and its surroundings. An example of a closed system is a stability study. Typically, a drug product is tested for its stability at a given temperature or temperatures in a temperature-controlled oven. The drug product is stored in a container closure system that protects it from outside moisture, oxygen, and other environment contaminants, but heat can flow into the container and reach equilibrium with the drug product. In this case, heat can flow into the closed container, but the product contents or matter cannot flow out of the container.

2.4.3 Open System

Both energy and matter can flow in and out of an open system. An example of an open system is an open beaker of drug solution on a hot plate. In this case, heat from the hot plate flows into the solution, and water can evaporate and be lost from the solution.

2.5 Macroscopic Properties and Intensive–Extensive Variables

Thermodynamics is based on the macroscopic properties of matter involving large numbers of molecules. The macroscopic properties of a system are easily measured by macroscopic variables such as temperature, volume, pressure, and composition. An intrinsic physical property or variable that is independent of the size of the system is called an *intensive* property or variable. Temperature, pressure, surface tension, density, melting point, viscosity, concentration, and average kinetic energy are intensive variables. Physical properties or variables that are additive and are quantity- or capacity-dependent are known as *extensive* properties or variables. Mass, moles, and volume are extensive variables.

An extensive variable can be converted to an intensive variable by dividing the extensive property by the mass or amount of material. Dividing volume (an extensive property) by grams gives *specific volume,* which is an intensive property. Likewise, dividing volume by the amount of moles of substance gives *molar volume,* which is an intensive property. In this chapter, you will also be briefly introduced to another extensive property called *internal energy* (U). Chapter 2 introduces other extensive forms of energy, such as entropy (S), enthalpy (H), and Gibbs free energy (G). Chemical potential (μ) will be introduced as an intensive property.

Work, which is a form of energy, is defined as the product of force and displacement or distance (Eq. 2.3), where force is an intensive variable and displacement is an extensive variable. Table 2.1 shows the relationship between energy and the product of specifically associated intensive and extensive variables.

2.6 Thermodynamic State Variables

The state of the system (see Section 2.4) can be defined completely by four measurable thermodynamic variables or *state variables*. These four observable state variables are temperature, pressure, volume, and composition.

2.6.1 Ideal Gas Law

The ideal gas law describes an ideal gas system by Eq. (2.8) and state variables

$$PV = nRT \tag{2.8}$$

where P = pressure, V = volume, n = number of moles, R = gas constant, and T = temperature in kelvin. A pure homogeneous gas or system can be defined by using two of the state variables since

Table 2.1 Relationship between Energy Form and Product of Intensive and Extensive Variables.

Energy form	Intensive (I) variable	Extensive (E) variable	Product I·E[a]
Thermal energy (heat)	Temperature (K)	ΔS (J K^{-1})	J
Work of expansion	Pressure (N m^{-2})	ΔV (m^3)	N m = J
Surface free energy	Surface tension (N m^{-1})	ΔA (m^2)	N m = J
Chemical energy	Chemical potential (J mol^{-1})	Moles (m)	J

[a] *Note:* dimensional analysis always results in J as the unit of energy Δ = change, S = entropy, V = volume, and A = area.

P, V, and T are not independent. Once the system is defined by these state variables, all other macroscopic properties, such as density, osmotic pressure, surface tension, viscosity, and others, are defined and set. At constant temperature, Eq. (2.8) has only two state variables that can change, P and V.

2.6.2 Dalton's Law

Dalton's Law states that in a mixture of gases confined to a rigid container with volume V maintained at a constant temperature, the total pressure exerted by all the gases equals the sum of all the *partial pressures* of the gases (Fermi, 1936). The *partial pressure* is defined as the pressure exerted by a single gas at the same temperature when it occupies the same volume of the mixture of gases. In other words, according to Dalton's Law, each gas behaves as an ideal gas and does not interact with each other and does not affect the pressure of the other gas.

2.6.3 Pharmaceutical Application of the Ideal Gas Law and Dalton's Law: Sterile Vial and Prefilled Syringe Manufacturing

Parenteral drug products are often sterilized in a steam *autoclave or sterilizer*. A typical steam sterilizer process employs steam at a temperature of 121.1 °C. Most commercial steam sterilizers measure chamber pressure in *bar*. Water vapor pressure at 121.1 °C is 2.0655 bar, or 206.55 kPa, or 2.0385 atm. The pressure buildup inside a vial or prefilled syringe is significant at 121.1 °C.

Example 2.1 Application of Ideal Gas Law: Increase in Pressure with Increase in Temperature

Let us look at steam sterilization in more detail and assume that a closed glass vial, a closed system, of an aqueous drug product is filled at 20.0 °C at 1.00 atm air pressure (101.325 kPa). At 20.0 °C, the vapor pressure of water is 2.3393 kPa. The total headspace pressure at 20.0 °C, according to Dalton's law, would be the sum of the air pressure (101.325 kPa) and the water vapor pressure (2.3393 kPa), giving a total vial headspace pressure of 103.664 kPa.

The total pressure in the vial headspace at 121.1 °C can be calculated using the ideal gas law and Dalton's law *to calculate the air pressure at* 121.1 °C and *knowing* the water vapor pressure at 121.1 °C.

The air pressure in the vial at 121.1 °C can be calculated from the ideal gas law, Eq. (2.8), where $P_1V_1 = nRT_1$ and $P_2V_2 = nRT_2$, where subscript 1 represents the pressure and volume at temperature 1, in this case, 20.0 °C. The subscript 2 represents the pressure and volume at a second temperature. In this example, T_2 is 121.1 °C. Let's assume that the vial liquid volume does not change significantly when heated from 20.0 to 121.1 °C.

Assuming the vial liquid volumes are equal, meaning that the headspace volume at each temperature stays constant, we can write the following equations for V_1 at 20.0 °C and V_2 at 121.1 °C.

$$V_1 = (nRT_1)/P_1 \text{ and } V_2 = (nRT_2)/P_2 \tag{2.9}$$

Since $V_1 = V_2$, we can place each of the functions of the right-hand sides of each equation equal to each other. Equations (2.9) can thus be rewritten to give Eq. (2.10).

$$(nRT_1)/P_1 = (nRT_2)/P_2 \tag{2.10}$$

P_2 can be determined by rearranging Eq. (2.10) and substituting the values for P_1, T_1, and T_2 which are known. First, we can divide both sides by nR, which are constants in this example. Then we can multiply both sides of the equation by P_2. and then divide both sides of the equation by T_1, giving Eq. (2.11).

$$P_2/P_1 = T_2/T_1 \tag{2.11}$$

Finally, Eq. (2.12) is obtained by multiplying both sides of Eq. (2.11) by P_1.

$$P_2 = P_1 \ (T_2/T_1) \tag{2.12}$$

We can confirm the correct dimension of the calculations by using unitary ratios and dimensional algebra.

$$P_2 = 101.325\ \text{kPa}\ (394.250\ \cancel{\text{K}}/293.15\ \cancel{\text{K}}) = \underline{136.27\ \text{kPa air pressure}} \tag{2.13}$$

Note: unless stated otherwise, all thermodynamic equations use Kelvin as the temperature scale, where K = °C + 273.150. Also note that P_2 has five SFs since the second temperature has five SFs, not six SFs, which is shown for the second temperature.

To find the total pressure the vial experiences, we use Dalton's law which states that the total pressure is equal to the sum of the individual pressures. In this example, the total vial pressure is the sum of the water vapor pressure and air pressure at 121.1 °C. At 121.1 °C, the water vapor pressure is 206.55 kPa (obtained from a water vapor pressure-temperature table), and the air pressure was determined earlier to be 136.27 kPa (Eq. 2.13). The total vial pressure at 121.1 °C is 342.82 kPa compared to 103.664 kPa at 20.0 °C.

This is an ~231% increase in headspace pressure, according to the calculation

$$[(342.82\ \cancel{\text{kPa}} - 103.664\ \cancel{\text{kPa}})/103.664\ \cancel{\text{kPa}}] \times 100\% = \underline{230.70\%} \tag{2.14}$$

Let us put the total headspace pressure into units we might be more familiar with.

One kPa is equivalent to 0.145 pounds per square inch (PSI). Therefore, using a unitary conversion ratio and dimensional analysis, the 342.82 kPa is equivalent to 49.7 PSI. Using a 0.145 PSI/1 kPa unitary conversion ratio gives the pressure in pounds PSI.

$$342.82\ \cancel{\text{kPa}}\ (0.145\ \text{PSI}/1\ \cancel{\text{kPa}}) = \underline{49.7\ \text{PSI}} \tag{2.15}$$

This pressure requires the vial rubber stopper to be held in place by an aluminum band. Otherwise, the rubber stopper would "pop out."

What about the case of a prefilled syringe where the barrel is typically not aluminum banded nor the needle opening plugged? How can one prevent the barrel from being ejected from the syringe and liquid being forced out the needle opening or end cap during steam sterilization? Before we tackle this problem, there is one additional complication that we have not discussed. We have not considered expansion of the liquid that occurs when the liquid is subjected to 121.1 °C. Liquid expansion is more critical with prefilled syringes because the headspace is smaller than vials.

One approach to address this problem is to create pressure inside the sterilizer that is equal to or slightly greater than the syringe's internal pressure. This requires a specially designed sterilizer called an *over-pressure sterilizer*. The calculation of the required sterilizer counter pressure must account for four factors: the expansion of the liquid at sterilizing temperatures, the decrease in the headspace caused by the liquid expansion, and the increase in air pressure and water vapor pressure resulting from the temperature increase.

2.6.4 Pharmaceutical Application of Dalton's Law: Determination of Total Gas Pressure of Pressurized Metered Dose Inhalers Formulated with a Mixture of Propellant Gases

Pressurized metered-dose inhalers have significantly improved the treatment of asthma, chronic obstructive pulmonary disease, and other lung diseases. Propellants play a key role in delivering the drug to the lung. The most common propellants are liquefied propellants like hydrofluoroalkanes, HFA-134a and HFA-227, and butane. Nonliquefied gases, carbon dioxide, and nitrogen have also been investigated. The advantage of using liquefied gases is that as long as there is some liquid in the inhaler canister, the exit gas pressure will remain constant. It is well known that the exit gas pressure is critical to the delivery of accurate and effective drug doses. Too high an exit pressure will cause the drug particles to impact the back of the throat. Too low an exit pressure can result in low shear at the exit nozzle, leading to unacceptably large aerosolized particles that cannot make their way to the lungs.

Dalton's law can be used to formulate optimized aerosol propellant exit pressures by mixing propellants that have different vapor pressures. Williams III and Liu (1998) showed that mixing HFA-134a and HFA-227 obeyed Dalton's law over a temperature range of 6–42 °C.

The partial vapor pressure of a single propellant is

$$P_i = P_i^{\circ} x_i \tag{2.16}$$

where P_i is the partial pressure of constituent i, P_i° the vapor pressure of the pure liquid ($x_i = 1$), and x_i is the mole fraction of constituent i in solution. For a two-component system, the total pressure (P_t) above the solution is the sum of the partial pressures of the two components

$$P_t = P_1 + P_2 \tag{2.17}$$

where P_t is the total headspace pressure, P_1 is the partial gas pressure of component 1, and P_2 is the partial gas pressure of component 2.

A 50 : 50 mixture of HFA-134a and HFA-227 at 20 °C would have a total pressure, $P_t = 479$ kPA. Using Dalton's law provides some flexibility in optimizing the exit pressure from a pressurized metered dose inhaler. Other excipients can also be added, such as butane or alcohol, to adjust the total exit pressure.

Example 2.2 Determination of the Total Pressure in a Metered Dosage Inhaler Using a Mixture of Hydrofluoroalkanes

HFA-134a has a vapor pressure of 559 kPa (81.1 PSI). At the same temperature and atmospheric pressure, HFA-227 has a vapor pressure of 399 kPa (57.9 PSI).

The total pressure of a 50 : 50 mixture of HFA-134a and HFA-227 is calculated using Dalton's law and Eqs. (2.16) and (2.17). Using dimensional algebra

$$P_t = P_1 + P_2 = 559\,\text{kPa}\,(0.5) + 399\,\text{kPa}\,(0.5) = \underline{479\,\text{kPa}}. \tag{2.18}$$

The product design scientist can obtain a specific pressurized metered-dose inhaler exit pressure by blending different propellants in specific ratios.

2.6.5 Pharmaceutical Application of Ideal Gas Law: Determination of a Pressurized Metered-Dose Inhaler when Stored at −20.0° Fahrenheit (−28.9 °C) in a Car Glove Compartment in the Middle of Winter

When metered-dose inhalers were first marketed, a number of product efficacy concerns were reported during the winter months from patients in the coldest regions of the US. At first, it was thought that the increased asthma-related problems were most likely due to the well-known fact that cold air can initiate bronchospasms in asthmatic patients. However, on further investigation, other factors were also postulated. It turned out that asthma patients commonly kept extra inhalers for emergency use in case their inhaler ran out of doses or they forgot to keep one with them. It was found that extra inhalers were commonly stored in car glove box compartments.

As mentioned in Section 2.6.4, pressurized metered-dose inhalers are designed to optimally function with a given propellant exit pressure. This exit pressure can change significantly at extreme low and high temperatures.

Example 2.3 Determination of the Exit Pressure of a Metered Dosage Inhaler when Stored at −20.0 °Fahrenheit (−28.9 °C) in a Car Glove Department
Let's assume that the inhaler was designed to have an exit pressure of 559 kPa at 20.0 °C.

What would the exit pressure be at −28.9 °C?

First, we convert −28.9 °C to Kelvin using K = °C + 273.150 or, in this case, 244 K. As mentioned in Section 2.6.4, HFAs generally meet the ideal gas assumptions, so we can substitute into Eq. (2.12) and calculate the pressurized metered-dose inhaler exit pressure at −28.9 °C.

$$P_2 = P_1(T_2/T_1) = 559\text{ kPa}\,(244\,\cancel{\text{K}}/293.2\,\cancel{\text{K}}) = \underline{466\text{ kPa}}.$$

Storing an extra inhaler for emergency purposes in a car glove compartment at −20.0 °C results in a <u>20.0% decrease</u> in its exit pressure [(Final pressure – initial pressure)/Final pressure × 100%].

The cause of the increased number of asthma-related problems was not fully resolved, but awareness of the effect of extreme temperature on exit pressure was increased.

2.7 Isothermal, Adiabatic, Isovolumetric, and Isobaric Thermodynamic State Processes

Isothermal processes are those conducted under constant temperature. Experimentally, isothermal processes are performed using a constant-temperature bath that acts as a temperature sink. An *adiabatic* process is one where there is no heat flow, which occurs in an insulated storage vessel such as a Dewar flask or thermos. An *isovolumetric*, also known as an *isometric* or *isochoric* process, is performed at constant volume, such as in a bomb calorimeter or other closed vessel that can withstand increased pressures such as a steam sterilizer. *Isobaric or isopiestic* means at constant pressure. Since atmospheric pressure is relatively constant at any given location and altitude, experiments carried out in open containers are performed under isobaric conditions.

Figure 2.1 shows a pressure versus volume graph at different temperatures indicating isothermal, adiabatic, isovolumetric (isochoric), and isobaric processes.

State 1 is an isothermal process starting at point "a" and following the constant temperature curve, T_3. The graph shows how pressure and volume change along the rectangular hyperbola

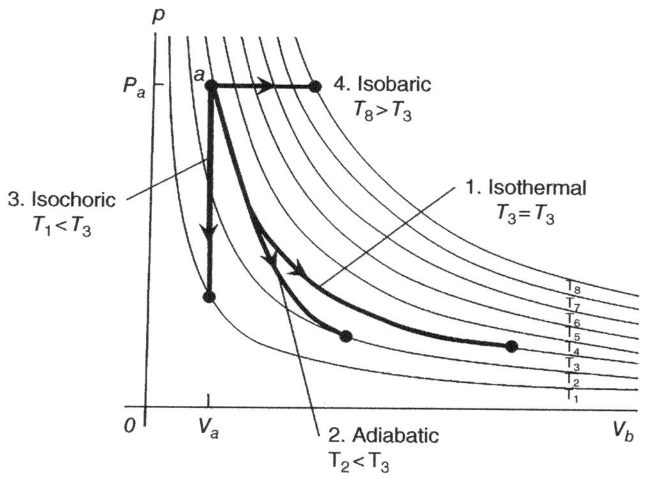

Figure 2.1 Pressure versus volume (PV) plot of isothermal, adiabatic, isovolumetric (isochoric), and isobaric processes. *Source:* Adapted from Young et al. (2011).

under constant temperature. For an ideal gas under constant temperature (isothermal), the product, PV, is constant. The work of isothermal expansion of an ideal gas can be written as:

$$-w = nRT \log (V_2/V_1) = nRT \log (P_1/P_2) \tag{2.17}$$

where subscripts 1 and 2 represent the initial and final conditions. A more detailed discussion regarding the work of isothermal expansion is provided in Section 2.10.1.

Isothermal processes occur where the system temperature can be controlled. Phase changes like melting, evaporation, and polymorph transitions are isothermal processes that usually occur at constant pressure.

State 2 is an adiabatic process. An adiabatic process requires an isolated system where the process occurs without transferring heat or mass to or from the surroundings. It is a nonisothermal process where the temperature changes from T_3 to T_2. Nuclear magnetic decoupling and demagnetization are adiabatic processes where no heat is lost or gained. Adiabatic calorimetry will be discussed in more detail.

State 3 illustrates an isochoric or isovolumetric process that results in a change in temperature of the system when the system pressure changes. The reverse is also true. That is, as the pressure changes, the temperature also changes. This is important in freeze-drying processes. The freeze-dryer has constant volume, so it is an isochoric process. Changes in pressure are used to provide rapid and accurate temperature changes within the freeze-dryer. This will be discussed in more detail later in this chapter.

State 4 is an isobaric process where the pressure of the system is held constant. Most lab experiments and manufacturing processes are carried out at constant pressure.

2.7.1 Pharmaceutical Application: Using an Adiabatic Calorimeter to Determine the Enthalpy of Solution of a Drug or Material of Interest

Discussion of this application offers the opportunity to introduce you to several important concepts that are covered in detail in Chapter 3, The Laws of Thermodynamics, and provides an opportunity to review some of the concepts already presented in this chapter.

In this section, we are interested in determining the *enthalpy of solution* of a drug. Enthalpy of solution is an energy of a reaction in which the drug dissolves and forms a solution with the solvent. The drug's change in internal energy is given by Eq. (2.19).

$$\Delta U = q + w \tag{2.19}$$

where ΔU is the internal energy of the system (in this case, the drug), q is the heat absorbed by the system, and w is the work done on the system. Calorimetry is the measurement of heat. For adiabatic solution calorimetry, the drug is considered the *system* because we are interested in the drug's heat of solution. The *solvent* in the adiabatic calorimetry functions as the surroundings. Constant pressure solution adiabatic calorimetry ("coffee-cup calorimetry") is performed at constant pressure, and Eq. (2.19) becomes

$$\Delta U = q_v \text{ and } \Delta H_{solution} = q_p \tag{2.20}$$

where $w = 0$ in Eq. (2.19), q_v is the heat measured, $\Delta H_{solution}$ is the drug's heat of solution, and q_p is the heat measured at constant pressure. Since we are interested in the $\Delta H_{solution}$, we use q_p. For adiabatic solution calorimetry, Eq. (2.21) is analogous to the energies discussed in Eq. (2.7), where:

$$q_{system} = -q_{surroundings} \tag{2.21}$$

Now that the theoretical background has been presented, we will discuss the adiabatic calorimetry experimental setup and design. A constant pressure adiabatic calorimeter may be designed using a Dewar flask that has an insulated lid and an opening for a thermometer. The calorimeter is designed to prevent heat flow into the Dewar flask, and the flask does not allow heat to escape the flask. When a known amount of water at a known temperature is added to a known amount of drug that is in the Dewar flask, the drug undergoes a dissolution/solubilization reaction or, simply, a solution reaction. This reaction can be an exothermic, endothermic, or athermic process. In the case of an exothermic process, the drug (the system) gives up heat to the solvent (the surroundings). This increases the solvent's kinetic energy. As noted in Eq. (2.1), kinetic energy is directly proportional to temperature. So, an exothermic reaction will lead to an increased solvent temperature, which is measured by the thermometer. An exothermic solution reaction is associated with a $-\Delta H_{solution}$. In an endothermic solution reaction, the drug system adsorbs heat from the surrounding solvent. This takes heat from the solvent, decreasing its kinetic energy, which leads to a decrease in the solvent's temperature. A $+ \Delta H_{solution}$ is associated with an endothermic solution reaction. An athermal solution reaction occurs with no or minimal heat flow. NaCl has a heat of solution of ~+3.9 kJ/mol, which is a modest endothermic solution reaction and is often considered athermal. The heats of solution for potassium nitrate and sodium hydroxide are ~+34.5 and −44.5 kJ/mol, respectively.

Example 2.4 Calculation of Ammonium Nitrate ($NH_4\ NO_3$) Enthalpy of Solution in kJ/mol

Ammonium nitrate ($NH_4\ NO_3$) is used in "instant ice packs" for first aid. $NH_4\ NO_3$ undergoes a strong endothermic reaction when it dissolves in water. The commercial product has two components: water in a plastic bag and $NH_4 NO_3$ in a container that can be broken to dissolve in the water. The temperature of the activated system reaches about 1.00 °C or 33.8 °F in 15 minutes.

To determine $NH_4\ NO_3$ enthalpy of solution in water, a Dewar flask is used as an adiabatic calorimeter. 3.88 g of finely triturated $NH_4\ NO_3$ is added to the Dewar flask. The temperature of 60 g of water is accurately measured and quickly added to the Dewar flask containing the $NH_4\ NO_3$ powder. The Dewar flask is gently shaken while the temperature of the system is measured.

The starting temperature of the water was 23.0 °C. The ending temperature of the solution was 18.4 °C.

The applicable equations to calculate $NH_4\,NO_3$'s heat of solution are provided. Equation (2.21) becomes

$$q_{reaction} = -q_{calorimeter} = -\left(m \cdot C_p \cdot \Delta T\right) \tag{2.22}$$

where m is the mass of the solution, C_p is the heat capacity of the solution, and ΔT is the temperature change (final temperature minus starting temperature). Let's assume that the heat capacity of the solution is the same as the heat capacity of water, 4.18 J/(g °C). The $NH_4\,NO_3$ enthalpy of solution can now be calculated by substituting into Eq. (2.22) and employing unitary ratios and dimensional analysis.

$$q_{reaction} = -\left[63.88\,\text{g}\left(4.18\,\text{J}\,\text{g}^{-1}\,{}^{\circ}\text{C}^{-1}\right)(18.4-23.0)\,{}^{\circ}\text{C}\right] = \underline{+\ \ 1.23\text{E}3\,\text{J}} \tag{2.23}$$

The amount of $NH_4\,NO_3$ undergoing dissolution is 3.88 g. The number of moles is calculated using unitary ratios and dimensional algebra.

$$3.88\,\text{g}\,(1\,\text{mol}\,NH_4NO_3/80.04\,\text{g}) = 4.84\text{E-}2\,\text{mol}\,NH_4NO_3 \tag{2.24}$$

The $NH_4\,NO_3$ enthalpy of solution in kJ/mol $NH_4\,NO_3$ is

$$[(1.23\text{E}3\,\text{J}/4.84\text{E-}2\,\text{mol}\,NH_4NO_3)](1\,\text{kJ}/1000\,\text{J}) = \underline{25.8\,\text{kJ/mol}} \tag{2.25}$$

2.7.2 Pharmaceutical Application: Freeze-Drying as an Isochoric or Isovolumetric Process

In the case of freeze-drying, pressure is considered the "*master lever*" for controlling the temperature inside a freeze drier and product vial. The critical temperature limit during freeze-drying is the product's eutectic melting temperature or collapse temperature. Eutectic melting and product collapse temperature are associated with two different physical systems. Eutectic melting is associated with a crystalline system. Product collapse occurs with an amorphous system. If the product temperature reaches the system's critical temperature, catastrophic product failure can occur due to product melting or product collapsing. Typically, during primary drying, the maximum upper product temperature is set at 2–3° below the eutectic temperature or 4–5° below the collapse temperature. If these "action limit" temperatures are exceeded, it is important to correct any temperature excursion quickly to avoid product failure. Minor adjustments in the freeze-dryer chamber pressure result in an almost instantaneous change in chamber and vial product temperature.

In comparison, trying to make a direct temperature adjustment is nearly impossible. In a large freeze-dryer, the mechanism of heat flow is primarily conduction and convection. At primary freeze-drying temperatures of −10 to −25 °C, heat flow by conduction is modest at best. Freeze-drying vacuum conditions are also a complicating factor. During primary freeze-drying, the vacuum is set to around ~10 to 40 mPa. At this vacuum, the number of molecules available to participate in convection heat flow is marginal.

Example 2.5 Calculate the Pressure Rise When a Freeze-Dryer Temperature Increases 1.00 °C

Let's calculate the temperature change that occurs in a freeze dryer under isovolumetric conditions when the pressure is changed. In this example, the eutectic product melting temperature is −10.0 °C. The *setpoints* for primary drying are −13.0 °C and 30.0 mPa. The *action temperature* is −12.0 °C. That is, the product temperature should not be greater than −12.0 °C. At this temperature, adjustment is required to lower the temperature to the set point.

When the temperature reaches −12.0 °C, what will be the observed pressure?

Equation (2.12), $P_2 = P_1\ (T_2/T_1)$, can be used to calculate this pressure change, where T is in kelvin, T_1 (the set point) = 260.2 K, T_2 (action temperature) = 261.2 K, P_1 = 30.0 mPa, and P_2 is the pressure at the action temperature:

$$P_2 = (T_2/T_1)\,P_1 = (261.2\,\cancel{\text{K}}/260.2\,\cancel{\text{K}})\ 30.0\,\text{mPa} = \underline{30.1\,\text{mPa}} \tag{2.26}$$

This pressure change of 0.1 mPa is well within the 0.25% sensitivity of capacitance manometers. Thus, the *action pressure* of an in-chamber capacitance manometer would be 30.1 mPa. Once the freeze dryer reaches this pressure, the pressure can be reduced to 30.0 mPa to immediately readjust the temperature to −13.0 °C.

2.8 Thermodynamic State Functions

The term *state function* is dependent on the state variables and current state of the system. A state function depends on the state of the system or matter that is defined by temperature, pressure, volume, and composition. By definition, a *state function* is a thermodynamic property whose value does not depend on the path taken to reach the *final state. Path functions* depend on the path taken to get to the final state.

The location (longitude and latitude) of a city can be thought of in an analogous way to a thermodynamic state function. There are many different driving routes one can take to get from New York City, NY (initial state) to Buies Creek, NC (final state). The location of Buies Creek is not dependent on the route driven or *path*. Therefore, the location of Buies Creek behaves in a similar way to thermodynamic state functions. Using this same analogy, the driving time and amount of gas will depend on the path driven. In this case, time and amount of gas do not behave like a state function. They are path functions. If we extend the analogy to driving back to New York City, the overall change in location is zero, no matter the path used to reach New York City. Typically, state functions are denoted by Latin/Roman upper-case or Greek lower-case letters, and path functions are denoted by Latin/Roman lower-case letters. The integral of a state function differential equation, along any path from the initial state to the final state, is the difference in the two states. This is illustrated for internal energy, *U*, by Eq. (2.27).

$$\int_{Initial\ (i)}^{Final\ (f)} dU = U_{final} - U_{initial} \tag{2.27}$$

The differential of the state function is called an *exact differential* and is represented by *d* (state function) like dU in Eq. (2.27). Temperature (T), pressure (P), volume (V), composition or mass, density (ρ), specific volume (ν), and energy [internal energy (U), enthalpy (H), Gibbs free energy (G), Helmholtz energy (A), and entropy (S)] are thermodynamic state functions. Heat (q) and work (w) are path functions, and their differentials are known as *inexact differentials* and are represented by $\delta(q)$ and $\delta(w)$, where δ indicates an inexact differential. These properties and their relationships will be discussed in more detail in later chapters.

2.9 Spontaneous Processes and Nonspontaneous Processes

A spontaneous process occurs naturally without help or intervention. As the volume of a container is expanded, the gas in the container will spontaneously fill the new volume. Heat always spontaneously flows from a material at a higher temperature to a material at a lower temperature. A thermodynamically unstable polymorphic form of a drug spontaneously converts to the more stable crystalline drug form. Polymorphic forms of a material have the same composition but have different crystalline structures. Thermodynamics can be used to predict the spontaneity of a process, but it cannot tell us how fast or necessarily when the process will occur. An interesting example of two polymorphic forms of carbon is diamond and graphite. It turns out that graphite is a more stable crystalline form of carbon. Diamond will spontaneously convert to graphite over time.

Example 2.6 Pharmaceutical Application of a Spontaneous Process: Iopanoic Acid Amorphous Drug Transformation to a Thermodynamically Stable Polymorphic Form
As mentioned earlier, polymorphs have the same composition but have different crystalline structures. Different drug polymorphs have different energies, which result in differences in physical–chemical properties. The higher energetic drug forms are less stable physically and chemically, are more soluble, have higher dissolution rates and bioavailability, and have a lower density, to note a few differences.

Once a polymorph is in solution, the physical–chemical differences disappear. The observed energetic differences are derived solely from the crystalline structure differences. Amorphous drugs also have the same composition, chemical makeup, and are not crystalline but are more glass-like. Amorphous materials have a higher energy, *escaping tendency* than high-energy polymorphic forms.

Stagner and Guillory (1979) studied the dissolution behavior of amorphous iopanoic acid, a high-energy polymorph, and the stable polymorph. They observed the spontaneous conversion of iopanoic acid's amorphous form to iopanoic acid's thermodynamically stable crystalline form. This conversion is shown in Figure 2.2.

In this figure, Form I is the low energy, stable crystalline form of iopanoic acid. Form II is the higher energy, less stable crystalline form. The amorphous iopanoic acid form is the highest energy and least stable iopanoic acid solid-state form. It is important to note that as a part of the dissolution experiment, excess solid was maintained throughout the dissolution process. It can be seen from this powder dissolution study that the higher the energy amorphous form, has a higher *escaping tendency* that leads to a higher dissolution rate and greater solubility.

The spontaneous transformation of the amorphous form can be observed. After about two hours, the concentration of the amorphous form starts to decrease exponentially and approaches the concentration of the stable Form I. A differential scanning calorimetry thermal profile of the excess solid in the amorphous dissolution vessel confirmed that amorphous material had converted to the stable Form I iopanoic acid.

The metastable Form II did not convert to the more stable Form I over the two study days.

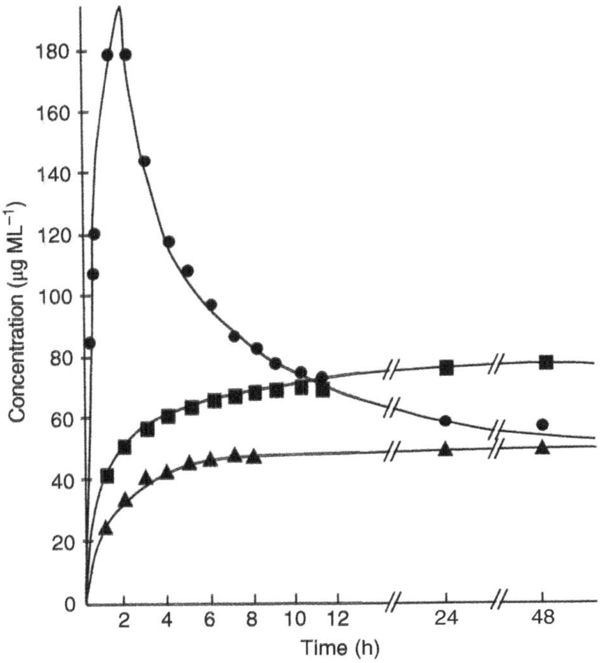

Figure 2.2 Spontaneous conversion of iopanoic acid amorphous form (closed circles) to stable Form I iopanoic acid (closed triangles) during powder dissolution. High energy polymorph Form II is shown as closed rectangles. *Source:* From Stagner and Guillory (1979)/with permission of Elsevier.

Example 2.7 Pharmaceutical Application of a Spontaneous Process: Catastrophic Spontaneous Conversion of Metastable Chloramphenicol to It's Stable Form

Serious clinical consequences of high-energy drug forms spontaneously converting to lower energy more stable drugs are well documented. One of the earliest and most devastating cases of clinical failure was reported for the antibiotic chloramphenicol palmitate (Aguiar et al., 1967). The stable, Form A polymorph was biologically inactive. The metastable Form B was developed as the commercial active pharmaceutical ingredient. Unexpectedly, numerous treatment failures were reported for chloramphenicol palmitate. Investigation into the failures revealed that chloramphenicol palmitate had unexpectedly and spontaneously converted from the metastable, active Form B to the stable, inactive Form A, leading to the clinical failures.

Example 2.8 Pharmaceutical Application of a Spontaneous Process: Spontaneous Conversion of Metastable Ritonavir to It's Stable Form

Thirty-one years after the catastrophic clinical failure of chloramphenicol, a similar account was documented. In this case, a higher energy form of ritonavir, an anti-AIDS compound, converted to a lower energy, less bioavailable form. This resulted in the product being taken off the market for an extended period of time. One of the biggest hurdles to solving the problem was the well-established fact that even trace amounts of a stable low energy form can initiate the spontaneous conversion of the higher energy form. This is analogous to adding a "seed" crystal to a supersaturated solution and having the crystals form spontaneously. The ritonavir lower energy form was found on the production line and other machines, which halted product manufacturing. Several partial remedies were found, but it was not until 2000, four years after the issue was first detected, that an acceptable product remedy received FDA approval (Ritonavir, Wikipedia, n.d.).

2.9.1 Author's Anecdote - Unexpected Polymorphic Conversion in Preparation of Toxicology Supplies

There are numerous accounts of stable lower energy, polymorphic forms contaminating not only surfaces but also the environmental air. I experienced this situation where environmental air was contaminated by a lower energy polymorph. In this case, the toxicology department called me and said they could no longer make up a toxicology solution at the original concentration I had formulated. For several months, the toxicology supplies had been prepared without issue. Suddenly, they could no longer prepare the solution at the original concentration without there being a significant portion of the drug precipitating. After some investigation, I hypothesized that the toxicology department had become contaminated with a lower energy polymorphic form. I also speculated that the department air was contaminated with the lower energy polymorph. I must admit, I had many doubters. To prove my hypothesis, I did two things. First, I invited one of the toxicologists to come directly to my lab first thing in the morning before going to the lab. When the toxicologist arrived, I prepared the solution in the original concentration without any difficulty or precipitation. The toxicologist was greatly surprised and marveled that the solution did not precipitate. Once the solution was prepared, the toxicologist and I went to the toxicology lab in another building. If the air in the toxicology lab was contaminated with the lower energy form, I posited that exposing the solution to the air in the toxicologist's lab would result in the solution precipitating. When I took the cap off the solution container, precipitation started almost immediately!

The pragmatic lesson here and in the above catastrophic examples is simple. Understand the thermodynamic principle(s) underpinning a pharmaceutical problem and carefully weigh the benefits versus the risks of making a decision that is not supported by thermodynamic principle(s). "Mother Nature" and thermodynamics hold the "trump" cards.

2.9.2 Nonspontaneous Processes

A nonspontaneous process does not occur naturally without help or intervention. Water does not flow uphill. A messy office does not naturally clean itself. Nonspontaneous processes require intervention that requires energy input, typically in the form of work. A motorized pump can push water uphill or against gravity. The messy office requires effort and energy to be cleaned up. So, nonspontaneous processes are possible to achieve, but energy of some form is required to help them happen.

2.10 Reversible Processes

A *reversible process* is one that is in equilibrium and can be changed or reversed at any moment by changing an independent variable by an infinitesimal amount. That is, the process is always at or near equilibrium. Examples of reversible processes are phase transitions such as melting and vaporization, physical adsorption, and solubility. A reversible process can be thought of as one that undergoes a quasistatic process that consists of a series of infinitely small shifts in the equilibrium. For an ideal gas, an infinitely small temperature change or pressure change will result in an infinitely small volume change to create a new equilibrium state. These infinitesimal changes are considered a quasistatic process. If an equal but opposite change in temperature or pressure is made on the system, the system will return to its original volume state.

Reversible systems can do the maximum amount of work for a given net change. The maximum amount of work can be obtained from the isothermal expansion of an ideal gas. In reality, no system is completely reversible without loss of energy. An example of a reversible process would be ice melting at zero degrees centigrade. Adding or taking away heat adjusts the equilibrium, and the ice will go from being more solid or more liquid.

2.10.1 Reversible Isothermal Expansion

It is possible to calculate the maximum work obtained from reversible isothermal expansion of an ideal gas. Figure 2.3 illustrates an isothermal idealized gas expansion process.

The gas is confined in a cylinder that is immersed in a constant-temperature bath. At the top of the cylinder is a weightless, frictionless piston that is covered with infinitesimally small weights (Figure 2.3a). If one of the weights is removed, the external pressure is decreased, and the volume inside the cylinder will expand infinitesimally. This quasistatic process can be repeated many times until the system arrives at a new final state or final volume in this case (Figure 2.3b). Figure 2.3 depicts an isolated system where there is no heat or material loss. The meticulous reader has probably noticed that Figure 2.3b has fewer gas molecules than Figure 2.3a. Good catch! Some artistic license was taken to emphasize the gas expansion that occurs with the volume increase.

This reversible process produces the maximum amount of work the system can perform. This amount of work is necessary to reversibly compress the gas to its original volume. If the process is not reversible and some of the energy of the system is lost to friction, then the system cannot return to its original volume because of the loss of energy.

Example 2.9 Calculate the Work Done as a Reversible Isothermal Expansion
Let's calculate how much work is done as a reversible isothermal expansion. In this example, one mole of ideal gas expands from 11.20 L to 22.40 L by a quasistatic reversible process, that is by infinitesimally small steps at 25 °C?

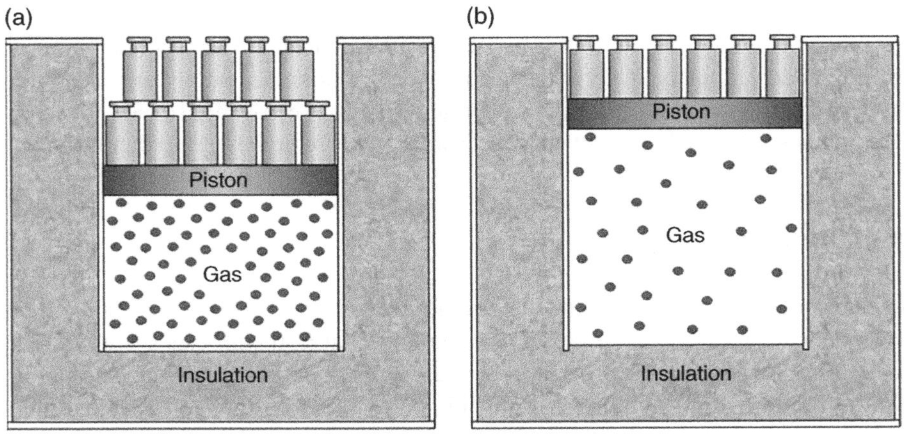

Figure 2.3 Reversible pressure-volume work of expansion under isothermal conditions. A-initial state and B-expanded state.

Note that work is being done by the system during the expansion process. Let us ask the question: "If work is being done by the system, does the system gain or lose energy?" By analogy, when you do work, do you gain or lose energy? Intuitively, for the gas to expand, the system needs to do work against its surroundings, and in doing work, the system would lose energy. The loss in the system's energy is symbolized by a negative sign $-w_{reversible}$ or loss in energy. The sign convention for heat and work will be discussed in more detail in Chapter 3.

In Table 2.1, isothermal reversible work of expansion, $-w_{reversible}$, is given by $P\Delta V$. The ideal gas law, $PV = nRT$, can be rearranged to $P = nRT/V$. Given that the work of expansion is $P\Delta V$, one can substitute for P and integrate over the infinitesimal changes in volume, dV, to give

$$-w_{reversible} = \int_{V1}^{V2} (nRT/V)\, dV \tag{2.28}$$

where V_1 and V_2 are the initial and final volumes, n is the number of moles of gas, R is the gas constant, T is the temperature in kelvin, and $-w_{reversible}$ is the energy lost by the system in doing work. The definite integral of dV/V (see integration rule 1.101) is $\ln V$ evaluated at the limits of V_1 and V_2. The work of expansion is (see logarithmic operation Eq. (1.47))

$$-w_{reversible} = nRT[\ln V_2 - \ln V_1] = nRT[\ln(V_2/V_1)] \tag{2.29}$$

Substituting the appropriate values into Eq. (2.29) gives:

$$-w_{reversible} = 1\,\cancel{\text{mol}}\,(8.314\,\text{J}\cdot\cancel{\text{K}}^{-1}\cdot\cancel{\text{mol}}^{-1})\,(298.15\,\cancel{\text{K}})(\ln 22.40/11.20) = \underline{1718\,\text{J}}$$

Therefore, the maximum work, $w_{reversible}$, performed by the system is $\underline{-1718\text{ J}}$.

It is important to note that dimensional analysis is critical to choosing the correct gas constant and gas constant units. Dimensional algebra also allows one to check for the dimensional consistency of an equation and associated property values. For Eq. (2.29), the final unit of joules is the correct unit for work and energy.

2.11 Irreversible Processes

An irreversible process cannot be reversed unless the surroundings are changed. Nonequilibrium thermodynamics is a branch of thermodynamics that studies processes like temperature and concentration gradients that occur at finite rates. Irreversible processes are the phenomena that nonequilibrium thermodynamics seeks to explain. De Groot and Mazur (1984) provide an introduction to nonequilibrium thermodynamics for those readers interested in this topic. This book focuses on spontaneous, nonspontaneous, and reversible processes.

References

Aguiar, A.J., Krc, J. Jr., Kinkel, A.W., and Samyn, J.C. (1967). Effect of polymorphism on the absorption of chloramphenicol from chloramphenicol palmitate. *J. Pharm. Sci.* 67: 847–853.

De Groot, S.R. and Mazur, P. (1984). *Nonequilibrium Thermodynamics*. New York: Dover.

Fermi, E. (1936). *Thermodynamics*. New York: Dover.

Holzer, S. (2010). *Physics II for Dummies*. Indianapolis, IN: Wiley Publishing Inc. https://www.dummies.com/education/science/physics/how-to-convert-potential-energy-into-kinetic-energy/ Accessed Feb 2021.

Ritonavir. Wikipedia. https://en.wikipedia.org/wiki/Ritonavir. Accessed Feb 2021.

Stagner, W.C. and Guillory, J.K. (1979). Physical characterization of solid forms of iopanoic acid. *J. Pharm. Sci.* 68: 1005–1009.

Williams, R.O. III and Liu, J. (1998). Influence of formulation additives on the vapor pressure of hydrofluoroalkane propellants. *Int. J. Pharm.* 166: 99–103.

Young, H.D., Freedman, R.A., and Ford, A.J. (2011). *University Physics with Modern Physics*, 13e. Boston: Addison Wesley.

Chapter 2 Problem Set

2.1 What is the average kinetic energy of helium, neon, argon, krypton, xenon, and radon at 25.0 °C?

2.2 Work is defined as force times distance. Work is a form of energy and has units of joules. Using dimension algebra, show that force times distance equals joules.

2.3 Define isolated, closed, and open systems.

2.4 List four state variables.

2.5 Define intrinsic and extrinsic variables. List several intensive and extensive variables.

2.6 Forms of energy are products of intensive and extensive variables. What are the intensive and extensive variables for thermal energy, work of expansion, surface energy, and chemical energy?

2.7 Calculate the enthalpy of solution in joules per mole for sodium hydroxide when 50 g of NaOH is completely dissolved in 1000 g of water at 20.0 °C. Assume there is no heat loss and the heat capacity of the solution is the same as water, 4.148 J g^{-1} $°C^{-1}$.

2.8 In Section 2.7.1, Pharmaceutical Application: Using an Adiabatic Calorimeter to Determine the Enthalpy of Solution of a Drug or Material of Interest, identify the boundary between the system and surroundings.

2.9 Describe a thermodynamic state function. Name several thermodynamic state functions.

2.10 Describe a thermodynamic path function. Name several thermodynamic state functions.

2.11 Define a thermodynamic spontaneous process. Give several examples of spontaneous processes.

3

The Laws of Thermodynamics

3.1 Introduction

This chapter is an introduction to the classic laws of thermodynamics. Thermodynamics is the branch of science that deals with the relationships between heat and other forms of energy. It describes how thermal energy is converted to and from other forms of energy and how it affects matter. Classical thermodynamics describes macroscopic properties of a system such as its temperature, mass, volume, pressure, heat, and work. It is applicable to all physical and biological systems.

Classical thermodynamics also deals with systems in equilibrium. It is not concerned about the time it takes to reach equilibrium. Nor is it concerned about the rate loss of a drug or the rate of formation of degradation products. The extent of an equilibrium reaction can be determined, as can the spontaneous nature of the reaction. Classical thermodynamics equilibrium assessment is independent of assumptions and knowledge about molecular structure or mechanisms of reactions.

Classical thermodynamics developed slowly over a period of about 200 years, from the 1700s to early 1900s. In 1715, Daniel Fahrenheit, a German scientist, established the Fahrenheit temperature scale. Anders Celsius, a Swedish scientist, developed the Celsius temperature scale in 1742. Sadi Carnot is known as the "father" of thermodynamics. In 1824, this French scientist developed the Carnot Cycle and showed that heat cannot be completely converted to work. German scientists Julius Mayer (1842), Rudolf Clausius (1850), and Hermann Helmholtz (1860) advanced the idea of conservation of energy, which is now known as the first law of thermodynamics. About the same time, Rudolf Clausius introduced a new thermodynamic property that he named *entropy*. This concept became the second law of thermodynamics. In about 1885, Austrian physicist Ludwig Boltzmann formulated a probabilistic and statistical definition of entropy This unified statistical thermodynamics and the classical entropy definition of Clausius. At about this same time, William Thomson, also known as Lord Kelvin of Britain, introduced the idea of "absolute zero temperature." In 1906, a German scientist, Walther Nernst, developed the third law of thermodynamics, for which he received the Nobel prize in 1920. The third law incorporated the ideas of entropy and absolute zero temperature. Finally, in 1935, Ralph Fowler, an English scientist, completes the four laws of classical thermodynamics by delineating the zeroth law. The zeroth law comes full circle and develops the notion of "hotness" and heat flow, employing temperature and temperature scales developed in the early to mid-1700s.

As an aside, the first Nobel Prizes were awarded in 1901. So many of the giants who first made the major contributions to thermodynamics were no longer living when the prize was first awarded.

Thermodynamics for Drug Product Design: A Guide for Pharmaceutical Scientists, First Edition. William Craig Stagner.

Farrington and Alberty (1975) and Connors and Mecozzi (2010) texts served as reference sources and were used to develop many of the discussions presented in this chapter. A brief description of the thermodynamic laws starts our introduction to the classic laws of thermodynamics.

The zeroth law defines the "equality of temperatures" and the "degree of hotness." When two systems are each in thermal equilibrium and have the same degree of hotness with a third system, the first two systems are in thermal equilibrium with the third system, which has the same degree of hotness.

The first law of thermodynamics is known as the law of conservation of energy. It states that energy can neither be created nor be destroyed, but it can flow from one object to another and can be converted to other forms of energy. The first law states that a change in a system's internal energy is equal to the sum of the heat added to the system from its surroundings and work done on the system by its surroundings.

The second law has two general components. The second law characterizes *disorder* as a quantity or property known as *entropy*. Entropy can be used to determine if a process will occur spontaneously in an isolated system. It states that spontaneous reactions go from a state of order to less order or more disorder. It also states that heat will spontaneously flow from a hot object to a colder object, not the reverse.

The third law of thermodynamics states that the disorder or entropy of a system approaches a constant minimum value as the temperature approaches absolute zero. The third law also states that the entropy of a perfect crystal is zero at 0 K. Imperfections in a crystal increase its disorder and entropy.

Other thermodynamic topics that are not covered in this book are: (1) statistical thermodynamics, which is another branch of thermodynamics that considers the individual state of particles such as atoms and molecules and their interactions; (2) nonequilibrium thermodynamics, which deals with systems having temperature and concentration gradients; and (3) irreversible processes, which cannot be reversed without producing a change in the surroundings.

3.2 The Zeroth Law of Thermodynamics: Thermal Equilibrium and Temperature

The "zeroth" law of thermodynamics was not established until after the first, second, and third laws were established. The zeroth law resulted from a need to develop the concepts of temperature and thermal equilibrium. The zeroth law describes heat flow and the establishment of an equilibrium thermal condition. If two closed systems having different degrees of *hotness* are brought in contact with each other, property changes take place in both systems. At thermal equilibrium, there are no further property changes. At thermal equilibrium, the hottest of both systems are the same, and they are at the same temperature.

The zeroth law can now be stated. Considering three systems, *A, B,* and *C*, it can be shown experimentally that if *A* is in thermal equilibrium with *B* and *B* is in equilibrium with C, then *A* and *C* are in thermal equilibrium.

As noted earlier, system hottest is related to system temperature. It follows then that if two systems are in thermal equilibrium and they have the same hotness, they will have the same temperature.

It is also understood that once the system reaches a point of thermal equilibrium, its other physical properties will remain unchanged as well. This law is the basis of conducting experiments under *isothermal* conditions. For example, immersing a glass beaker containing a solution in a constant-temperature water bath is a direct application of the zeroth law.

Incidentally, the United States Pharmacopeia has five temperature classifications. These terms are cold, cool, room, warm, and excessive heat. Any temperature not exceeding 8 °C is classified as a cold temperature. For example, the refrigerator's temperature is usually maintained at 4 °C. A cool temperature is defined to be between 8 °C and 15 °C. Temperatures between 15 °C and 30 °C reflect room temperature readings. In reference textbooks, room temperature is often stated to be 20 °C or 25 °C. A warm temperature is classified as temperatures between 30 °C and 40 °C. Any temperature that exceeds 40 °C is labeled as excessive heat.

3.3 The First Law of Thermodynamics: Conservation of Energy and Internal Energy as a State Function

The statement of the first law of thermodynamics is an expression of conservation of energy and matter during a process or energy transfer. Energy is not created or destroyed during any process, energy flow, or transfer, but rather it is simply transformed from one form into another (Bartell, 2001; DeMeo, 2001). In addition, according to Einstein's Equation ($E = mc^2$; m = mass; c = speed of light in vacuum), matter itself is a form of energy (Bartell, 2001). Therefore, in an isolated system where there is no heat or material flow from the system to its surroundings or vice versa, the total energy of the system is constant. It is important to emphasize that the first law does not consider spontaneity and the direction by which the process proceeds.

As noted in Section 2.4 and brought forward here for convenience, the First Law also states that the overall change in the energy of the universe is zero. The conservation of energy is described by Eq. (2.6).

$$\Delta E_{Universe} = \Delta E_{System} + \Delta E_{Surroundings} = 0 \tag{2.6}$$

As noted in Chapter 2, on a cosmic scale, $\Delta E_{Universe}$ is the change of energy in the universe, ΔE_{System} is the change in the energy of a system in the universe, and $\Delta E_{Surroundings}$ is a change in the energy of everything outside the system.

It also follows that:

$$\Delta E_{System} = -\Delta E_{Surroundings} \tag{2.7}$$

Internal energy is a thermodynamic state function and an extensive property. That is, the change in internal energy only depends on its initial and final condition. Internal energy is also dependent on the extent of the system, including its number of moles, mass, and volume. The total *internal energy* (U) of a system can be thought of as the sum of the energies of all the individual particles in the system. As discussed in Chapter 2, Eq. (2.1), an ideal monatomic gas undergoes elastic collisions with no loss in energy or momentum, and the total internal energy of the system results from its average kinetic energy. The total energy of other bodies or systems can include not only kinetic energy but also vibrational, rotational, and electronic energy; bond energy; potential energy; mechanical energy; and others. If a system is made of n particles, then the total internal energy of the system can be estimated by:

$$U = \sum n_i \varepsilon_i \tag{3.1}$$

$$n = \sum n_i \tag{3.2}$$

where U is the total internal energy of a system, n_i is the number of particles of the ith component, and ε_i is the energy of the ith component. In an open system, exchanges between the system and its environmental surroundings can occur in terms of molecular or energy exchange, with both exchanges being relatively small. When these exchanges occur, the *change* in the internal energy can be estimated from Novak (2003):

$$dU = \sum \varepsilon_i dn_i + \sum n_i d\varepsilon_i \quad (3.3)$$

The combined Equations of Eqs. (3.1)–(3.3) represent the first law of thermodynamics in its molecular format.

In general, in an open system, the statement for the first law of thermodynamics is defined as:

$$dU = \delta q + \delta w + \mu dn \quad (3.4)$$

where dU is the internal energy change of the system, δq is the differential heat absorbed by the system, δw is the differential work done on the system, μ is the chemical potential, n is the moles of material, and μdn is the differential energy provided to the system by the addition of one particle. In a closed system, there is no exchange of matter in the system and dn is zero. Note that the values δq and δw are dependent on the path the system takes, and therefore their cyclic integral is not zero (see Section 2.8). Whereas internal energy is a state function and its cyclic integral is zero. For an isolated or closed system, Eq. (3.4) can be rewritten

$$\Delta U = q + w \quad (3.5)$$

Equation (3.5) is applicable only to *isolated and closed systems* as opposed to *open systems* where matter can flow in and out of the system (see Section 2.4 for a discussion of systems). Equation (3.5) tells us that the change in internal energy of a system is fully described by only two forms of energy, heat and work. The sign conventions for internal energy depends if heat is being added or taken away from the system and whether work is being performed by the system or on the system. The author has chosen the Eq. (3.5) sign convention because it is generally more intuitive and understandable. A practical way to think about Eq. (3.5) is:

1) If heat is absorbed by the system from its surroundings, the system gains energy, and q is positive;
2) If heat is given up by the system to its surroundings, the system loses energy, and q is negative;
3) If work is done by the system, the system loses energy, and w is negative;
4) If work is done on the system, the system gains energy, and w is positive.

Adiabatic work, where there is no heat exchanged with the surroundings, is given by Eq. (3.6). The negative sign tells us that the work is done by the system. In this case, the system expends energy, and energy is lost.

$$\Delta U = -w = P\Delta V \quad (3.6)$$

q in Eq. (3.5) is zero and the new state has a decreased internal energy. Since heat cannot be exchanged, the temperature of the system will decrease (see Section 2.7 and adiabatic work). This is consistent with Eq. (3.7), where the average kinetic energy (KE_{ave}) of a monatomic ideal gas system is proportional to the system's temperature.

$$KE_{ave} = U = 3/2(nRT) \quad (3.7)$$

Equation (3.7) shows that internal energy is equal to its average kinetic energy. Equation (3.7) shows the important relationship between internal energy and temperature. The internal energy for a diatomic ideal gas is

$$U = 5/2(nRT) \tag{3.8}$$

Using the ideal gas law equation, $PV = nRT$, and substituting PV for nRT in Eq. (3.7), it can be shown that the total internal energy of a monatomic system is related by

$$U = 3/2(PV) \tag{3.9}$$

Work of expansion is given by $P\Delta V$. For reversible isothermal expansion of an ideal gas, the internal energy of the final state is the same as the initial state, and $\Delta U = 0$. If $P = 0$ (expansion occurs in a vacuum), then no work is performed by the system, and $w = 0$. This leads to the notion that the change in internal energy of an ideal gas is independent of its volume under isothermal conditions and is written as the partial differential:

$$\left(\frac{\partial U}{\partial V}\right)_T = 0 \tag{3.10}$$

The change in internal energy of an ideal gas undergoing isothermal expansion of an ideal gas can be written as

$$\Delta U = 0 = q - w \tag{3.11}$$

where $q = w$, which tells us that the isothermal work done by the system is equal to the heat absorbed by the gas. Note the *work term* is negative because the system is expanding against its surroundings. In doing so, some of the energy in the system is expended, resulting in a decrease in system energy. Using the ideal gas law, $PV = nRT$, at constant temperature, P is inversely proportional to V, and it is reasonable to consider that the internal energy is independent of the pressure of the system. This is given by the partial derivative:

$$\left(\frac{\partial U}{\partial P}\right)_T = 0 \tag{3.12}$$

3.3.1 Example of Change in Internal Energy for a Liquid Undergoing Vaporization

Example 3.1 Conditions and Determination of Change in Internal Energy
In this example, we are interested in the overall change in the internal energy of one mole of water when it vaporizes at 1 atm pressure and 100 °C. The water is placed in a cylinder that has a theoretical weightless and frictionless piston. The whole apparatus is placed in a large heat reservoir at 100 °C. It is also assumed that the water vapor behaves as an ideal gas.

There are two components to the vaporization process. First, the water must absorb enough heat from the reservoir to create enough energy to separate the water molecules to allow them to vaporize. The second step of the vaporization process is the expansion of the water vapor at 100 °C. Being an ideal gas, the *increase* in the water vapor volume can be calculated using the ideal gas law.

Two heat flow processes occur in this example. The first heat flow is associated with the heat of vaporization, q, or enthalpy of vaporization, which will be discussed in Section 3.3.2. The second heat flow is the work of the expansion of the water vapor, w. The change in the internal water energy per mole can be written:

$$\Delta U/\text{mol} = q/\text{mol} + w/\text{mol} \tag{3.13}$$

Calculation of the Heat of Vaporization

It takes 539.7 cal g^{-1} to vaporize water at 100 °C or 373.15 K. The molecular weight of water is 18.02 g mol^{-1}. Using dimensional analysis (see Section 1.2), the heat of vaporization per mole of water is:

$$q/\text{mol} = \left(18.02\ \cancel{\text{g}}\,\text{mol}^{-1}\right)\left(539.7\ \text{cal}\ \cancel{\text{g}}^{-1}\right) = 9.725\text{E3 cal mol}^{-1} \text{ or } \underline{9.725\ \text{kcal mol}^{-1}} \tag{3.14}$$

This heat of vaporization is positive since the system, water, absorbs this amount of heat, and the water's internal energy is increased. Also note that the four SFs are correct.

Calculation of the Work of Expansion

Using the historical Standard Temperature and Pressure (STP) conditions, the water conditions are temperature 0 °C or 273.15 K, pressure 1 atm, and water vapor volume 22.41 L mol^{-1}. The work of expansion of the water vapor is

$$-w/\text{mol} = P(\Delta V/\text{mol}) \tag{3.15}$$

The work term is negative because the water vapor system expends energy for the gas to expand against the frictionless and weightless piston. The water vapor loses energy in this case, and the water's internal energy is decreased in this step.

We can use the ideal gas law to determine the ΔV, the increase in water vapor volume.

$$P_1 V_1 = nRT_1 \text{ and } P_2 V_2 = nRT_2 \tag{3.16}$$

P_1 and P_2 are equal to 1 atm. We solve for P_1 and P_2, which gives

$$P_1 = (nRT_1)/V_1 \text{ and } P_2 = (nRT_2)/V_2 \tag{3.17}$$

The right-hand side of both equations in (3.17) can be set equal to each other

$$(\cancel{nR}T_1)/V_1 = (\cancel{nR}T_2)/V_2 \tag{3.18}$$

Solving Eq. (3.18) for the V_2, which is the increase in water vapor volume from 22.41 L mol^{-1} at STP to 373.15 K is

$$V_2 = V_1 (T_2/T_1) \tag{3.19}$$

Substituting the values for V_1, T_2, and T_1 into Eq. (3.18) gives V_2/mol. The water vapor at 0 °C was given as 22.41 L mol^{-1}. This is substituted for V_1.

V_2/mol = (22.41 L mol^{-1}) (373.15 $\cancel{K}$/273.15 $\cancel{K}$) = $\underline{30.61\ \text{L mol}^{-1}}$ $\underline{\text{water vapor}}$ (3.19) at 100 °C.

Substituting V_2/mol and P = 1 atm into Eq. (3.15) gives:

$$-w/\text{mol} = (1\ \text{atm})\ \left(30.61\ \text{L mol}^{-1}\right) = 30.61\ \text{L atm mol}^{-1} \tag{3.20}$$

$$w/\text{mol} = \underline{-30.61\ \text{L atm mol}^{-1}} \tag{3.21}$$

which is the work of expansion.

From the ideal gas law, the work of expansion for one mole of gas ($n = 1$) is $P\Delta V = RT$. So, RT can be substituted for $P\Delta V$ in Eq. (3.15).

$$-w/\text{mol} = RT \tag{3.22}$$

Substituting the values into Eq. (3.22) gives

$$w/\text{mol} = RT = \left(0.08205\,\text{L atm}\,\cancel{\text{K}^{-1}}\,\text{mol}^{-1}\right)\ 373.15\ \cancel{\text{K}} = \underline{-30.61\,\text{L atm mol}^{-1}} \tag{3.23}$$

This value is consistent with the value in (3.21).

NOTE: To calculate the work of expansion in terms of cal or kcal, we use the "molar gas constant," 1.987 cal K^{-1} mol^{-1}, for R.

$$w/\text{mol} = RT = \left(1.987\,\text{cal}\,\cancel{\text{K}^{-1}}\,\text{mol}^{-1}\right)\ (373.15\ \cancel{\text{K}}) = \underline{-741.4\,\text{cal mol}^{-1}}$$
$$\text{or} - 7.414\text{E} - 1\ \text{kcal mol}^{-1} \quad \supset \tag{3.25}$$

Equation (3.13), $\Delta U/mol = q/mol + w/mol$, describes the change in the internal energy of a liquid undergoing vaporization. From Eq. (3.14), the heat of vaporization (q/mol) of one mole of water is 9.725 kcal mol^{-1}. The work of expansion from Eq. (3.25) is −7.414E-1 kcal mol^{-1}. Substituting the heat of vaporization and work of expansion into Eq. (3.13) gives the change in the internal energy of one mole of water that undergoes vaporiztion.

$$\Delta U/\text{mol} = q/\text{mol} + w/\text{mol} = 9.725\,\text{kcal mol}^{-1} + \left(-7.414\text{E-1}\,\text{kcal mol}^{-1}\right) = \underline{8.984\,\text{kcal mol}^{1}} \tag{3.26}$$

The total change in internal energy can be converted to J mol^{-1} using the unitary conversion quotient 4.184E3 J · $kcal^{-1}$ and dimensional analysis (see Section 1.2.5).

$$\left(8.984\ \cancel{\text{kcal}}\,\text{mol}^{-1}\right)\left(4.184\text{E}3\,\text{J}\,\cancel{\text{kcal}^{-1}}\right) = \underline{3.759\text{E}4\,\text{J mol}^{-1}} \tag{3.27}$$

Note: The historic STP conditions were changed by the International Union of Pure and Applied Chemistry in 1982 from temperature (273.15 K) and pressure (1 ATM or 101.325 kPa), and 1 mol of an ideal gas occupies ~22.4 L to temperature (273.15 K) and pressure (1 bar or 100 kPa) and mole of ideal gas occupies ~22.7 L.

3.3.2 Enthalpy: A Thermodynamic State Function

Enthalpy is a thermodynamic state function that is closely related to internal energy, U. Enthalpy is the internal energy measured under constant pressure or isobaric conditions when only pressure-volume work is done. Like internal energy, it is also an extensive property that depends on the extent of the system. More specifically, it is the heat measured under constant pressure. Most experiments are performed in a laboratory at atmospheric pressure, which is generally considered a constant pressure. Therefore, most laboratory heat measurements are enthalpy heat measurements. Common enthalpies important to pharmaceutical product design scientists are enthalpy of fusion, enthalpy of solution, enthalpy of vaporization, enthalpy of sublimation, polymorphic phase transformation enthalpies, glass or amorphous transition enthalpies, and others. Heat and enthalpy are often used synonymously, such as in the terms of heat of fusion or enthalpy of fusion. The use of *enthalpy* informs the reader that the heat measurement was done under constant pressure conditions.

Enthalpy of transformation is also known as *latent heat or latent enthalpy*. Latent heat occurs when a substance, a thermodynamic system, absorbs or releases heat without the substance's temperature changing. Examples are enthalpy of fusion and enthalpy of vaporization. For example, when a substance melts, the substance temperature remains the same during melting. This adsorbed heat is considered a *hidden* or *latent* form of energy because the temperature during the process is constant. A substance's *sensible heat,* on the other hand, is *sensed* because the energy absorbed or released results in a temperature change of the substance.

Enthalpy is related to internal energy. As mentioned earlier, enthalpy is heat or energy measured under constant pressure or isobaric conditions when only pressure-volume work is done. Substituting $P\Delta V$ for w in Eq. (3.5) gives $\Delta U = q - P\Delta V$. We can designate ΔU using initial (i) and final (f) states, giving

$$U_f - U_i = q_p - P(V_f - V_i) = q_p - PV_f + PV_i \tag{3.28}$$

Consolidating the initial and final terms and solving for q_p q_{oph} (heat under isobaric or constant pressure conditions) results in:

$$q_p = (U_f + PV_f) - (U_i + PV_i) \tag{3.29}$$

As can be seen, u_p is the difference of two quantities involving a state function, U_f and two state variables, P and V. As such, under constant pressure conditions and only pressure-volume work, q_p can be defined as a new state function called *enthalpy, H*.

$$H = U + PV \tag{3.30}$$

and

$$\Delta H = \Delta U + P\Delta V = q_p \tag{3.31}$$

where the heat of the system, $q_p = \Delta H$ under isobaric conditions involving only $P\Delta V$ work.

As noted earlier, there are a number of enthalpic processes or reactions that involve only $P\Delta V$ work under isobaric conditions. These processes are described as exothermic and endothermic processes. Melting and vaporization or evaporation are *endothermic* physical processes. To melt a crystalline substance, it requires heat or energy to disrupt and overcome the lattice structure of the material. When a system, or in this case, a crystalline drug absorbs heat, ΔH is positive, and the process is called an *endothermic process*, that is, heat is absorbed by the system. The process of evaporation requires heat or energy to overcome the intermolecular liquid bonds so that the molecules can escape the liquid and become vapor. Therefore, heat of vaporization is also an endothermic process. Crystallization and vapor condensation are *exothermic* processes because heat or energy is released by the system.

Example 3.2 Pharmaceutical Application Exploiting Vapor Condensation: Steam Sterilization

Steam sterilization is the most common technology used to produce sterile parenteral products. An autoclave is used to achieve a typical pharmaceutical steam sterilization cycle of 121 °C for 15 minutes. The water vapor pressure in the closed autoclave is 206.5 kPa at this temperature. The latent heat or enthalpy of vaporization is ~525.6 kcal mol^{-1} under these conditions.

How much energy does 1 kg of steam release when it condenses onto cooler vials present in the autoclave?

Since the latent heat of vaporization is an endothermic reaction, the superheated water absorbs 525.5 kcal mol^{-1} of energy. This "hidden" energy is released to the vials when the water vapor, steam, condenses on the cooler vials as an exothermic reaction. We can solve the above question using this heat of condensation, quotient ratios, and dimensional algebra.

$$1\,\cancel{\text{kg steam}}\,\left(1\text{E}3\,\text{g}\cdot\cancel{\text{kg}}^{-1}\right)\left[\cancel{1\,\text{mol}\,\text{H}_2\text{O}}\cdot(18.02\,\cancel{\text{g}})^{-1}\right]\left(525.5\,\text{kcal}\cdot\cancel{\text{mol}}^{-}\,\cancel{\text{H}_2\text{O}}^{-1}\right)=\underline{2.916\text{E}4\,\text{kcal}} \tag{3.32}$$

Note the answer has four SFs (see Section 1.7). One kg of steam at 121 °C and 206.5 kPa pressure releases 2.916E4 kcal of sterilizing heat. For perspective, the general recommended daily dietary caloric guideline is about 2,000 kcal day^{-1}. It would take 14.5 days to consume the number of calories of heat released by 1 kg of steam.

3.3.3 Enthalpy and Hess's Law

Hess's Law states that the total enthalpy change of a reaction is independent of whether the reaction takes place in one step or any number of steps. This law is corollary to the fact that enthalpy is a state function and only depends on the initial and final state of the system. Using Hess's law, the enthalpies of individual reaction steps can be algebraically combined, which are equivalent to the overall heat of reaction. All the reaction steps must be at the same experimental conditions, and all reaction equations must balance. Hess's law makes it possible to determine the overall enthalpy of a reaction when it is difficult or impossible to determine the enthalpy from a single reaction.

In Chapter 6, Drug Solubility Equilibrium, we will use Hess's law to break down the heat of solution into three steps. The ability to evoke Hess's law to explain the drug solution process by breaking it down into dissolution, solubilization, and mixing processes illuminates seemingly contradictory enthalpy measurements.

Example 3.3 Pharmaceutical Application: Using Hess's Law to Determine the Neutralization Enthalpy When NaOH Is Neutralized with HCl

Although NaOH-HCl neutralization enthalpy can be measured in one step by calorimetry measurements, the purpose of this example is to show how Hess's law can be applied to other measurements that are more difficult or impossible to achieve.

Direct Measure of NaOH-HCl Neutralization Reaction Enthalpy

In this example, equal molar quantities of NaOH solution and HCl solution are added together. The reaction equation is given in Eq. (3.33).

$$\text{NaOH}_{(aq)} + \text{HCl}_{(aq)} \rightarrow \text{NaCl}_{(aq)} + \textit{heat}\left(-57.46\,\text{kcal mol}^{-1}\right) + \text{H}_2\text{O} \tag{3.33}$$

where *heat* is on the right side of the equation because *heat* is a product of the reaction. The minus sign indicates the reaction is an exothermic reaction where heat is released or lost from the NaOH-HCl system. Adiabatic calorimetry *heat* of neutralization experiments shown in Eq. (3.33) reaction have been reported to be −57.46 kcal mol^{-1}.

Implementing Hess's Law to Determine NaOH-HCl Neutralization Reaction Enthalpy

The two reaction steps used to determine the *NaOH-HCl* neutralization reaction enthalpy are provided. Both reactions shown in Eqs. (3.34) and (3.35) are exothermic reactions, so the *heat* is a product and exothermic. The reported enthalpy values for reactions Eq. (3.34) and Eq. (3.35) are also provided.

Step 1:

$$NaOH_{(s)} \rightarrow NaOH_{(aq)} + \textit{heat} \ \left(-44.50\,\text{kcal mol}^{-1}\right) \tag{3.34}$$

Step 2:

$$NaOH_{(s)} + HCl_{(aq)} \rightarrow NaCl_{(aq)} + \textit{heat} \ \left(-102.1\,\text{kcal mol}^{-1}\right) \tag{3.35}$$

We can reverse Step 1 Eq. (3.34), which becomes

$$NaOH_{(aq)} + \textit{heat} \ \left(+44.50\,\text{kcal mol}^{-1}\right) \rightarrow NaOH_{(s)} \tag{3.36}$$

Reversing Eq. (3.34), shows the precipitation reaction for $NaOH_{(s)}$. For the case of NaOH precipitation, the reaction is an endothermic, so the *heat* term becomes a reactant, not a product, as shown in Eq. (3.36). The *heat* also has a "+" sign to show the endothermic nature of the precipitation process.

Now we can employ Hess's law and use algebraic cancellation for Eqs. (3.35) and (3.36) to get the equivalent reactants and products of Eq. (3.33) as shown below in Eq. (3.38)

$$\cancel{NaOH_{(s)}} + HCl_{(aq)} \rightarrow NaCl_{(aq)} + \textit{heat} \ \left(-102.1\,\text{kcal mol}^{-1}\right) \tag{3.35}$$

$$NaOH_{(aq)} + \textit{heat} \ \left(+44.50\,\text{kcal mol}^{-1}\right) \rightarrow \cancel{NaOH_{(s)}} \tag{3.36}$$

Since the $NaOH_{(s)}$ term is on both sides of Eq. (3.36), it can be cancelled, as indicated by the strikethrough. We sum the *heats* from Eqs. (3.35) and (3.36), equals

$$\begin{array}{l} -102.1\,\text{kcal mol}^{-1} \\ \underline{+44.50\,\text{kcal mol}^{-1}} \\ \underline{-57.6\,\text{kcal mol}^{-1}} \end{array} \tag{3.37}$$

$$NaOH_{(aq)} + HCl_{(aq)} \rightarrow NaCl_{(aq)} + \textit{heat} \ \left(-57.46\,\text{kcal mol}^{-1}\right) \tag{3.38}$$

This example demonstrates the use of Hess's law. Equations (3.33) and (3.38) are equivalent even though the paths to get to these equations were different. Individual reaction steps and their enthalpies were algebraically combined and shown to be equivalent to the overall heat reaction. The underlying principle of Hess's law is based on enthalpy being a state function. As a state function, enthalpic processes are not path-dependent.

3.3.4 Pharmaceutical Application: Thermochemistry Using Differential Scanning Calorimetry

Thermochemistry is the branch of chemistry that is concerned with the absorption and evolution of heat involved with various chemical and physical processes. The most common thermochemical physical processes pharmaceutical scientists are interested in are phase transitions such as vaporization/condensation, melting/freezing, sublimation, heats of solution/precipitation, and heat of dilution. Precise enthalpies and temperatures of transition can be determined by differential scanning calorimetry (DSC).

DSC measures heat flow and the difference in heat flow between the sample and reference as a function of time and temperature. It can be used to measure enthalpies of melting, crystallization, vaporization, and so on. It can be used to calculate compound purity and assess complex thermal phenomena such as multiple thermotropic events. A DSC thermogram plots *specific* power consumption in $J\,g^{-1}\,s^{-1}$ (heat flow per gram) or *specific* watts ($1\text{ watt g}^{-1} = 1\,J\,g^{-1}\,s^{-1}$) versus furnace temperature. A DSC cell cross-section is illustrated in Figure 3.1.

Since the DSC is run under constant pressure, the heat flow is a measure of the sample's/system's enthalpy. As shown in Figure 3.1, there are two thermoelectric discs. The reference disc is

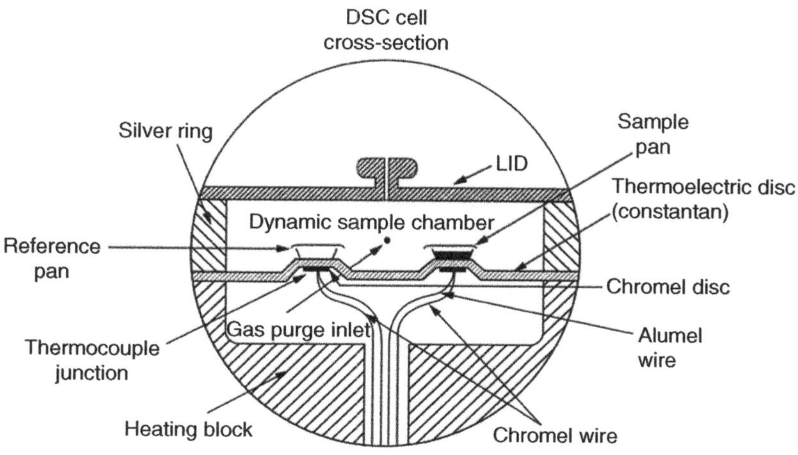

Figure 3.1 DSC cell cross-section. Adapted from Baine RL.

composed of chromel, and the sample disc is made of constantan. These discs form a thermocouple that generates a voltage directly related to the temperature difference between the reference and sample discs. Preheated inert gas purges the sample chamber. The sample heating rate can be set depending on the thermal event that is of interest and the type of thermal study. Heating rates up to 200 °C min^{-1} can be used. For isothermal studies, the heating rate can be set to 0 °C min^{-1}. Nitrogen is the most frequently used inert purge gas, which is typically set at a 40–50 mL min^{-1} flow rate. Helium is often used for subambient,−40 °C, temperature studies because it contains less moisture than nitrogen. Aluminum pans, which can be open to the environment or hermetically sealed, are most often used to hold the sample and serve as the reference pan with no sample.

Example 3.4 Pharmaceutical Thermochemistry Application: Using Temperature and Enthalpy to Determine the Ideal Solubility of Drugs

Figures 3.2 and 3.3 are DSC thermograms of lidocaine and prilocaine.

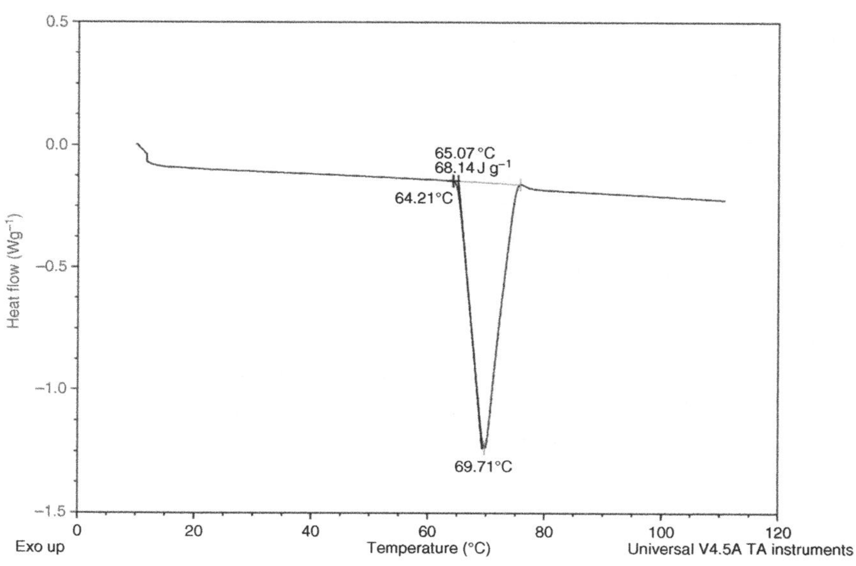

Figure 3.2 DSC of lidocaine melting point. The sample was heated at 5.0 °C min^{-1} in an open aluminum pan with nitrogen purge at 50 mL min^{-1}. Endotherm is "down." Heat flow in watts per gram is equivalent to J s^{-1} g^{-1}.

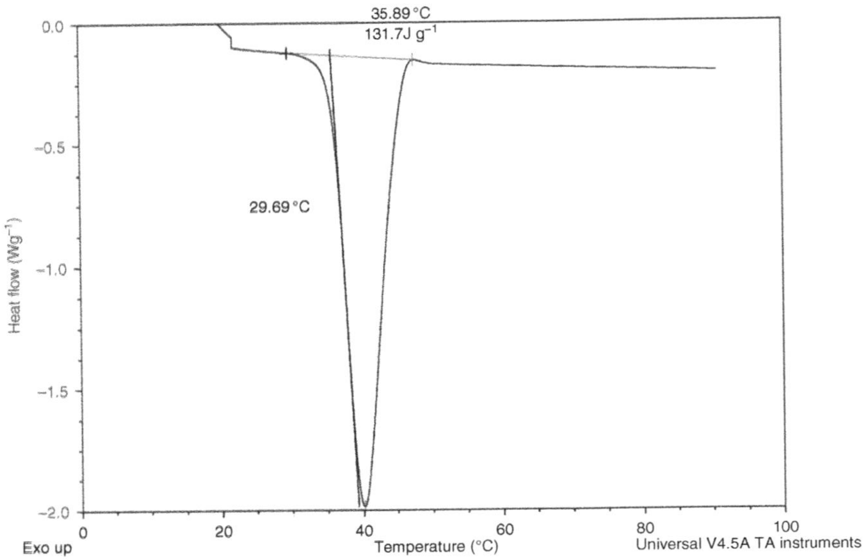

Figure 3.3 DSC of prilocaine melting point. The sample was heated at 5.0 °C min^{-1} in an open aluminum pan with nitrogen purge at 50 mL min^{-1}. Heat flow in watts per gram is equivalent to J s^{-1} g^{-1}. Endotherm is represented as "down."

As seen in these figures, melting is shown as an endothermic process. The heat flow is measured as J s^{-1} g^{-1} (specific enthalpy of fusion per s) or watts g^{-1}. The *extrapolated onset of the melting point method* is used to report the melting points. This method has been shown to be less sensitive to differences in heating rate and sample quantity. The melting process is completed at the peak temperature. The DSC melting points (mp) for lidocaine and prilocaine are 65.05 °C and 39.85 °C, respectively. The specific enthalpy of fusion, ΔH_{sf}, is the area under the melting curve divided by the sample weight. Lidocaine's ΔH_{sf}, is 68.14 J g^{-1}. Prilocaine's ΔH_{sf}, is 131.73 J g^{-1}. ΔH_f is known as the molar heat of fusion and can be calculated from the specific heat of fusion by knowing the material's molecular weight.

Both the mp and ΔH_{sf} are related to the strength of the drug's intermolecular bonds and its organizational structure. A higher mp indicates the need for increased kinetic energy to disrupt the drug's intermolecular bonds and start the solid-to-liquid transition. A drug's higher fusion enthalpy means more energy needs to be adsorbed to break the crystalline lattice organization. Equation (3.58) shows that heat capacity, C_p, is directly proportional to ΔH. Heat capacity is discussed in more detail later in Sections 3.3.5–3.3.7. For this current discussion, it is important to know that C_p is the incremental amount of heat required to raise the temperature of an object 1 K. C_p is also a measure of molecular mobility. Comparing lidocaine's and prilocaine's melting point temperatures suggests that lidocaine requires more kinetic energy to disrupt its intermolecular bonds than prilocaine. That is, lidocaine seems to have stronger intermolecular bonds than prilocaine. On the other hand, prilocaine's almost twofold greater ΔH_{sf} and related C_p suggest prilocaine requires more energy and greater molecular mobility to overcome its crystalline organizational structure than lidocaine. Mercury 4.2.0 was used to generate the crystal packing structure of lidocaine and prilocaine. The crystal organization structure of prilocaine indicates that it has a continuous

H-bonding network along the C-plane, which leads to a smaller cell volume and denser crystalline packing compared to lidocaine. This denser packing and higher H-bonding structural network most likely explains prilocaine's much larger ΔH_{sf}. This illustrates the interplay between the melting point and enthalpy of fusion. The higher lidocaine mp indicates that more energy is required to break intermolecular interactions and initiate melting. The higher ΔH_{sf} of prilocaine suggests that once melting is initiated, a greater amount of energy is required to overcome the lattice organizational structure to complete the melting.

The melting point and ΔH_f can be used to estimate mole fraction solubility without performing more time-consuming solubility studies. Estimating the rank order solubility of development compounds in a time efficient manner can be important in lead drug candidate selection. A more detailed discussion of solubility will be presented in Chapter 6, Drug Solubility Equilibrium.

For now, the following equation can be used to determine the ideal solubility of a drug using the drug's melting point and molar enthalpy of fusion.

$$-\ln X_D = \frac{\Delta H_f}{R}\left[\frac{T_0 - T}{T_0\, T}\right] \tag{3.39}$$

where X_D is the mole fraction solubility of the drug, ΔH_f is the drug's molar enthalpy of fusion, T_0 is the drug's melting point, and T is the temperature of the solvent.

This ideal solubility equation is based on the concepts of an ideal solution, which will be discussed in more detail in Section 6.8. Some of the important ideal solution assumptions for this discussion are provided. The intermolecular forces between the solute and solvent are considered to be the same as those between the solute molecules and solvent molecules themselves. Another key ideal solution assumption is that the enthalpy of mixing is zero.

To use Eq. (3.39) the lidocaine-specific enthalpy of fusion ΔH_{sf}, obtained by DSC, needs to be converted to the molar heat of fusion, ΔH_f. Using a unitary conversion ratio of the lidocaine molecular weight and dimensional analysis gives ΔH_f.

$$\Delta H_f = 68.14\,\mathrm{J}\,\cancel{\mathrm{g}^{-1}}\left(234.34\ \cancel{\mathrm{g}}\cdot\mathrm{mol}^{-1}\right) = \underline{1.597\mathrm{E}4\,\mathrm{J\,mol}^{-1}} \tag{3.40}$$

Let's assume that we are interested in lidocaine's ideal solubility at 25.00 °C. We can now substitute in Eq. (3.39) with the appropriate quantities to determine the lidocaine ideal solubility.

$$-\ln X_D = \left(1.597\mathrm{E}4\ \cancel{\mathrm{J}}\ \cancel{\mathrm{mol}^{-1}}\right)\cdot\left(8.314\ \cancel{\mathrm{J}}\ \cancel{\mathrm{K}^{-1}}\cancel{\mathrm{mol}^{-1}}\right)^{-1}$$
$$\times\left[(338.20\,\mathrm{K} - 298.15\,\cancel{\mathrm{K}})\cdot(338.20\,\cancel{\mathrm{K}}\cdot 298.15\,\cancel{\mathrm{K}})\right]^{-1} = \underline{0.7629}$$

Multiply both sides by −1 and then take the antilog of both sides.

$$X_D = \text{Antilog}\ (-0.7626) = \underline{0.4663\ \text{mole fraction lidocaine}} \tag{3.41}$$

The ideal % w/w solubility of lidocaine (L) in water (W) can be determined as follows:

Step 1. Using a unity ratio and dimensional analysis, the weight of 0.4463 mol of L is

$$0.4463\ \cancel{\mathrm{mol}}\ \mathrm{L}\cdot\left(234.34\ \mathrm{g}\cdot\cancel{\mathrm{mol}^{-1}}\right) = \underline{104.6\,\mathrm{g\,L}} \tag{3.42}$$

Step 2. Given: the moles of W are 1 −0.4463 mol L or 0.5537 mol W. One mol of W weighs 15.999 g. The g of W is calculated using a unity ratio and dimensional analysis.

$$0.5537\,\mathrm{mol\,W}\cdot\left(15.999\,\mathrm{g\,mol}^{-1}\right) = \underline{8.858\,\mathrm{g\,W}}$$

Step 3. w/w % is the weight of solute divided by the weight of solution times 100%.

$$w/w\%L = [104.6\,g\,L/(8.858\,g\,W + 104.6\,g\,L)] \times 100\% = \underline{92.16\,w/w\%.} \tag{3.43}$$

The ideal solubility of L in W at 25.00 °C is 92.16 w/w%. The USP reports L to be poorly soluble in W. The USP defines poorly soluble as 1 part drug solubilized in 30–100 parts solvent. Assuming 1 part of L is soluble in 30 parts W, it can be shown that L's solubility in W is only ~3.2% w/w. A semi-quantitative determination of L's solubility in ethanol was 62.5% w/w. L appears to be acting more ideally in ethanol than in water. For an ideal solution, the energy input required to break the solute–solute and solvent–solvent intermolecular interactions is balanced by the energy released during the solute–solvent interactions. In this case, L shows a very large negative solubility deviation from the calculated ideal L-W solubility. This implies that L is significantly more nonpolar than W, which results in a nonideal solution. Positive and negative deviations from ideal solubility are discussed in more detail in Section 6.10.

Example 3.5 Pharmaceutical Thermochemistry Application: Characterizing the Highly Complex Thermal Behavior of Two Grades of Magnesium Stearate and Their Liquid Crystal Forms

Magnesium stearate (MgSt) is the most common excipient used in oral solid dosage forms (Perrault et al., 2011). Historically, MgSt has been shown to have significant thermal behavior variability. Recently, a DSC study (Haware et al., 2018) explored MgSt's highly complex thermal behavior, which may help explain some of the historical thermal variability reported. DSC and temperature-dependent small-angle X-ray diffraction confirmed that MgSt lacks a conventional melting point and undergoes six reversible thermal transitions. Four of the thermal events were attributed to reversible liquid crystal transitions.

MgSt's complex thermal behavior is shown in Figure 3.4. The thermotropic difference between two different MgSt grades is also shown.

The DSC experimental conditions for the above study were as follows: DSC Q200 TA Instruments; sample size ~5.5 mg; open Tzero aluminum pan; heating from −20 to 250 °C at 10 °C min^{-1}; cooling from 250 °C to −40 °C min^{-1} at 5 °C min^{-1}; and heat-cool-heat-cool-heat (HCHCH) cycle under nitrogen purge at 40 mL min^{-1}. Peak temperatures were reported because of the difficulty in creating flat baselines that are required for extrapolated onset thermal event temperature determinations.

Figure 3.4 shows the first heating, and the remaining CHCH cycles are highly reproducible. The first heating in Blue exhibited six highly reproducible thermal events for each grade (Table 3.1). The six thermotropic events included five endothermic events and one exothermic event. The remaining CHCH cycles showed six reproducible exothermic/endothermic conjugate pair peaks. Four of the conjugate pairs were attributed to different liquid crystal transitions.

This example demonstrates the use of a thermochemistry analytical method, DSC, to characterize a pharmaceutical material's thermotropic behavior. Unique thermotropic behavior can be used to develop quality control acceptance criteria. A material's thermotropic properties may also be highly correlated to its performance (Haware et al., 2014, 2015, 2016). The results of the above example can be used to establish a *quality control specification* for MgSt. This specification can be used in selecting material vendors. It can also be used as an incoming quality control test that is used to release the material for use in the manufacturing of a drug dosage form.

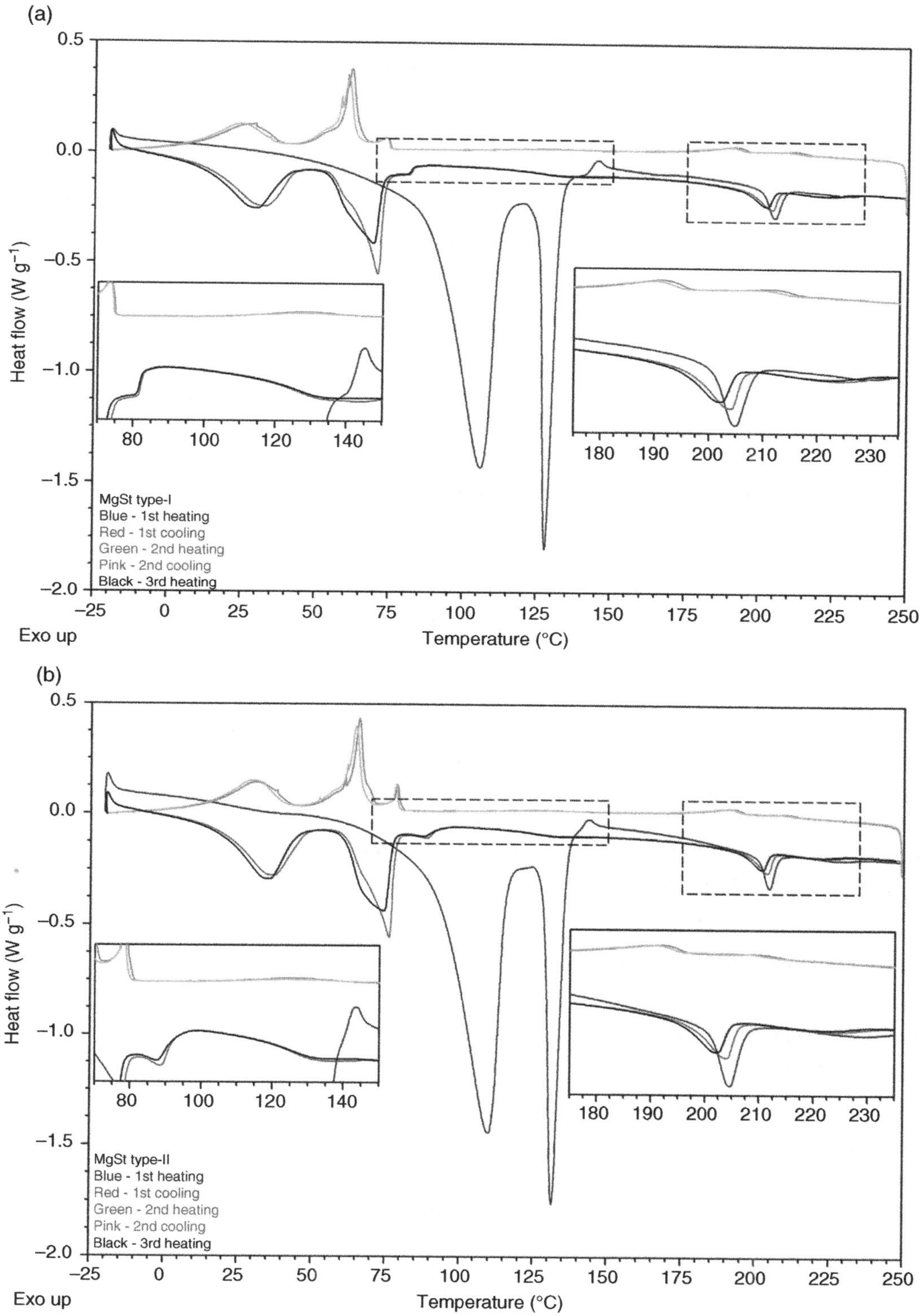

Figure 3.4 Magnesium stearate (MgSt), Type-I (a) (MgSt, NF Kosher Passover HyQual™) and Type-II (b) (MgSt, NF Vegetable Grade Impalpable Powder HyQual™). Blue- first heating; red- first cooling; green- second heating; pink- second cooling; and black- third heating. Endotherm is down. Heat flow in watts per gram is equivalent to joules per gram per second. *Source:* From Haware et al. (2018)/with permission of Elsevier. (Color figure will be available in the e-product.)

Table 3.1 Highly Reproducible MgSt Thermal Events on the First Differential Scanning Calorimetry Heating

Thermal events	Type-1 MgSt, °C	Type-II MgSt, °C
Endo	24.87 ± 0.21	33.25 ± 0.36
Endo	106.20 ± 0.06	109.76 ± 0.5
Endo	127.91 ± 0.06	131.18 ± 0.14
Exo	144.64 ± 0.95	143.13 ± 0.36
Endo	204.85 ± 0.29	204.54 ± 0.01
Endo	228.23 ± 0.46	227.82 ± 0.05

Type-I (MgSt, NF Kosher Passover HyQual™) and Type-II (MgSt, NF Vegetable Grade Impalpable Powder HyQual™). $N = 3$.
Source: Haware et al. (2018)/with permission of Elsevier.

3.3.5 Heat Capacity: A Thermodynamic State Function

As discussed in the temperature Section 2.2, the more heat or kinetic energy an object has, the hotter it gets. All substances increase in temperature when heated, but the magnitude of the temperature change for a given amount of heat/energy varies from substance to substance. Heat capacity, *C*, is the incremental amount of heat required to raise the temperature of an object 1 K and is given in Eq. (3.44).

$$C = dq/dT \tag{3.44}$$

The heat capacity can also be thought of as the ratio of heat absorbed to temperature change. The greater the heat capacity, the greater the amount of heat required to produce a one-degree temperature change. Heat capacity is also a measure of a material's molecular mobility. The larger the heat capacity, the greater a material's molecular mobility. Heat experiments are either carried out at constant volume, which is the case for a constant volume calorimeter, or at constant pressure, which is the most common experimental case.

3.3.6 Heat Capacity at Constant Volume

Heat capacity at constant volume can be derived from the first law of thermodynamics and Eq. (3.5), which is provided here for convenience, $\Delta U = q + \text{w}$. When *P-V* work is done by the system, this equation becomes

$$\Delta U = q - P\Delta V \text{ and } dU = dq - PdV \tag{3.45}$$

The negative sign indicates that work is done by the system on its surrounding, and the system gives up energy to the surroundings.

Solving for dq gives

$$dq = dU + PdV \tag{3.46}$$

The changes in internal energy for a chemically inert system are a function of temperature change (kinetic energy) and volume change (see Eq. 3.47). The partial differential (see

Section 1.12) for the change in internal energy as a function of temperature and volume can be written

$$dU = (\partial U/\partial T)_V\, dT + (\partial U/\partial V)_T\, dV \tag{3.47}$$

Substituting Eq. (3.47) into Eq. (3.46), gives

$$dq = (\partial U/\partial T)_V\, dT + (\partial U/\partial V)_T\, dV + PdV \tag{3.48}$$

Rearranging Eq. (3.48), results in Eq. (3.49).

$$dq = (\partial U/\partial T)_V\, dT + \left[(\partial U/\partial V)_T + P\right]\, dV \tag{3.49}$$

If the volume is constant, $dV = 0$ and

$$dq_v = (\partial U/\partial T)_V\, dT \tag{3.50}$$

where dq_v indicates the heat is measured at constant volume.

Dividing both sides by dT, gives

$$dq_v/dT = (\partial U/\partial T)_V \tag{3.51}$$

Equation (3.44), $C = dq/dT$ becomes Eq. (3.52) when the volume is held constant.

$$C_v = dq_v/dT = (\partial U/\partial T)_V \tag{3.52}$$

Equation (3.52) is the heat capacity at constant volume and gives the relationship of C_v to internal energy.

3.3.7 Heat Capacity at Constant Pressure

In Section 3.3.2, enthalpy as a thermodynamic state function that is closely related to internal energy was discussed. For now, we will assume that for a chemically inert closed system, the change in enthalpy is caused by a change in temperature at constant pressure and a change in pressure at constant temperature. The partial differential of the enthalpy change provides

$$dH = (\partial H/\partial T)_P\, dT + (\partial H/\partial P)_T\, dP \tag{3.53}$$

At constant pressure, the second term in Eq. (3.53) goes to zero. Therefore,

$$dH = (\partial H/\partial T)_P\, dT + 0 \tag{3.54}$$

and

$$dH = dq_p = (\partial H/\partial T)_P dT \text{ at constant pressure} \tag{3.55}$$

where dq_p indicates the heat is measured at constant pressure.

Dividing both sides by dT gives

$$dH/dT = dq_p/dT = (\partial H/\partial T)_P = C_P \tag{3.56}$$

This is the equivalent form of the definition of heat capacity provided in Eq. (3.44).

Note: the heat capacity at constant pressure, C_p, is always larger than C_V because pressure-volume work is done by the system when a system is heated at constant pressure. If C_p is constant over the temperature range of interest:

$$\int dH = \int C_p dT \tag{3.57}$$

and

$$\Delta H = C_p \left(T_{final} - T_{initial}\right) \tag{3.58}$$

Heat capacity can be expressed on a per gram or per mole basis. These heat capacities are known as *specific heat capacity* and *molar heat capacity*, respectively. Specific heat capacity and molar heat capacity are intensive physical properties.

On a molecular level, heat capacity is a measure of molecular motion. Heat capacity is a thermodynamic property of the material and is a state function compared to heat flow, which is dependent on path. Substituting Eq. (3.58), $\Delta H = C_P\,\Delta T$, into Eq. (3.31), $\Delta H = \Delta U + P\Delta V$, gives

$$C_P\,\Delta T = \Delta U + P\Delta V \tag{3.59}$$

which relates the material's molar internal energy change and change in molar volume to its heat capacity. For example, a polymer below its glass transition temperature will have a lower heat capacity, molar volume, enthalpy, internal energy, molecular mobility, higher viscosity, and be more brittle than above its glass transition temperature.

Heat capacity can also be used to understand complexation and binding mechanisms. For example, the binding of several drugs to two forms of cyclodextrin (β and γ) was studied by Todorova and Schwarz (2007). The change in C_P for the binding process of flurbiprofen (an anti-inflammatory agent) to β-cyclodextrin and γ-cyclodextrin was found to be -362 ± 48 J mol^{-1} K^{-1} and -25.1 ± 9.2 J mol^{-1} K^{-1}, respectively. Water, as the solvent, played a key role in the binding to cyclodextrin. From these data, the authors concluded that the structure of water was reorganized when the drug bound to the β form and not to the γ form of cyclodextrin.

3.3.8 Pharmaceutical Application: Thermochemistry Using Heat Capacity and Modulated Temperature Differential Scanning Calorimetry

Modulated temperature DSC (MTDSC) or simply modulated DSC (MDSC) is especially useful to explore the nature of thermotropic events that are hidden or undetected by conventional DSC. It is also employed to increase the sensitivity and resolution of conventional DSC. MDSC also directly measures the heat capacity of material compared to DSC, which measures heat flow. The same DSC cell is used for MDSC, but the programing involves superimposing a modulating heating–cooling cycle in conjunction with a constant linear ramp heating rate. The MDSC working equation is

$$dH/dt = C_p(dT/dt) + f(T,t) \tag{3.60}$$

where dH/dt is the heat flow signal or *total heat flow*, which corresponds to the same thermogram as conventional DSC, C_p is the isobaric heat capacity, dT/dt is the constant linear ramp heating or cooling rate, and $f(T,t)$ is the heat flow as a function of temperature and time. The *total heat flow* is the sum of what is called the *reversing heat flow* component that is associated with the $C_p\,(dT/dt)$ term and a *nonreversing heat flow* component, which is associated with $f(T,t)$ term. Nonreversing heat flow is determined indirectly by subtracting the reversing heat flow from the total heat flow.

The sample reversing heat flow is attributed to the *reversing* of the heat capacity as it mirrors the heating–cooling temperature modulation cycle. If the heat capacity mirrors the temperature modulation cycle and there are no latent heat contributions, the measured heat capacity is the equilibrium C_p. In general, reversing heat flows are due to heat capacity dependent transformation events like melting, glass transitions, liquid crystal phase transitions, and enantiotropic and sometimes monotropic polymorphic phase transitions.

Nonreversing heat flow is associated with a sample whose C_p does not mirror the temperature modulation cycle. This inability of the heat capacity to follow the modulation cycle is due to slow processes or nonequilibrium reactions that react slowly to temperature modulations. Nonreversing heat flows are generally related to kinetic (time)-related heats like chemical reactions, *cold* crystallization, crystallization, desolvation, dehydration, denaturation, and enthalpic recovery. Enthalpic recovery will be highlighted later. Cold crystallization is an exothermic crystallization process that occurs when a glassy or amorphous solid is heated and crystallizes. This phenomenon will also be discussed in more detail later.

Typically, the heating rate for MDSC is much slower than DSC. It is not uncommon to employ a 1.0 °C min^{-1} heating rate. The modulation cycle is generally set to have 4–5 modulations at half peak-width of an enthalpic event to achieve adequate deconvolution between reversing and non-reversing signals.

Example 3.6 Pharmaceutical Application: Using MDSC and C_p to Uncover Hidden Thermal Behavior

Figure 3.5 is a MDSC thermogram that displays commonly observed thermal events for an *amorphous* drug. Amorphous drugs or materials in general are noncrystalline substances that possess no or little long-range order. Amorphous drugs are thermodynamically unstable and have a natural tendency to transform to a more stable crystalline form. In addition, these drugs are more chemically unstable than their more stable crystalline counterpart. Having said that, amorphous drugs

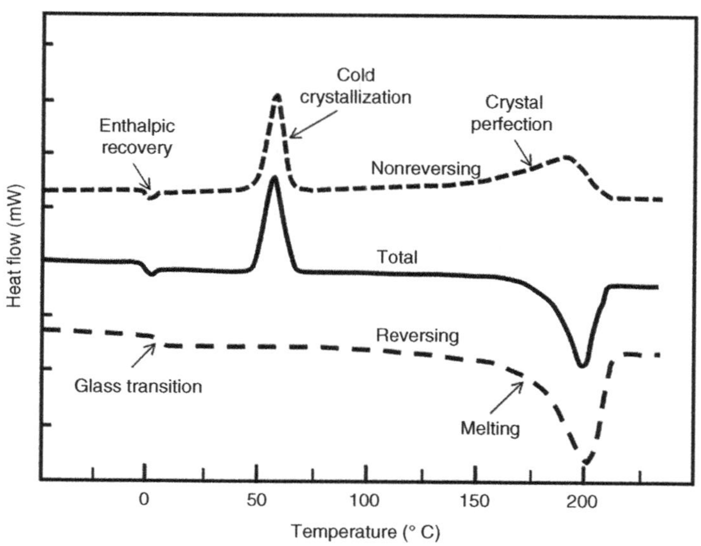

Figure 3.5 Typical MDSC thermogram of an amorphous drug substance. Exotherm is up. *Source:* Adapted from Newman and Zografi (2020).

have several pharmaceutically important attributes that make them attractive pharmaceutical active ingredients (API) when compared to their crystalline forms. Amorphous APIs have higher thermodynamic *activity* or *escaping tendency*, so they have higher solubility, dissolution rate, bioavailability, and others. It is important to point out that once the solid amorphous material is in solution, the solute will have a tendency to transform to the more stable form. When this occurs the solute's thermodynamic activity will be the same as the stable form.

The MDSC total heat flow thermogram is equivalent to a conventional DSC thermogram. The reversing and nonreversing heat flows are also plotted in Figure 3.5. From Eq. (3.58), $\Delta H = Cp\ \Delta T$, shows that heat capacity is directly proportional to enthalpy.

The total heat flow curve exhibits three thermal events. The first event around 0 °C shows an initial sigmoidal shaped change in heat flow indicative of a heat capacity change associated with a glass/amorphous solid phase transition. On closer inspection, there is a very slight endothermic peak at the end of the sigmoidal glass transition event. The reversing and nonreversing heat flow curves deconvolute the total heat flow of this thermal event. With a little imagination, one can see that the reversing heat flow curve shows a step-function change in C_p that is typical of glass transition phase changes. The glass transition temperature can be assessed by determining the temperature at the inflection point of the glass transition sigmoid curve. In this case, the sigmoid curve is more apparent in the total heat flow thermogram than the reversing curve.

The nonreversing heat curve separates the glass transition phase change from the slight endothermic peak seen in the total heat flow curve. The nonreversing curve shows a small separate endothermic peak, which is attributed to the amorphous powder's *enthalpic recovery* at the glass transition temperature. As mentioned earlier, glassy and amorphous materials are higher energy forms than their crystalline counterparts. Over time, at temperatures below its glass transition temperature, amorphous substances will naturally rearrange into a more compact, lower-energy form. This process is called *structural relaxation* and is associated with the aging of the amorphous material. During a glass transition phase change, this lower energy structurally relaxed material will absorb heat at the glass transition temperature to regain the energy it lost during structural relaxation aging. This process of a structural relaxed amorphous aged sample regaining lost energy is called *enthalpic recovery. Enthalpic recovery* is an endothermic process. The magnitude of the enthalpic recovery peak is a measure of the extent of aging. On occasions when significant loss of energy occurs over time, a single, relatively large DSC endothermic enthalpic recovery peak can obscure the sigmoidal glass transition curve. In this example, the reversing and nonreversing heat flow curves have deconvoluted these two processes, a glass transition and an enthalpic recovery, that are "hidden/obscured" in the total thermogram.

The second total heat flow transition is an exotherm that occurs around 45–50 °C. This thermal event is totally accounted for as a nonreversing heat flow behavior. This kinetic heat flow is credited to *cold crystallization*. Cold crystallization refers to a crystallization process that happens during heating a glass or amorphous powder above its glass transition temperature. Cold crystallization occurs above the drug's glass transition point and below its crystalline melting point. In some ways this terminology seems counterintuitive. It seems the term "heat crystallization" describes the process better. Not to belabor the point, but for clarification, *cold crystallization* is opposed to a crystallization process that occurs from the melted material on cooling.

The third total heat flow peak also has hidden thermal events that cannot be resolved by DSC. This total heat flow endotherm would most likely be interpreted as the melting point of the drug using conventional DSC. This interpretation would be correct, but it would not be complete. The MDSC nonreversing heat flow uncovers an additional time dependent thermal behavior associated with the drug crystal lattice imperfections rearranging to form a more perfect crystal. This process is like *heat annealing* material to obtain a more stable material form.

An example is annealing steel to harden it by heat treatment. The lack of a sharp melting point seen in the total heat flow and reversing heat flow thermograms supports the idea that the drug has a reasonable level of impurities. In this case, the nonreversing thermogram suggests that crystal imperfections are acting as a source of drug crystal impurity. At a temperature near the drug melting point, the crystal imperfections gain enough mobility to organize into a more perfect crystal structure. As seen with cold crystallization, this crystallization perfection process is also exothermic and gives up heat.

Example 3.7 Pharmaceutical Application: Using Thermochemistry to Understand Product Design Risks

Thermochemistry characterization of drugs is used to understand potential product design risks. Nanodrug delivery systems have gained significant interest and importance. Most nanodrug systems are amorphous or have some amorphous content. Figure 3.5 illustrates the use of MDSC in characterizing the thermal behavior of an amorphous drug. The glass transition occurs at about 0 °C and cold crystallization occurs around 45–50 °C. These thermal transitions would be considered critical product design risks since pharmaceutical processing operations may increase to temperatures above the critical glass transition and cold crystallization temperatures.

For example, the drug in Example 3.6 would not be a good candidate for a hot melt extrusion process, which routinely operates at temperatures higher than 100 °C and typically reaches temperatures of 150–160 °C. At these temperatures, the amorphous drug would be converted to a crystalline material, which would not have the amorphous pharmaceutical advantages of an amorphous nanodrug particle.

Drugs with low critical temperature transitions, like the API mentioned in this example, high-energy polymorphs, and low-melting drugs are not good candidates for manufacturing tablets. Interparticulate consolidation friction, particle fragmentation, shear localization during plastic deformation, particle bonding, and ejection friction can lead to exothermic processes that result in increased tablet temperatures. For example, at tablet production speeds, tablet temperatures have been shown to increase about 20-25 °C reaching temperatures 40–50 °C or more (Zavaliangos et al., 2008).

There have been numerous reports describing the effect of compaction on polymorphic transitions due to temperature rise and compression stresses (Ibrahim et al., 1977; Ghan and Lalla, 1992; Otsuka, 1993; and Wildfong et al., 2007).

Mao et al. (2016) reviewed the melting points of 3581 globally approved and discontinued drugs. Approximately 2% or 70 drugs had melting points between 25 °C and 55 °C, which is within the range of typical tablet compression temperature rises. 158 drugs had melting points between 25 °C and 60 °C.

The potential risk of low-temperature transitions requires careful consideration. Even partial transformation, such as partial melting or softening, would most likely lead to failed tablet batches.

3.3.9 Other Pharmaceutical Applications Using Heat Capacity

Heat capacity can also be used to understand complexation and binding mechanisms. For example, the binding of several drugs to two forms of cyclodextrin (β and γ) was studied by Todorova and Schwarz (2007). The change in C_p for the binding process of flurbiprofen, an anti-inflammatory agent, to β-cyclodextrin and γ-cyclodextrin was found to be -362 ± 48 J mol^{-1} K^{-1} and -25.1 ± 9.2 J mol^{-1} K^{-1}, respectively. Water, as the solvent, played a key role in the binding to cyclodextrin.

The decrease in C_p suggests a rearrangement in water structure during the binding reaction. From these data, the authors concluded that the structure of water was reorganized when the drug bound to the β form and not to the γ form of cyclodextrin.

3.4 Second Law of Thermodynamics: Entropy Is Unavailable to Do Useful Work and Does Not Decrease – Entropy Is a State Function

3.4.1 What is Entropy?

The terms heat, heat flow, and work are familiar to us through common daily experiences. We heat our homes, we experience a skillet becoming hot from the heat flowing from the stovetop, and we know it takes work to build a brick wall. Entropy, on the other hand, is not a term frequently used in our general daily living.

So, what is *entropy*? Basically, entropy is a measure of disorder. The more disordered a system is, the higher its entropy. We know what disorder looks like in daily living, but we rarely use the term entropy to describe it. An orderly room or office naturally tends to become messy and disordered, but we don't usually say the room's entropy increased.

Second Law is often defined in terms of different key entropy concepts. Clausius has been attributed to describing entropy in terms of disorder or randomness. He defined the Second Law by noting that in any natural process, the total amount of entropy as a measure of disorder or randomness which in the universe always increases or stays the same. It never decreases.

3.4.2 Entropy: A Statistical and Molecular Interpretation

As stated in Chapter 2, Section 2.1, "thermodynamic concepts are based on the macroscopic properties of material involving large numbers of molecules. It provides the foundation for understanding the effects of macroscopic properties such as temperature, pressure, volume, and concentration on equilibrium. Thermodynamics is not dependent on understanding the molecular or quantum chemistry properties of the material or system."

However, it is often helpful to think in terms of molecules to develop an intuitive feeling for various thermodynamic functions. Rudolf Clausius (1865) developed the macroscopic concept of entropy in terms of a spontaneous change. Clausius stated that for an irreversible process in an isolated system a spontaneous change proceeds in the direction of increasing entropy. Clausius' classic thermodynamic entropy will be discussed in detail later in this chapter. Boltzmann (1870s) provided a statistical interpretation of entropy, linking it to the number of microstates that correspond to a macroscopic state. He established the connection between classical macroscopic thermodynamic entropy and statistical entropy associated with particle physics.

The work of Connors and Mecozzi (2010) has been used as a guide to this development of the statistical and molecular interpretation of entropy.

Because the systems we are interested in have large numbers of molecules, Avogadro's number of 6.02214076E23 molecules per mole, statistical thermodynamics has been developed to describe the most probable state the molecules exhibit. A molecule is associated with a certain *microstate*. Each microstate has an associated entropy. Entropy, S, is an extensive property, and as such, the total entropy of the system, S_T, is the sum of its individual microstates, $S_1 + S_2 + S_n$ where S_1, S_2, and S_n are the entropies of the individual microstate.

$$S_T = S_1 + S_2 + S_n \tag{3.61}$$

Using statistical concepts and nomenclature, let Ω represent the probability of a molecule being in a particular energy state or microstate. Let Ω_1 and Ω_2 be equally probable microstate arrangements of one part of the system and a second part of the system. The entropy can be given as a function of Ω.

$$S = f(\Omega) \tag{3.62}$$

The number of the equally probable arrangements for the whole systems is $\Omega_1\,\Omega_2$.

$$S_T = f(\Omega_1\,\Omega_2) = f(\Omega_1) + f(\Omega_2) \tag{3.63}$$

Boltzmann concluded that S and Ω have a logarithmic relationship shown in Eq. (3.64).

$$S_T = k \ln \Omega \tag{3.64}$$

where S_T is the total entropy of the system, k is the Boltzmann constant, which is the molar gas constant divided by Avogadro's number to give the Boltzmann constant, which is the gas constant per molecule, and Ω is the total number of microstates of equal probability. Ω represents the total number of possible distinguishable microstates that are associated with a given macrostate. The basis of statistical thermodynamics is the *equal a priori probability postulate*, which states that for an isolated system in equilibrium, all accessible microstates have an equal probability of occurring.

The next example, follows that of Denbigh (1966) and Connors and Mecozzi (2010). It is a simplified analogy illustrating how microstate configurations might be thought about. This example shows that even though all microstates are equally probable, at equilibrium the state of the system is most likely to be one that has the most microstates.

Figure 3.6 shows the initial configuration of the crystals A and B. The crystals are placed side-by-side to allow the atoms from each crystal to freely diffuse under ideal conditions. The ideal diffusion conditions assume that atoms of either crystal are exactly the same size, have the same atom-atom interactions, and form an ideal solid solution.

This initial atomic arrangement, crystal A, has only one distinguishable microstate. Likewise, the initial configuration of crystal B has only one distinguishable microstate. This is denoted as

$$\Omega_{4:0} = 1 \text{ and } \Omega_{0:4} = 1 \tag{3.65}$$

where $\Omega_{4:0}$ signifies that all four atoms of A are on the left side of the dividing plane of the A and B crystals. $\Omega_{0:4}$ indicates that all four atoms of B are on the right side of the dividing plane between A and B crystals.

Suppose that one A atom diffuses to one of the four lattice points of B and one B atom moves to one of the lattice points of A. This would result in $4 \times 4 = 16$ distinct arrangements. This arrangement is noted as

$$\Omega_{3:1} = 16 \text{ and } \Omega_{1:3} = 16 \tag{3.66}$$

Now consider that two atoms of A diffuse into crystal B and two atoms of B move across the dividing plane into crystal A. This gives 4×3 arrangements on the left side and 4×3 arrangements on the right side. However, only $(4 \times 3)/2$ configurations are distinguishable on each side of the plane. This organization is described as by $[(4 \times 3)/2] \times [(4 \times 3)/2] = 36$ microstates. This is designated by Eq. (3.67)

$$\Omega_{2:2} = 36 \tag{3.67}$$

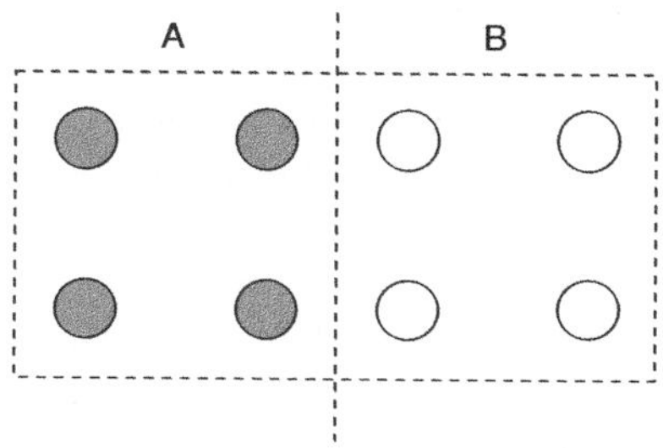

Figure 3.6 Two idealized crystals, A and B, with A atoms in the dark circles and B atoms in the light circles.

where there are 36 distinguishable configurations. Once the two crystals come to equilibrium, the atoms will have 70 ($1 + 1 + 16 + 16 + 36 = 70$) *distinguishable arrangements* or microstates. There are also five *distinct configurations* or macrostates or observable system states. Each configuration has equal energy and equal probability of existing. This also tells us that the probability of finding the crystals in their initial state is 1 in 70, and the probability of finding two A atoms in B and two B atoms in A is 36 in 70. This says that $\Omega_{2:2} = 36$ is the most probable configuration or macrostate.

From Eq. (3.64), we can calculate the entropy of the $\Omega_{2:2} = 36$ macrostate.

$$S_T = k \ln 36$$

Given: S_T is the total entropy of the $\Omega_{2:2}$ macrostate. k is the Boltzmann constant, 1.38E-23 J/K

$$S_T = k \ln 36$$

$$S_T = 1.38\text{E-}23\,\text{J/K}\,(3.583) = \underline{4.945\text{E-}23\,\text{J/K}} \tag{3.68}$$

As shown in this example, an *absolute entropy value* can be calculated for entropy. In addition, the Third Law states that the entropy of a perfect crystal at 0 K is zero. This is important. This sets a *zero point* for the entropy scale. At 0 K, a perfect crystal has one microstate and one macrostate. Internal energy and enthalpy have no zero point or absolute scale. Internal energy and enthalpy are expressed as changes in U and H. Entropy ranges from zero to a maximum value at a state of equilibrium. As stated above a perfect crystal has only one microstate and microstate at absolute zero. In this case, Eq. (3.64) becomes

$$S_T = k \ln \ \Omega = k \ln \ 1 = 0 \tag{3.69}$$

Equation (3.69) shows the entropy of a perfect crystal is zero at absolute zero.

3.4.3 Entropy: State Function That Explains Spontaneous Reactions

Section 2.9 in the last chapter discussed spontaneous reactions. For many years, scientists tried to predict spontaneous reactions that occurred naturally without intervention or help. Enthalpy was first thought to be the single property that would predict a spontaneous reaction. It was speculated that exothermic reactions would always be spontaneous events. Hypothesizing that enthalpy, and specifically exothermic heats, held the key to prediction created another dilemma because many endothermic reactions are spontaneous. Example 2.3 in Chapter 2 showed that ammonium nitrate spontaneously undergoes an endothermic solution process.

Rudolf Clausius (1865) created a new thermodynamic state function that explained spontaneous heat flow in only one direction. Clausius is quoted as saying, "Heat can never pass from a colder to a warmer body without some other change, connected therewith, occurring at the same time." He named the new function *entropy, S*. Clausius also observed that in any energy conversion process, some energy will always be lost as unusable heat. This unusable heat increases the entropy of the system and surroundings, making that portion of the total energy less available to do useful work. Clausius is also attributed to noting that an irreversible spontaneous change that occurs in an isolated system, will always proceed in the direction of increasing entropy. Entropy is a thermodynamic property of all material. The greater the number of possible system microstates, the greater the disorder and the higher the entropy. Clausius defined the classical change in entropy for a reversible system as

$$\Delta S = \Delta q_{rev}/T \tag{3.70}$$

where ΔS is the entropy change for a process, Δq_{rev} is the change in the amount of heat absorbed or released by the system during a reversible change, and T is temperature in kelvin. It is important to reflect on Eq. (3.70). It tells us that while entropy is closely related to energy, it is not considered an energy term itself. The units associated with entropy, J/K, clearly distinguish it from work, heat, or energy that have the unit of J.

As stated previously, work, heat, and energy depend on the process path, but it can be shown that heat for a reversible process divided by temperature in Kelvin is a state function. *It is important to note that entropy is a state function.*

The statement of the Second Law of Thermodynamics also describes the direction by which a process follows. As stated earlier, for an irreversible spontaneous and natural processes in an isolated system, its direction always proceeds toward a state of "*more randomness*" or an *increase* in *entropy, S*. For an isolated system, a reversible process at equilibrium, $\Delta S_{universe} = 0$ (Manninen, 2004; Flowers et al. 2023). The second law can be written:

$$\Delta S_{universe} = \Delta S_{system} + \Delta S_{surroundings} \tag{3.71}$$

The Second Law states that in any natural process, the total amount of entropy in the universe always increases or stays the same. Entropy never decreases. This is captured in Eq. (3.71a).

$$\Delta S_{\text{universe}} \geq 0 \tag{3.71a}$$

where:

$$\Delta S_{\text{universe}} > 0 \text{ for an isolated irreversible spontaneous process} \tag{3.72}$$

$$\Delta S_{\text{universe}} = 0 \text{ for an isolated reversible equilibrium process at its maximum} \tag{3.73}$$

$$\Delta S_{\text{universe}} < 0 \text{ is impossible for an isolated irreversible spontaneous process} \tag{3.74}$$

Equations (3.72) and (3.73) can be summarized by the following differential inequality equation:

$$dS \geq dq/T \tag{3.75}$$

where $dS > dq/T$ describes an irreversible spontaneous process and $dS = dq/T$ describes a reversible process.

In general, the entropy of a gas or vapor is greater than the entropy of a liquid, which is greater than the entropy of a solid.

$$S_{gas} > S_{liquid} > S_{solid} \tag{3.76}$$

3.4.4 Entropy: The Effect of Volume and Temperature Changes

Entropy increases with moles, volume, and temperature increases at constant pressure.

Effect of Volume Change on Entropy Change for a Reversible Process:

Starting with the entropy Eq. (3.77) and the internal energy Eq. (3.11), the relationship between entropy change and volume change will be derived. As a preview, this relationship is shown here in Eq. (3.83), $\Delta S = nR \ln V_{final}/V_{initial}$.

For a reversible process at constant pressure, Eq. (3.70) becomes

$$\Delta S = \Delta q_{rev,P}/T = \Delta H/T \tag{3.77}$$

Equation (3.77) can be derived from the first law, Eq. (3.11), which is restated below, where the change in internal energy of an ideal gas undergoing reversible isothermal expansion can be written as

$$\Delta U = 0 = q_{rev} - w \tag{3.11a}$$

In this case of isothermal expansion $q = w$, where $P\Delta V = w$. It follows that

$$q_{rev} = P\Delta V \tag{3.78}$$

At a given pressure, $P = nRT/V$. Substituting for P into Eq. (3.78) above gives

$$q_{rev} = nRT(\Delta V/V) \tag{3.79}$$

Dividing both sides by T gives

$$q_{rev}/T = nR(\Delta V/V) = \Delta S \text{ (from Eq.3.79)} \tag{3.80}$$

The differential form of Eq. (3.80) is

$$dS = dq_{rev}/T = dH/T = (nR/V)dV \tag{3.81}$$

If we integrate both sides of Eq. (3.81) considering an infinitesimal volume change, we get

$$dS = 1/T\int_{initial}^{final} dq_{rev} = nR\int_{initial}^{final} dV/V \tag{3.82}$$

$$S_{final} - S_{initial} = \Delta S = q_{rev}/T = \Delta H/T = nR \ \ln V_{final}/V_{initial} \tag{3.83}$$

Equation (3.83) shows the logarithm effect of a volume change on the entropy change for a system undergoing reversible *P-V* work at constant pressure. Equation (3.83) also shows that for this system at constant pressure, the enthalpy change can be used to calculate the change in entropy. As the volume increases, this increases the number of potential microstates, allowing for more disorder or randomness and greater entropy.

Effect of Temperature Change on Entropy Change for a Reversible Process: It is also possible to calculate the increase in entropy as a system's temperature is increased under constant pressure. From Eq. (3.56), $dq_p = C_pdT$ and substituting for dq_p in Eq. (3.70), $dS = dq_{rev}/T$ gives

$$dS = C_p\, dT/T \tag{3.84}$$

If C_p is independent temperature, integration of Eq. (3.84) gives

$$S_{final} - S_{initial} = \Delta S = C_p \ln T_{final}/T_{intial} \tag{3.85}$$

Equation (3.85) gives the entropy change for a system where the temperature is changed at constant pressure. Equation (3.85) shows that entropy increases with an increase in temperature, which is consistent with the notion that entropy increases as the randomness or disorder increases and the number of potential microstates increases.

3.4.5 Entropy of Mixing Ideal Gases

The mixing entropy change that occurs when two ideal gases are brought into contact with each other under constant temperature and pressure conditions is the same that results when an ideal gas is allowed to expand. Equation (3.83) is used to calculate the entropy of mixing, ΔS_{mixing}.

$$\Delta S_{mixing,1} = n_1R \ \ln V_{final}/V_{initial} = n_1R \ln{(V_1 + V_2)}/V_1 \tag{3.86}$$

and

$$\Delta S_{mixing,2} = n_2\, R \ln V_{final}/V_{initial} = n_2\, R \ln (V_1 + V_2)/V_2 \tag{3.87}$$

where $\Delta S_{mixing,1}$ and $\Delta S_{mixing,2}$ are the entropies of mixing for gas 1 and gas 2, n_1 and n_2 are moles of gas 1 and 2, V_1 and V_2 are the initial volumes of gas 1 and gas 2, and $V_1 + V_2$ is the final volume of the gas mixture.

Often times it is more meaningful to discuss the entropy of mixing in terms of mole fraction of a particular gas versus using gas volume. The next section derives the ideal gas mixing equations in terms of mole fraction.

The mole fractions of the two gases are

$$X_1 = n_1/(n_1 + n_2) \quad \text{and} \quad X_2 = n_2/(n_1 + n_2) \tag{3.88}$$

At constant temperature and pressure, the ideal gas law, $PV = nRT$, can be used to convert Eq. (3.88) into volume terms.

$$V_1 = (n_1 RT)/P \quad \text{and} \quad V_2 = (n_2 RT)/P \tag{3.89}$$

Mole fraction in terms volumes is

$$X_1 = V_1/(V_1 + V_2) \quad \text{and} \quad X_2 = V_2/(V_1 + V_2) \tag{3.90}$$

Equation (3.90) can be shown to be equivalent to Eq. (3.88) by substituting the volume values in Eq. (3.89) for V_1 and V_2 in Eq. (3.90).

$$X_1 = n_1 RT/P/[n_1 RT/P + n_2 RT/P] \tag{3.90a}$$

Now, collecting common terms and then cancelling common terms gives

$$X_1 = n_1 \cancel{RT/P}/[\cancel{RT/P}(n_1 + n_2)] = n_1/(n_1 + n_2) \tag{3.91}$$

which is the same as Eq. (3.88). The same algebraic manipulations lead to $X_2 = V_2/(V_1 + V_2)$.

From Eqs. (3.86) and (3.87), the entropies of mixing for gas 1 and gas 2 are

$$\Delta S_{mixing,1} = n_1 R \ln V_{final}/V_{initial} = n_1\, R \ln (V_1 + V_2)/V_1 \tag{3.86}$$

$$\Delta S_{mixing,2} = n_2\, R \ln V_{final}/V_{initial} = n_2\, R \ln (V_1 + V_2)/V_1 \tag{3.87}$$

Using exponential rule 1.34, Eqs. (3.86) and (3.87) can be transformed to

$$\Delta S_{mixing,1} = n_1\, R \ln\, [V_1/(V_1 + V_2)]^{-1} \tag{3.92}$$

$$\Delta S_{mixing,2} = n_2\, R \ln\, [V_2/(V_1 + V_2)]^{-1} \tag{3.93}$$

From logarithm rule 1.48, Eqs. (3.92) and (3.93) can be rewritten as

$$\Delta S_{mixing,1} = -R\, n_1 \ln\, [V_1/(V_1 + V_2)] \tag{3.94}$$

$$\Delta S_{mixing,2} = -R\, n_2 \ln\, [V_2/(V_1 + V_2)] \tag{3.95}$$

Since $X_1 = V_1/(V_1 + V_2)$ and $X_2 = V_2/(V_1 + V_2)$ from Eq. (3.90), we can substitute the respective mole fractions into Eqs. (3.94) and (3.95).

$$\Delta S_{mixing,1} = -R\, n_1 \ln X_1 \tag{3.96}$$

$$\Delta S_{mixing,2} = -R\, n_2 \ln X_2 \tag{3.97}$$

As the two gas molecules behave ideally, the entropy of mixing of both gases is the sum of the individual mixing entropies of each gas. Note that the $\Delta S_{mixing} = -R\, n \ln X$ term, has an overall positive value. Mole fraction is less than 1, and the natural log of a decimal number is negative. Therefore, the negative log value times the negative sign in the above equation gives a positive value. This means the entropy of mixing is spontaneous.

$$\Delta S_{mixing,1\&2} = -R\ (n_1\ \ln X_1 + n_2 \ln X_2) \tag{3.98}$$

3.4.6 Determination of Entropy Changes for Phase Transitions

Phase transitions are isothermal processes. The melting of a material is a phase change where, at the melting point, the solid and liquid phases coexist in equilibrium. During this equilibrium melting phase, heat is absorbed by the system, yet the temperature is constant. The absorbed heat/energy is required to overcome solid's intermolecular cohesive forces. This absorbed energy increases the disorder of the solid molecular structure, resulting in an entropy increase. This increase in entropy results in no useful work. The melting entropy change is a measure of the increase in the amount of disorder and opportunity for the liquid to access more microstates.

Entropy changes for constant pressure phases transitions like melting, evaporation, sublimation, polymorphic transitions, and liquid crystal transitions, can all be determined by Eq. (3.77) for phase transitions. $\Delta S_{transition} = \Delta q_{rev,P}/T_{transiton} = \Delta H_{transition}/T_{transition}$ (from Eq. (3.31), where $\Delta H = q_{rev}$)

Example 3.8 Pharmaceutical Application: Calculate the Entropy Change for the Fusion of Lidocaine and Prilocaine and Interpret the Meaning of Their Entropy Changes

Example 3.4 provided the enthalpies of fusion for two local anesthetics, lidocaine (68.14 J g^{-1}) and prilocaine (131.73 J g^{-1}). Lidocaine's melting point was reported to be 65.05 °C. Prilocaine's melting point was 39.85 °C.

1) Calculate the entropy of fusion for lidocaine using Eq. (3.77) and its enthalpy of fusion and melting point.

$$\Delta S_{fusion} = \Delta H_{fusion}/T_{melting\ \ point}$$

$$\Delta S_{fusion} = \left(68.14\,\mathrm{J\,g^{-1}}\right)/338.2\,\mathrm{K} = \underline{2.015\mathrm{E}-1\,\mathrm{J\,g^{-1}\,K^{-1}}} \tag{3.99}$$

2) Calculate the entropy of fusion for prilocaine.

$$\Delta S_{fusion} = \left(131.73\,\mathrm{J\,g^{-1}}\right)/313.0\,\mathrm{K} = \underline{4.209\mathrm{E}-1\,\mathrm{J\,g^{-1}\,K^{-1}}} \tag{3.100}$$

Generally, the entropy change for a phase transition is referred to as the *entropy for that specific phase change*. In this melting case, the entropy change is called the *entropy of fusion*, with the understanding that the value is the change in entropy that is being calculated. The same is true when discussing *enthalpies of fusion*, even though the enthalpy value is the change in enthalpy that occurs during melting.

Prilocaine's entropy of fusion is almost two times greater than lidocaine's. Prilocaine's entropy of fusion tells us that the prilocaine crystal lattice required about two times more disorder or microstates to melt. The increase in microstates is most likely in the form of increased mobility in its structure before melting could occur.

These observations are consistent with the crystalline structure of these two similar compounds. As noted earlier, Mercury 4.2.0 was used to generate the crystal packing structure of lidocaine and prilocaine. Prilocaine was shown to have a denser packing and higher H-bonding network. Prilocaine has a continuous H-bonding network along the C-plane, which leads to a smaller cell volume and denser crystalline packing compared to lidocaine. This would indicate stronger intermolecular interactions, which would require more absorbed energy to break the greater level of H-bonding and create more disorder and mobility in the continuous H-bond network and denser crystalline structure to cause melting.

3.4.7 Other Reversible Processes

Heating of a substance at constant pressure is a reversible process. Equation (3.84) describes this process ($dS = C_p\, dT/T$). Note the change in entropy is directly proportional to the heat capacity of the material. As we mentioned in (3.37), heat capacity is a thermodynamic state function. On a molecular level, heat capacity is a measure of molecular motion or mobility. As the molecular mobility increases, the system becomes more disordered by having increased potential for microstates and entropy increases. A substance can be heated or cooled in infinitesimal increments, and the heat capacity and entropy change accordingly.

Another reversible process is the expansion of an ideal gas. This was discussed extensively in Section 3.4.5.

3.4.8 Factors Affecting Entropy

Several factors affect entropy. As noted in Section 3.4.2 Entropy: A Statistical and Molecular Interpretation, an *absolute entropy value* can be determined. Factors that affect entropy are outlined in Table 3.2.

Table 3.2 Factors Affecting the Entropy of a System

Factor	Discussion
Physical state: $S_{gas} > S_{liquid} > S_{solid}$	At 25 °C Gas entropies range from 120 (He) to 190 (H_2O) J mol^{-1} K^{-1} Liquid entropies range from 70 (H_2O) to 200 (cyclohexane) J mol^{-1} K^{-1} Solid entropies range from 2 (diamond) to 60 (MgF_2) J mol^{-1} K^{-1}
Temperature	*S* increases with increase in temperature. The is more kinetic energy and more microstates
Mixing	Mixing results in an increase in *S*. This was shown to be the case in Section 3.4.5, Entropy of Mixing Ideal Gases
Formation of a solution	Solution formation is another case of mixing. Typically, there is increased positional disorder and entropy increases. Entropy can decrease if there are strong solute-solvent interactions that can cause ordering
Atomic size	Alkali metals Li, Na, K, Rb, and Cs have increasing entropy with increasing molecular size and weight ($S \sim$ 30, 50, 60, 70, and 85 J mol^{-1} K^{-1}, respectively)
Molecular complexity	Increases in possible atom locations leads to increases in entropy. A monoatomic gas can only exhibit translational motion. Diatomic gases can undergo translational, rotation, and vibrational motion such as bending, symmetrical stretching, and asymmetric stretching As the number of atoms in a molecule increase the entropy of the molecule increases. Likewise, as the structure becomes more complex the entropy increases
Concentration	As the pressure or concentration is decreased, the entropy increases For a gas, as the pressure is decreased, the volume increases, and there is increased disorder or opportunities for increased microstates. As the concentration decreases, there is an increase in the possibility to occupy more positions
Number of reaction products	Entropy increases when the number of products is greater than the number of reactants

Example 3.9 When One Mole of NaCl Is Dissolved in Water, Predict Which Form of NaCl Has the Highest Entropy: Solid or Solution

$$NaCl\ (s) + H_2O \rightarrow NaCl\ (aq)$$

The dissolved NaCl (aq) form of NaCl has the higher entropy because solution formation requires that the salt mixes with the water solvent. NaCl has much more freedom and availability of microstates in the water.

Example 3.10 Which Molecule Has the Highest Entropy, O_2 (g) or O_3 (g)?
O_3 (g) has higher entropy than O_2 (g) because it is a larger molecule, and it has a more complex structure.

Example 3.11 Is the ΔS_{system} for the Formation of Lidocaine Hydrochloride Salt from the Free Lidocaine Base Positive or Negative?

Lidocaine (free base) + Hydrochloric Acid → Lidocaine HCL

The ΔS_{system} is negative because there are fewer products than reactants.

3.4.9 Determination and Prediction of ΔS Sign and Relative Value

Table 3.3 lists the possible scenarios resulting in spontaneous and nonspontaneous processes.

3.4.10 Entropy's Shortcomings for Prediction of Spontaneous Processes

$\Delta S_{\text{universe}}$ needs to be known to be able to predict the spontaneity of a process. Calculating $\Delta S_{Surroudings}$ can be complex. The calculation is not only complex, but it can be impractical because of its difficulty. $\Delta S_{Surroudings}$ is also very temperature-dependent. A process may be spontaneous at one temperature and not at another. ΔS_{System} can be negative for a spontaneous process, if $\Delta S_{Surroudings}$ is positive enough to make the total entropy positive.

Having said this, the importance Clausius' discovery of entropy should not be diminished. The discovery and understanding of a second thermodynamic law and state function, entropy, had a profound impact on the development of the Gibbs free energy by Josiah Willard Gibbs in the 1890's. The Gibbs free energy equation and theory unified the theories of the first and second laws. The Gibbs free energy will be discussed in detail in Chapter 4.

Table 3.3 Entropy of the Universe as a Predictor of a Spontaneous Process

$\Delta S_{Universe} > 0$	Spontaneous process[a]
$\Delta S_{Universe} < 0$	Nonspontaneous process[b]
$\Delta S_{Universe} = 0$	Reversible process[c]

[a] Also an irreversible process.
[b] Process is spontaneous in the opposite direction.
[c] System is at equilibrium

3.5 Third Law of Thermodynamics: Entropy of a System Approaches a Constant Value as Its Temperature Approaches Absolute Zero

The development of the third law of thermodynamics occurred during the early 1900's. One construction of the third law states that at absolute zero, the system is in a state of absolute minimum energy and exhibits a constant entropy value. Unlike internal energy, U, or enthalpy, H, which are expressed as differences in internal energy or enthalpy, entropy can be expressed as an *absolute* value. At absolute zero, the entropy of the system is expressed by the quantity S_0, the lowest possible system energy state. Thus, the absolute entropy of a system above 0 K is greater than zero.

The third law also supports the notion that all atomic motion ceases at absolute zero.

Another development of the third law of thermodynamics states that S_0 is equal to zero for a perfect crystal at absolute zero where there is no crystal disorder or imperfections. This definition of the third law is consistent with statistical thermodynamics. In the case of a perfect crystal at absolute zero, each atom is in its ground state, which is the same state for each atom. The number of microstates, Ω, equals 1. From Eq. (3.64)

$$S = k\ln\ \Omega$$

$$S_0 = k\ln\ 1 = 0$$

which is in agreement with the classical definition of the third law and defines the lowest value for entropy to be zero. As noted in Table 3.2, diamond's low entropy of 2 indicates how little motion there is in the diamond crystal lattice and how close it is to being a perfect crystal.

In some cases, the lowest energy state may contain some disorder. An example would be an amorphous or glassy solid. As the temperature is lowered, the amorphous or glassy disordered state may become "frozen" in a higher energy state than its perfect crystalline counterpart. This equilibrium constant value at absolute zero is called *residual entropy*.

Equation (3.85), $S_{final} - S_{initial} = \Delta S = C_p \ln T_{final}/T_{intial}$ can be used to calculate the absolute entropy of a substance. The absolute entropy value can be determined for a system by integrating from absolute zero to a temperature of interest where:

$$S_T = S_0 + Cp \int_0^T d\ln T \tag{3.101}$$

System entropy can be determined experimentally at constant pressure by plotting C_p versus ln T. The experimental curve is extrapolated to absolute zero, and the area under the curve equals the entropy of the system at the final temperature. As noted in Section 3.3.8 Pharmaceutical Application: Thermochemistry Using Heat Capacity and Modulate Differential Scanning Calorimetry, C_p can be measured over the desired temperature range.

References

Bartell, L.S. (2001). Apparent paradoxes and instructive puzzles in physical pharmacy. *J. Chem. Educ.* 78 (8): 1067–1069.

Baine, R.L. Technical NoteTA081. The Case for a Generic Definition of Differential Scanning Calorimetry. https://www.tainstruments.com/pdf/literature/TA081.pdf (accessed April 2023).

Boltzmann, L.E. https://en.wikipedia.org/wiki/Ludwig_Boltzmann (accessed February 2025).

Carnot, N.L.S. https://en.wikipedia.org/wiki/Nicolas_L%C3%A9onard_Sadi_Carnot (accessed February 2025).

Celsius, A. https://en.wikipedia.org/wiki/Anders_Celsius (accessed February 2025).

Clausius, R.J.E. https://en.wikipedia.org/wiki/Rudolf_Clausius (accessed February 2025).

Connors, K.A. and Mecozzi, S. (2010). Thermodynamics of Pharmaceutical Systems: An Introduction to Theory and Applications, 2nde. Hoboken, NJ: John Wiley & Sons.

DeMeo, S. (2001). Making assumptions explicit: how the law of conservation of matter can explain empirical formula problems. *J. Chem. Educ.* 78 (8): 1050–1052.

de Nevers, N. (1991). A simple heat of crystallization experiment. *Chem. Eng. Educ.* 154–156.

Denbigh, K. (1966). The Principles of Chemical Equilibrium, 2nde. London: Cambridge University Press.

Fahrenheit, D.G. https://en.wikipedia.org/wiki/Daniel_Gabriel_Fahrenheit (accessed February 2025).

Farrington, D. and Alberty, R.A. (1975). Physical Chemistry, 4e. New York: John Wiley & Sons.

Flowers, P., Theopold, K., and Langley, R. (2023). Criteria for Spontaneous Change: The Second Law of Thermodynamics. Chemistry Libretexts. https://chem.libretexts.org/Bookshelves/General_Chemistry/Map%3A_General_Chemistry_(Petrucci_et_al.)/19%3A_Spontaneous_Change%3A_Entropy_and_Gibbs_Energy/19.4%3A_Criteria_for_Spontaneous_Change%3A_The_Second_Law_of_Thermodynamics.

Fowler, R.H. https://en.wikipedia.org/wiki/Ralph_H._Fowler (accessed February 2025).

Ghan, G.A. and Lalla, J.K. (1992). Effect of compression forces on piroxicam polymorphs. *J. Pharm. Pharmacol.* 44 (8): 678–681.

Haware, R.V., Shivagari, R., Johnson, P.R. et al. (2014). Application of Multivariate Methods to Evaluate the Functionality of Bovine- and Vegetable-Derived Magnesium Stearate. *J. Pharm. Sci.* 103 (5): 1466–1477.

Haware RV, Kancharla, JP., Aishwarya UK, Staton S, Gupta, MR, Al-Achi A. and Stagner WC. Physico-Mechanical Properties of Coprocessed Excipient MicroceLac® 100 by DM^3 Approach." *Pharm. Res.* 2015;32(11): 3618-3635. DOI https://doi.org/10.1007/s11095-015-1721-3.

Haware RV, Dave VS, Kakarala B, Delaney S, Staton S, Munson, E., Gupta, MR, Stagner WC. "Vegetable-Derived Magnesium Stearate Functionality Evaluation by DM^3 Approach." *Eur. J. Pharm. Sci.*2016;89: 115-124. DOI: https://doi.org/10.1016/j.ejps.2016.04.019.

Haware, R., Vinjamuri, B.P., Sarkar, A. et al. (2018). Deciphering Magnesium Stearate Thermotropic Behavior. *Int. J. Pharm.* 548: 314–325. https://doi.org/10.1016/j.ijpharm.2018.07.001.

Helmenstine, T. (2024). Heat of Fusion Example Problem: Melting Ice. ThoughtCo. https://www.thoughtco.com/heat-of-fusion-melting-ice-problem-609498 (Accessed February 2025).

Hess Law: The Use of Hess Law to Determine the Reaction Heat When Sodium Solution is Mixed with Diluted Hydrochloric Acid. https://gradesfixer.com/free-essay-examples/the-use-of-hess-law-to-determine-the-reaction-of-heat-when-sodium-solution-is-mixed-with-diluted-hydrochloric-acid/ (accessed January 2022).

Ibrahim, H.G., Pisano, F., and Bruno, A. (1977). Polymorphism of Phenylbutazone: Properties and compression behavior of crystals. *J. Pharm. Sci.* 66 (5): 669–673.

Manninen, A.H. (2004). Is a calorie really a calorie? Metabolic advantage of low-carbohydrate diets. *J. Int. Soc. Sports Nutr.* 1 (2): 21–26.

Mao, F., Kong, Q., Ni, W. et al. (2016). Melting point distribution analysis of globally approved and discontinued drugs: a research for improving the chance of success of drug design and discovery. *ChemistryOpen* 5: 357–368.

Mayer, J.R. https://en.wikipedia.org/wiki/Julius_von_Mayer (accessed February 2025).

Nernst, W. https://en.wikipedia.org/wiki/Walther_Nernst (accessed February 2025).

Newman A and Zografi G. Commentary: Considerations in the Measurement of Glass Transition Temperatures of Pharmaceutical Amorphous Solids. *AAPS PharmSciTech* 2020;21(1). DOI: https://doi.org/10.1208/s12249-019-1562-1.

Otsuka, M. (1993). Effects of environmental temperature and compression energy of polymorphic transformation during tableting. *Drug Dev. Ind. Pharm.* 19 (17-18): 2241–2269.

Perrault, M., Bertrand, F., and Chaouki, J. (2011). An experimental investigation of the effect of the amount of lubricant on tablet properties. *Drug Dev. Ind. Pharm.* 37: 234–242.

Thermodynamics: Entropy, Free Energy, and the Direction of Chemical Reactions. http://srjcstaff.santarosa.edu/~oraola/CHEM1BLECT/ch.%2020/ch20_lecture_6e_final.pdf (accessed January 2022).

Thomson, W. (Lord Kelvin). https://en.wikipedia.org/wiki/Lord_Kelvin (accessed February 2025).

Todorova, A.N. and Schwarz, P.F. (2007). The role of water in the thermodynamics of drug binding to cyclodextrin. *J. Chem. Thermodyn.* 39 (7): 1038–1048.

Wildfong, P.L.D., Morris, K.R., Anderson, C.A., and Short, S.M. (2007). Demonstration of a shear-based solid-state phase transformation in small molecular organic system: Chlorpropamide. *J. Pharm. Sci.* 96 (5): 1100–1113.

Zavaliangos, G.S., Cunningham, J., and Winstead, D. (2008). Temperature evolution during compaction of pharmaceutical powders. *J. Pharm. Sci.* 97 (8): 678–681.

Chapter 3 Problem Set

3.1 What is the work of reversible isothermal expansion of an ideal gas when carried out under constant temperature?

3.2 What is the reversible work of expansion of one mole of ideal gas at 25 °C when the gas expands from 5 to 25 L?

3.3 Hand warmers can provide comfort during winter outdoor activities. These reusable hand warmers can be made using a 50% w/w percent sodium acetate and supersaturated water solution. A thin metal disc inside the solution plastic bag container can be clicked to initiate the supersaturated solution to undergo precipitation (de Nevers, 1991).
Given:

a) Assume 82.03 g NaAcetate (one mole) is dissolved in 82 g water.
b) The crystallization reaction at constant pressure is

$$CH_3CO_2Na + 3\,H_2O\ \ (l) \rightarrow CH_3CO_2Na \cdot 3\,H_2O\,(cryst)$$

c) $CH_3CO_2Na \cdot 3H_2O\ \ \Delta H_{crystallization} = -30.74\,kJ\,mol^{-1}$

$$= -30{,}744\ J\ \cancel{mol}^{-1} \cdot \left(\cancel{mol} \cdot 82.03\,g^{-1}\right) = -374.8\,J\,g^{-1}$$

d) At the final temperature:
- 56.44 g $CH_3CO_2Na \cdot 3\,H_2O$ (cryst) in liquid mixture
- The solution portion of the liquid mixture contains 59.59 g water and 47.97 g dissolved CH_3CO_2Na (aq)

e) Heat capacities are

$$CH_3CO_2Na \cdot 3\,H_2O\,(cryst)\,2.40\,J\,g^{-1}\,K^{-1};\ water\ 4.184\,J\,g^{-1}\,K^{-1};\ CH_3CO_2Na\ 2.58\,J\,g^{-1}\,K^{-1}$$

A) Is crystallization an exothermic or endothermic process?
Crystallization is an exothermic process. The system releases heat.
B) What is the final temperature if the handwarmer is activated at 25.0 °C?

3.4 Explain the different features of the temperature modulated differential scanning calorimetry thermogram.

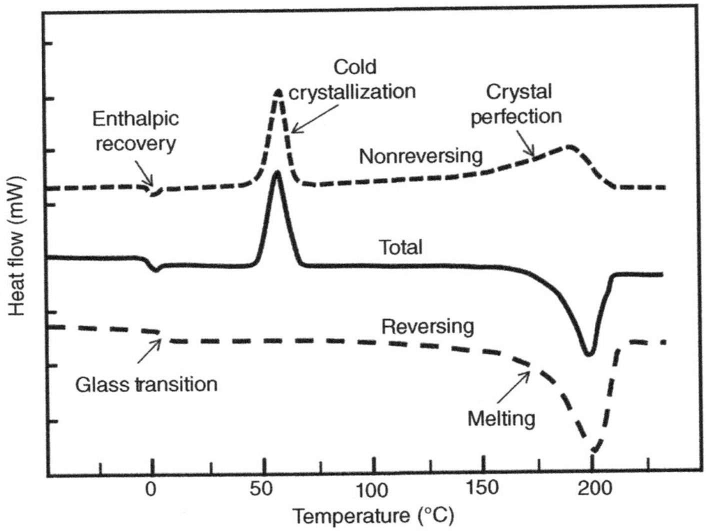

3.5 How much energy is required to melt 25 g of ice (Helmenstine, 2024).
Given:

a) The heat required for a phase transition is $q = m\ \Delta H_f$ where q is heat/energy, m is mass, and ΔH_f is the enthalpy or heat of fusion.
b) ΔH_f of ice is 334 J g^{-1}

3.6 When 25 g of ice is melted what is the entropy of the system?
Given:

a) ΔH_f of ice is 334 J g^{-1}
b) $S_{system} = \Delta H_f / T_{transition\ temperature}$

3.7 Which molecule has the greatest entropy? Explain.
a) 1 mol SO_2 or 1 mol SO_3
b) water at 5 °C or water at 25 °C
c) He or Ne
d) water or ethanol

3.8 Is entropy positive or negative for each reaction? Explain.
a) $N_{2(g)} + 3H_{2(g)} \rightarrow 2NH_{3(g)}$
b) $H_{2(g)} + I_{2(s)} \rightarrow 2HI_{(g)}$
c) $Ag^{+}{}_{(aq)} + Cl^{-}{}_{(aq)} \rightarrow AgCl_{(s)}$
d) Ice $\rightarrow$ Water Vapor

4

Gibbs Free Energy

4.1 Introduction

In the 1870s, Josiah Willard Gibbs from Yale University unified and extended the first and second laws of thermodynamics. He developed a *system-oriented* concept that allowed the *prediction of spontaneity* considering only the *energy status* of a *system* under *constant temperature and pressure.* His theory described the interplay between internal energy and enthalpy, and the entropy of a system. It also related work to internal energy and enthalpy, and entropy. This unified theory extended to multiple phases, phase transitions, and surfaces and interfacial energy. The concept of work was also expanded to multiple mixtures, chemical reactions, diffusion, equilibrium, adsorption, and other seemingly disparate physical-chemical phenomena.

This new unifying thermodynamic state function became known as the Gibbs free energy function or Gibbs free energy or simply the Gibbs energy. Gibbs free energy is the thermodynamic potential or capacity of a system's "available energy," as described by Gibbs that could perform "useful work." *Gibbs free energy is the maximum amount of nonPV expansion work available in a closed system under constant temperature and pressure.* NonPV work includes electrical mechanical, chemical, and others forms of work.

In 1882, Hermann Helmholtz described what became known as the Helmholtz free energy. The Helmholtz free energy is the maximum amount of work including PV work (expansion work) and other forms of work generated in a closed system at constant temperature and volume (Daniels and Alberty, 1975).

Since most pharmaceutical processes occur under constant temperature and atmospheric pressure, Gibbs free energy is more applicable for our ongoing discussions.

The Gibbs function is as follows:

$$G = H - TS \tag{4.1}$$

where G is the Gibbs free energy, H is enthalpy, T is the temperature in degree Kelvin, and S is the entropy of the system.

Gibbs free energy is not an absolute value as entropy. Gibbs free energy is expressed as a relative value or relative change like internal energy and enthalpy. The Gibbs function for a system becomes

$$\Delta G_{system} = \Delta H_{system} - T\Delta S_{system} \tag{4.2}$$

It is important to note that the change in Gibbs free energy can be calculated from the experimental values ΔH and $T\Delta S$.

Thermodynamics for Drug Product Design: A Guide for Pharmaceutical Scientists, First Edition. William Craig Stagner.

4.2 Relationship Between Gibbs Free Energy and Internal Energy

This discussion follows that of Connors and Mecozzi (2010) and Daniels and Alberty (1975).

Gibbs free energy is also related to the internal energy of a system knowing:

$$H = U + PV \tag{4.3}$$

where H is enthalpy, U is internal energy, P is pressure, and V is the volume of the system. Substituting for H in Eq. (4.1) gives:

$$G = U + PV - TS \tag{4.4}$$

Using the partial derivative Eq. (1.88) $d(uv) = udv + vdu$, the differential of Eq. (4.4) is:

$$dG = dU + PdV + Vdp - TdS - SdT \tag{4.5}$$

Equation (4.5) can be simplified to give what is known as Gibbs free energy *fundamental equation*. Simplifying Eq. (4.5) begins by showing that $dU = TdS - PdV$. This is accomplished by using the following equations first introduced in Chapter 3.

$$dU = dq \ - PdV \tag{3.45}$$

$$dq = TdS \tag{3.70}$$

TdS from Eq. (3.70) is substituted for dq into Eq. (3.45) to give:

$$dU = TdS - PdV \tag{4.5a}$$

Now we substitute $TdS - PdV$ for dU into Eq. (4.5) and cancel common terms to give:

$$dG = \cancel{TdS} - \cancel{PdV} + \cancel{PdV} + Vdp - \cancel{TdS} - SdT \tag{4.6}$$

or

$$dG = VdP - SdT \tag{4.7}$$

which is the Gibbs free energy *fundamental equation* for a closed system. This differential equation relates the change in Gibbs free energy to the change in pressure and temperature for a reversible process in a closed system. The volume and entropy terms are known as the *fundamental equation* coefficients. At constant temperature, dG becomes Eq. (4.8a).

$$dG = VdP \tag{4.8a}$$

At constant pressure, dG is Eq. (4.8b):

$$dG = \ - SdT \tag{4.8b}$$

A plot of the Gibbs free energy change versus temperature change gives a slope $-S$.

The partial derivatives (see Section 2.12.3) of Eq. (4.7) are:

$$(\partial G/\partial P)_T = V \tag{4.9a}$$

and

$$(\partial G/\partial T)_P = -S \tag{4.9b}$$

4.3 Gibbs Free Energy to Predict Spontaneous Processes

The second law of thermodynamics states that a spontaneous process will occur if the change in entropy of the universe is greater than zero. This was discussed in Section 3.4.3 and captured in Eq. (3.71a) and restated here as Eq. (4.10):

$$\Delta S_{universe} = \Delta S_{system} + \Delta S_{surroundings} > 0 \tag{4.10}$$

Using entropy to determine a spontaneous process has several major shortcomings. First, scientists are generally interested in understanding or predicting if the system that is being studied will undergo a spontaneous reaction. To make this prediction, the scientist needs to determine the entropy change for the system and the surroundings. The assumptions, use of Hess's law, and the amount of data required to make these calculations are onerous. In addition, each data value used in the calculations has a certain degree of error associated with the value. These errors are propagated throughout the calculations as discussed in Section 1.7.3 "Propagation of Error of Combined Measurements and Calculations." If the difference between the entropy changes of the system and surroundings is small, the propagated errors associated with their calculations may result in an incorrect value used to assess spontaneity. Another major shortcoming of using entropy to predict spontaneity is the common difficulty in determining $\Delta S_{surroundings.}$

Gibbs free energy, on the other hand, allows the prediction of a spontaneous process of the system using experimental data from the system. This was a major breakthrough in thermodynamics.

It is possible to start with Eq. (4.10) from the second law and derive the Gibbs free energy function using principles from the first and second law. This foundational derivation shows how the first and second laws are unified in the Gibbs free energy function.

First, we need to show that $\Delta S_{surroundings} = -\Delta H_{system}/T$. According to the first law of thermodynamics and the conservation of energy, any heat lost by the system will be gained by its surroundings. It is also true that that any heat gained by the system will be lost by the surroundings. This is represented by Eq. (4.10a):

$$-q_{system} = q_{surroundings} \tag{4.10a}$$

The enthalpy of the system at constant pressure is equal to the heat absorbed or released by the system (see Eq. 4.10b):

$$\Delta H_{system} = q_{system} \tag{4.10b}$$

Now, substitute ΔH_{system} for q_{system} in Eq. (4.10a) to give Eq. (4.10c):

$$-\Delta H_{system} = q_{surroundings} \tag{4.10c}$$

Divide both sides of Eq. (4.10c) by T to provide Eq. (4.10d):

$$-\Delta H_{system}/T = q_{surroundings}/T \tag{4.10d}$$

The entropy change of the surroundings is described by Eq. (4.10e) which follows the Clausius classic Eq. (3.70):

$$\Delta S_{surroundings} = q_{surroundings}/T \tag{4.10e}$$

Now, substitute $-\Delta H_{system}/T$ from Eq. (4.10d) for $q_{surroundings}/T$ in Eq. (4.10e) to become (4.10f).

$$\Delta S_{surroundings} = -\Delta H_{system}/T \tag{4.10f}$$

The remainder of the derivation follows that of Lower (2023) and Chang (1998). First, $-\Delta H_{system}/T$ is substituted into Eq. (4.10) for $\Delta S_{surroundings}$ which leads to Eq. (4.11):

$$\Delta S_{universe} = \Delta S_{system} - \Delta H_{system}/T > 0 \tag{4.11}$$

Equation (4.11) is then multiplied by T.

$$T\Delta S_{universe} = T\Delta S_{system} - \left(\Delta H_{system}/\not{T}\right)\ \not{T} > 0 \tag{4.12}$$

Multiply Eq. (4.12) by −1. Multiplying by −1 also results in changing the > sign to a < sign where 1 > 0 and − 1 < 0. This gives:

$$-T\Delta S_{universe} = -T\Delta S_{system} + \Delta H_{system} < 0 \tag{4.13}$$

Equation (4.2), $\Delta G_{system} = \Delta H_{system} - T\Delta S_{system}$, and Eq. (4.13) is $-T\Delta S_{\text{universe}} = \Delta H_{system} - T\Delta S_{\text{system}} < 0$. So, $\Delta G_{system} = -T\Delta S_{\text{universe}}$ which is shown in Eq. (4.14)

$$\Delta G_{system} = -T\Delta S_{universe} \tag{4.14}$$

ΔG_{system} from Eq. (4.2) is now substituted for $-T\Delta S_{\text{universe}}$ in Eq. (4.13) and results in Eq. (4.15):

$$\Delta G_{system} = \Delta H_{system} - T\Delta S_{system} < 0 \tag{4.15}$$

Equation (4.15) states that if a system releases useable energy to do work, ΔG_{system} is negative and the system will undergo a spontaneous reaction or process.

For a reversible reaction A ⇆ B, there are three possible outcomes:

> $\Delta G_{system} < 0$, the above reaction will spontaneously proceed from A to B;
> $\Delta G_{system} > 0$, the reaction is not spontaneous from A → B, but will spontaneously proceed from B → A; and
> $\Delta G_{system} = 0$, the reaction is at equilibrium.

ΔG_{system} is dependent on the sign and magnitude of ΔH_{system} and $-T\Delta S_{system}$ terms. A $-\Delta H_{system}$ promotes a spontaneous process, while a $+\Delta H_{system}$ counters spontaneity. The sign of the second term, $-T\Delta S_{system}$, is affected by the sign of ΔS_{system}. As a reminder, a positive ΔS_{system} means the system's entropy is increasing and results in the $-T\Delta S_{system}$ term being negative. This negative term contributes to the process being spontaneous. A negative ΔS_{system} means the system's entropy is decreasing. This negative value causes the $-T\Delta S_{system}$ term to be positive, which opposes a spontaneous process.

Table 4.1 and Figure 4.1a–d summarize the above comments for a reversible process. It is important to note that these conditions assume that ΔH_{system} and ΔS_{system} are independent of temperature over the temperature range of interest, even though ΔH and ΔS are affected by temperature to an extent as discussed in Chapter 3.

For those who find visual examples more informative, the above relationships are also illustrated in Figure 4.1a–d which is adapted from Lower (2023). The general format for all the figures is stated next. The *x*-axis and *y*-axis are temperatures in degree Kelvin and energy in kJ, respectively. At absolute zero, the term $T\Delta S_{system}$ is zero, so the value of this term always starts at 0 kJ and zero K. The slope of the $T\Delta S_{system}$ *line* from 0 kJ depends on if the value of ΔS_{sustem} is greater than zero or less than zero. It is important to be aware of a subtle but important difference in how the entropy related, $[-T\Delta S_{system}]$ term, is represented for discussion in Table 4.1 and Figure 4.1a–d. In Table 4.1, the ΔS_{system} sign affects the whole $[-T\Delta S_{system}]$ *term sign* as indicated in the Table 4.1 note "b." For graphical purposes, Figure 4.1a–d uses the ΔS_{system} sign to indicate the slope of the *descriptive* $T\Delta S_{system}$ line which represents the change in the disorder of the system. If ΔS_{system} is greater than

Table 4.1 Effects of ΔH_{system}, ΔS_{system}, and $-T\Delta S_{system}$ on the Sign of ΔG_{system}[a] for a Reversible Process.

ΔH_{system} Sign	ΔS_{system} Sign	$[-T\Delta S_{system}]$ *Term* Sign[b]	ΔG_{system} Sign	Spontaneity Description
− Exothermic	+	−	−	ΔG[c] is always negative. Process proceeds spontaneously at ALL temperatures
− Exothermic	−	+	(+/−)[d]	ΔG is negative at low temperatures and proceeds spontaneously. ΔG is positive at high temperatures and the forward reaction is not spontaneous. The reverse reaction is spontaneous at high temperatures.
+ Endothermic	+	−	(+/−)[d]	ΔG is negative at high temperatures and is spontaneous. ΔG is positive at low temperatures. The reverse reaction is spontaneous at low temperatures.
+ Endothermic	−	+	+	ΔG is always positive, and the forward reaction is never spontaneous at any temperature. The reverse reaction is spontaneous at ALL temperatures

[a] $\Delta G_{system} = \Delta H_{system} - T\Delta S_{system}$;
[b] Note the term $[-T\Delta S_{system}]$ sign is negative if ΔS_{system}, is positive and positive if ΔS_{system}, is negative;
[c] ΔG represents the system Gibbs free energy change; and
[d] Sign and spontaneity are temperature dependent.

zero, the *descriptive* $T\Delta S_{system}$ line will have a positive slope starting at 0 K and 0 kJ, indicating that the system's disorder or number of microstates is increasing. If, on the other hand, ΔS_{system} is less than zero, the *descriptive* $T\Delta S_{system}$ line will have a negative slope starting at 0 K and 0 kJ, representing a system whose disorder or number of microstates is decreasing. The ΔG_{system} lines are directional, originating from the $T\Delta S_{system}$ line. The ΔG_{system} "arrow direction" indicates if ΔG_{system} is increasing and greater than zero or decreasing and less than zero. An upward ΔG_{system} arrow means that the Gibbs free energy is greater than zero and the length of the arrow indicates the relative magnitude of the positive value. A downward ΔG_{system} arrow, on the other hand, means that the Gibbs free energy is less than zero and the length of the arrow indicates the relative magnitude of the negative value. As a reminder, for a reversible process or reaction, $\Delta G_{system} > 0$ represents a process or reaction that is not spontaneous. When $\Delta G_{system} < 0$, the process or reaction is spontaneous. When the $T\Delta S_{system}$ line crosses the ΔH_{system} line, $\Delta G_{system} = 0$ and the system is at equilibrium.

For the visual case example (Figure 4.1a), $\Delta S_{system} > 0$ and the descriptive $T\Delta S_{system}$ line is shown to have a positive slope, indicating that the system's disorder is increasing. For Figure 4.1a, the sign of the ΔH_{system} is negative. The ΔH line is located below 0 kJ on the graph indicating an exothermic reaction. The ΔG_{system} arrows point down from the $T\Delta S_{system}$ line, indicating that ΔG_{system} is negative and spontaneous over all temperatures. The figure also shows that as the temperature increases, the magnitude of ΔG_{system} becomes more negative.

Figure 4.1b shows the effect of ΔH_{system} and ΔS_{system} sign on ΔG_{system} when $\Delta H_{system} < 0$ and $\Delta S_{system} < 0$.

In this case, since $\Delta S_{system} < 0$, the slope of the $T\Delta S_{system}$ descriptive line is negative. ΔH_{system} is negative, indicating an exothermic reaction. At low temperatures, the ΔG_{system} arrows point down, indicating that ΔG_{system} is negative from 0 K to the point where the $T\Delta S_{system}$ line crosses ΔH_{system}. The process is spontaneous over this temperature range. As the $T\Delta S_{system}$ line approaches the "crossover point," ΔG_{system} becomes less negative. At the crossover point, ΔG_{system} equals zero and the system is at equilibrium. As the temperature increases beyond the crossover point, ΔG_{system} is positive. This is indicated by the upward pointing arrows. At temperatures above the equilibrium

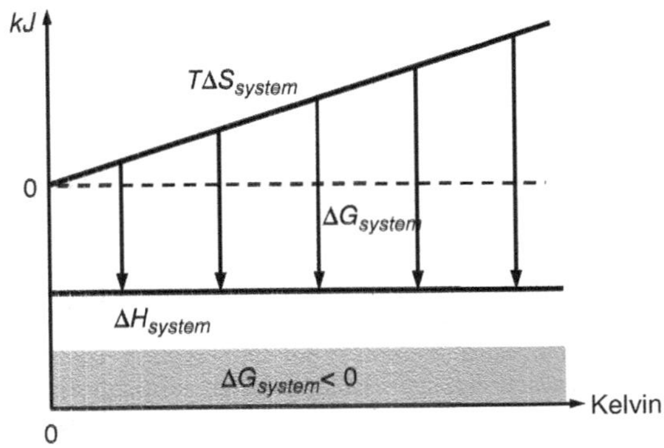

Figure 4.1a Effect of ΔH_{system} and ΔS_{system} sign on ΔG_{system} when $\Delta H_{system} < 0$ and $\Delta S_{\text{system}} > 0$. *Source:* Adapted from Lower (2023).

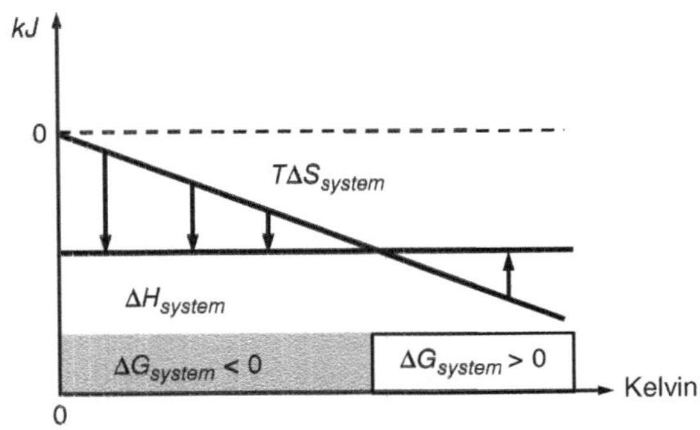

Figure 4.1b Effect of ΔH_{system} and ΔS_{system} sign on ΔG_{system} when $\Delta H_{system} < 0$ and $\Delta S_{\text{system}} < 0$. *Source:* Adapted from Lower (2023).

point, the process is no longer spontaneous. From 0 K to the crossover point, the process is said to be "enthalpic-driven" or dominated. At temperatures beyond the crossover point, the process is said to be "entropy-driven" or dominated.

Figure 4.1c depicts the effect of ΔH_{system} and ΔS_{system} sign on ΔG_{system} when $\Delta H_{system} > 0$ and $\Delta S_{system} > 0$.

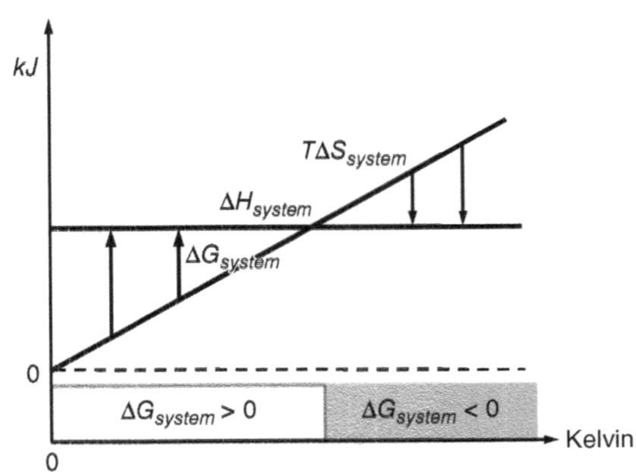

Figure 4.1c Effect of ΔH_{system} and ΔS_{system} sign on ΔG_{system} when $\Delta H_{system} > 0$ and $\Delta S_{\text{system}} > 0$. *Source:* Adapted from Lower (2023).

In this case, $\Delta S_{system} > 0$, and the slope of the system disorder descriptive $T\Delta S_{system}$ line is positive. ΔH_{system} is also positive, indicating an endothermic reaction. At low temperatures, the ΔG_{system} arrows point upward, indicating that ΔG_{system} is positive from 0 K to the point where the $T\Delta S_{system}$ line crosses ΔH_{system}. The process is not spontaneous over this temperature range. As the $T\Delta S_{system}$ descriptive line approaches the "crossover point," ΔG_{system} becomes less positive. At the crossover point, ΔG_{system} equals zero and the system is at equilibrium. As the temperature increases beyond the crossover point, ΔG_{system} is negative. This is indicated by the downward pointing arrows. At temperatures above the equilibrium point, the reaction becomes spontaneous. From 0 K to the crossover point, the process is said to be "enthalpic-driven" or dominated. At temperatures beyond the crossover point, the process is said to be "entropy-driven" or dominated. The higher temperatures in the $T\Delta S_{system}$ term are the primary factor in making the entropy-driven process dominate.

Figure 4.1d illustrates the effect of ΔH_{system} and ΔS_{system} sign on ΔG_{system} when $\Delta H_{system} > 0$ and $\Delta S_{system} < 0$.

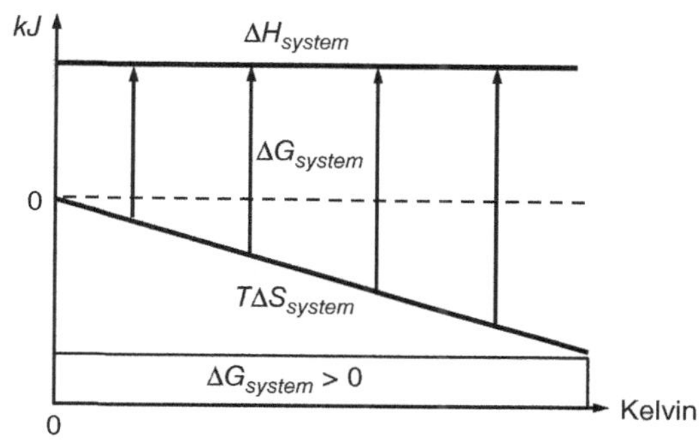

Figure 4.1d Effect of ΔH_{system} and ΔS_{system} sign on ΔG_{system} when $\Delta H_{system} > 0$ and $\Delta S_{system} < 0$. *Source:* Adapted from Lower (2023).

In this case, $\Delta S_{system} < 0$, and the slope of the $T\Delta S_{system}$ line is negative, indicating the system's disorder or microstates in decreasing. ΔH_{system} is positive, indicating an endothermic reaction. The ΔG_{system} arrows point up, indicating that ΔG_{system} is positive, and the reaction is nonspontaneous over all temperatures. The figure also shows that as the temperature increases, the magnitude of ΔG_{system} becomes more positive.

4.4 Temperature Influence on Gibbs Free Energy

The influence of temperature on Gibbs free energy is given by the Gibbs-Helmholtz equation. The derivation starts with the Gibbs function Eq. (4.1) $G = H - TS$, which gives Eq. (4.16):

$$G = H - TS \tag{4.16}$$

From Eq. (4.9b), $(\partial G/\partial T)_P = -S$, $(\partial G/\partial T)_P$ is substituted for $-S$ in Eq. (4.16)

$$G = H + T\,(\partial G/\partial T)_P \tag{4.17}$$

Both sides of the Eq. (4.17) are divided by T^2.

$$G/T^2 = H/T^2 + [\not{T}\ (\partial G/\partial T)_P]/(\not{T}\cdot T) \tag{4.18}$$

Equation (4.18) is rearranged to give the Gibbs free energy terms on the right-hand side and the enthalpy term on the left-hand side of the equation.

$$-H/T^2 = -G/T^2 + 1/T\,(\partial G/\partial T)_P \qquad (4.19)$$

It can be shown that the right-hand term of Eq. (4.19) is the same as taking to derivative of $[(G/T)/dT]$. Using the general derivative Eq. (1.90),

$$d[(G/T)/dT] = -G/T^2 + 1/T\,(dG/dT) \qquad (4.20)$$

When the derivative of Eq. (4.20) is taken at constant pressure, the partial derivative is

$$(\partial(G/T))\,\partial T]_P = -G/T^2 + 1/T\,(\partial G/\partial T)_P \qquad (4.20a)$$

The right-hand side of Eq. (4.20a) and (4.19) are the same. So, $(\partial(G/T))\,\partial T]_P)$ from Eq. (4.20a) can be substituted for the right-hand side of (4.19) giving:

$$-H/T^2 = (\partial(G/T))\,\partial T]_P \qquad (4.21)$$

This is the Gibbs-Helmholtz equation. The incremental form of Eq. (4.21) is:

$$-\Delta H/T^2 = (\partial(\Delta G/T))\,\partial T]_P \qquad (4.22)$$

4.5 Pressure Influence on Gibbs Free Energy

Equation (4.8a), $dG = VdP$ (4.8a), describes the effect of pressure on Gibbs free energy at constant temperature, which is given as Eq. (4.23):

$$dG = VdP \qquad (4.23)$$

Using the ideal gas law, $PV = nRT$, and $V = nRT/P$, and substituting V into Eq. (4.23) gives:

$$dG = (nRT/P)\,dP \qquad (4.24)$$

where $dP/P = d\ln P$ (see Eq. 1.85). Equation (4.24) becomes:

$$dG = nRT\,(d\ln P) \qquad (4.25)$$

Integrating Eq. (4.25) over the limits of P_1 to P_2 where P_1 is *1 mol of gas at 1 atm and 25 °C* standard temperature and pressure (STP) leads to Eq. (4.26).

$$G_2 - G^\circ = 1RT\ln P_2/1 \qquad (4.26)$$

G_2 is the Gibbs free energy at P_2, and G° is the standard Gibbs free energy. A more general form of Eq. (4.26) is:

$$G_2 = G^\circ + RT\ln P_2 \text{ or } G = G^\circ + RT\ln P \qquad (4.27)$$

For a reversible process, the incremental change in the Gibbs free energy change is written as follows:

$$\boldsymbol{\Delta G = \Delta G^\circ + RT\ln Q} \qquad (4.28)$$

where ΔG is the *actual* or *in situ* Gibbs free energy change, ΔG° is the standard Gibbs free energy change, R is the gas constant, T is temperature in Kelvin, and Q is the *reaction quotient*. Q measures the relative concentrations of products and reactants present in a reaction at any given point in time. It is a snapshot of the reaction's progress. The term "Q" is replaced with the equilibrium constant $K_{equilibrium}$ when equilibrium has been reached. The reaction quotient is the ratio of the "multiplication product" of the reaction *product* concentrations raised to the power of their reaction coefficients and the "multiplication product" of the reactant concentrations raised to the power of their reaction coefficients. The presence of Q in Eq. (4.28) tells the reader that the reaction is not at equilibrium. Q will be discussed in more detail in the next section. During a nonequilibrium reaction, ΔG changes throughout the course of the reaction and can be calculated from experimental values from ΔH and ΔS. ΔG is called the *actual* or *in situ* ΔG, because it changes as the reversible reaction approaches equilibrium. ΔG^o is a constant and does not change during the process.

When the reversible process reaches equilibrium, $\Delta G = 0$, and $Q = K$. Equation (4.28) becomes:

$$\boldsymbol{\Delta G^\circ = -RT \ln K} \tag{4.29}$$

where K is the equilibrium quotient or equilibrium constant for the reversible reaction that is at equilibrium. This generalized form of the change in Gibbs free energy equation has many applications that will be presented throughout the remainder of this text.

4.6 Standard Gibbs Free Energy of Reaction

The standard Gibbs free energy of reaction ($\Delta G^o{}_{reaction}$) is the change in Gibbs free energy for a reaction that takes place under standard state conditions. The reactants in their standard states are converted to products in their standard states. Table 4.2 provides the generally accepted conventions for standard state conditions for standard thermodynamic data. Note: in the context of standard thermodynamic values including $\Delta G^o{}_{reaction}$, the temperature and pressure are different than current STP conditions for an ideal gas that was changed by the International Union of Pure and Applied Chemistry in 1982 (see Example 3.3.1).

A brief review of the law of mass balance, the law of mass action, and equilibrium is useful at this point. The law of mass balance, also known as the law of material balance, is an accounting of material entering and leaving an *open* system. Fundamental to the balance is the first law of thermodynamics and conservation of mass principle. That is, matter cannot disappear or be created. The application of the law of mass balance for chemical reactions is balancing the total number of atoms and ions involved as reactants with the total number of atoms and ions formed as products.

Table 4.2 General Conventions for Standard States[a].

State of Matter	Standard State at 25 °C and 1 atm
Elements	$\Delta G^o{}_{formation} = 0$ for the most stable allotropic form
Gas	Pure gas (uniformly fixed composition)
Liquid	Pure liquid (uniformly fixed composition)
Solid	Pure solid (physical stable form of uniform composition)
Aqueous solution	1 molar concentration
Nonaqueous solution	X_2 (solute) = 1
Solvent	Pure liquid (uniformly fixed composition)

[a] Janscó and Fenby (1983).

The law of mass action states that the rate of any given chemical reaction is proportional to the product of the activities (or concentrations) of the reactants raised to the power of their stoichiometric amounts. Equation (4.30) is an example of a reversible reaction that will tend in the direction of equilibrium.

$$aA + bB \rightleftharpoons cC + dD \tag{4.30}$$

$$\text{Rate} \propto [A]^a B]^b \tag{4.31}$$

$$\text{Rate}_f = k_f[A]^a B]^b \tag{4.32}$$

$$\text{Rate}_r = k_r[C]^c[D]^d \tag{4.33}$$

where a, b, c, and d are the stoichiometric coefficients of the reactants A and B and products C and D, whose molar concentrations are [A], [B], [C], and [D]. Rate_f and Rate_r are the forward rate of the reaction and reverse reaction rate, respectively. k_f and k_r are the forward and reverse rate constants. For the reverse rate process, C and D become the reactant species and vice versa for A and B.

The *reaction quotient, Q*, is associated with the law of mass action and is defined as:

$$Q = [\text{Products}]/[\text{Reactants}] = [C]^c[D]^d/[A]^a[B]^b \tag{4.34}$$

Eventually, reversible reactions will reach an equilibrium where the reactant and product concentrations remain constant unless the equilibrium system is perturbed in some way. Q is a measure of how close or far the reaction is from equilibrium. Q also can be used to determine whether the reaction is approaching the equilibrium state from the reactant or product side.

At equilibrium, the forward rate and reverse rates are equal. Therefore,

$$\text{Rate}_f = \text{Rate}_r \text{ and } k_f[A]^a[B]^b = k_r[C]^c[D]^d \tag{4.35}$$

The equilibrium constant, $K_{equilibrium}$, of any reaction is equal to the value of the forward rate constant divided by the value of the reverse rate constant. $K_{equilibrium}$ is the concentration of the products raised to their stoichiometric coeffficients divided by the concentration of reactants raised to their stoichiometric coefficients. Rearrangement of Eq. (4.35) gives:

$$K_{equilibrium} = k_f/k_r = \left\{[C]^c[D]^d\right\}/\left\{[A]^a[B]^b\right\} \tag{4.36}$$

where the right-hand side of the equation is Q and the product and reactant concentrations remain unchanged.

With this background, $\Delta G^o{}_{reaction}$ at STP of 25 °C and 1 atm pressure is determined by the generalized equation:

$$\Delta G^o_{reaction} = \sum n\Delta G^o_f\,(\text{products}) - \sum m\Delta G^o_f(\text{reactants}) \tag{4.37}$$

where n and m are stoichiometric coefficients and $\Delta G^o{}_{reaction}$ is the standard Gibbs free energy of the reaction or *product formation*.

For the reaction scheme Eq. (4.30), aA + bB $\leftrightarrows$ cC + dD, $\Delta G^o{}_{reaction}$ can be written:

$$\Delta G^o{}_{reaction} = \left[c\Delta G^o{}_f(C) + d\,\Delta G^o{}_f(D)\right] - \left[a\Delta G^o{}_f(A) + b\Delta G^o{}_f(B)\right] \tag{4.38}$$

where $\Delta G^o{}_f$ is the standard Gibbs free energy of formation of compounds A, B, C, and D. That is, $\Delta G^o{}_f$ is the calculated Gibbs free energy change to form 1 mol of the compound from its elements in their standard states. These $\Delta G^o{}_f$'s are then substituted into Eq. (4.38) to determine the $\Delta G^o{}_{reaction}$.

There are several ways to calculate $\Delta G^o{}_f$. Tables are available that give $\Delta H^o{}_f$, $\Delta S^o{}_f$, and $\Delta G^o{}_f$ for some compounds. If the standard Gibbs free energy of formation of interest can be found in a table, this is the most direct method of calculating the $\Delta G^o{}_{reaction}$. The standard Gibbs free energies can then be calculated directly by substituting for the appropriate $\Delta G^o{}_f$ s into Eq. (4.38).

If the $\Delta H^o{}_f$ and $\Delta S^o{}_f$ are only available for certain compounds. These can be used to calculate the $\Delta G^o{}_f$ using the general form of the Gibbs free energy Eq. (4.2), $\Delta G = \Delta H - T\Delta S$.

A third and more complicated method to calculate $\Delta G^o{}_f$ is to employ Hess's law. This method is only used if the first two approaches do not have the necessary standard $\Delta H^o{}_f$, $\Delta S^o{}_f$, and $\Delta G^o{}_f$ values available.

Example 4.1 Determination of Standard Gibbs Free Energy Change for the Reaction of Methane with Oxygen to Produce Carbon Dioxide and Water. The Operating Reaction is

$$CH_{4(g)} + 2O_{2(g)} \rightarrow CO_{2(g)} + 2H_2O_{(l)} \tag{4.39}$$

Note that the law of mass balance has been obeyed. In this case, $\Delta G^o{}_{reaction}$ can be calculated from Eq. (4.39).

$$\begin{aligned} \Delta G^o{}_{reaction} = {} & \left[1\text{ mol } \Delta G^o{}_f\left(CO_{2(g)}\right) + 2\text{ mol } \Delta G^o{}_f\left(H_2O_{(0)}\right)\right] \\ & - \left[1\text{ mol } \Delta G^o{}_f\left(CH_{4(g)}\right) + 2\text{ mol } \Delta G^o{}_f\left(O_{2(g)}\right)\right] \end{aligned} \tag{4.40}$$

The following values were found in a table of standard Gibbs free energy of formation:

$$CO_{2(g)} = -394.4\text{ kJ mol}^{-1}$$

$$H_2O_{(l)} = -237.2\text{ kJ mol}^{-1}$$

$$CH_{4(g)} = -50.8\text{ kJ mol}^{-1}$$

$$O_{2(g)} = 0\text{ kJ mol}^{-1}$$

These values are substituted in Eq. (4.40), and the $\Delta G^o{}_{reaction}$ is calculated.

$$\Delta G^o{}_{reaction} = -818\text{ kJ}$$

The negative $\Delta G^o{}_{reaction}$ value indicates that the reaction proceeds spontaneously.

4.7 Gibbs Free Energy, Chemical Equilibrium, and Extent of Reaction

Consider a reversible reaction where the solution reactants are at STP conditions. That is, the reactant concentrations are one molar at 25 °C and 1 atm pressure. Once the reaction starts, the standard-state concentrations change. At equilibrium, the concentration of the products and reactants no longer changes. Equation (4.28) describes the change in Gibbs free energy for the reaction as it proceeds toward equilibrium.

$$\Delta G = \Delta G^o + RT\ln Q \tag{4.28}$$

As noted previously, ΔG is the *actual* or *in situ* Gibbs free energy change that occurs during the reaction. It constantly changes until equilibrium is reached as shown in Figure 4.2a and b. At equilibrium,

$$Q = K \tag{4.41}$$

where K is the *equilibrium constant*. At equilibrium, G, Gibbs free energy, is minimized and $\Delta G = 0$. Substituting 0 into the above Eq. (4.28) for ΔG gives:

$$0 = \Delta G^o + RT \ln K \tag{4.42}$$

$$\Delta G^o = -RT \ln K \tag{4.43}$$

or

$$-\Delta G^o = \ln K \tag{4.44}$$

As K increases, ΔG^o becomes more negative. Also at equilibrium, the system can no longer do additional work. This fact will be discussed in more detail in a later section. ΔG^o and K are constants at a given temperature and pressure, just as the reactant and product concentrations are constant.

Equation (4.44) is characterized by three cases. These are shown in Table 4.3.

Figure 4.2a shows a Gibbs free energy curve as a function of the extent of a reversible reaction. When the extent of the reaction is 0, the reaction mixture is 100% reactants A and B (see Eq. 4.30, $aA + bB \leftrightarrows cC + dD$). When the extent of the reaction is 1, the reaction mixture is 100% products C and D. As indicated in Figure 4.2a, $G^o{}_{products}$ (location 2) < $G^o{}_{reactants}$ (location 1). In this case, $\Delta G^o{}_{reaction} = (G^o{}_{products} - G^o{}_{reactants})$ is less than 0 (see double arrowed line). A negative $\Delta G^o{}_{reaction}$ favors the extent of the reaction moving "forward" toward more products which is shown in

Table 4.3 Three Cases Showing the Relationship between the Equilibrium Constant, K, ΔG^o, and Extent of Reaction.

K	Ln K	ΔG^o	State of Products and Reactants at Equilibrium
> 1	+	−	Extent of reaction favors products
= 1	0	0	Product and reactant concentrations are equal
< 1	−	+	Extent of reaction favors reactants

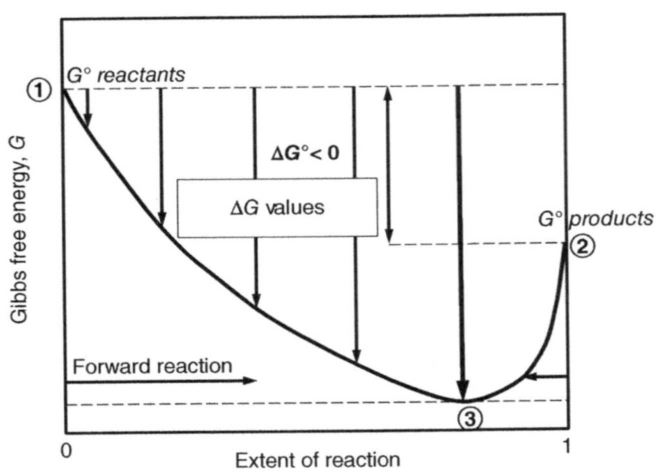

Figure 4.2a Relationship between the extent of a reversible reaction, and G, $\Delta G_{reaction}$, $G^o{}_{reactants}$, $G^o{}_{products}$, and $\Delta G^o{}_{reaction}$ when $\Delta G^o{}_{reaction} < 0$. *Source:* Adapted from Lower (2023).

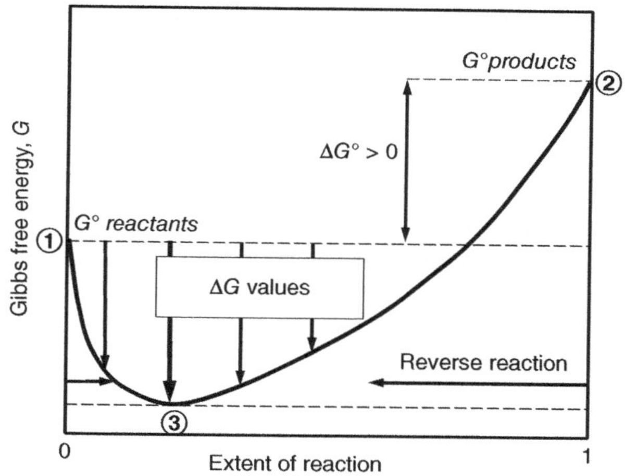

Figure 4.2b Relationship between the extent of a reversible reaction, and G, $G^o_{reactants}$, $G^o_{products}$, and $\Delta G^o_{reaction}$ when $\Delta G^o_{reaction} < 0$. *Source:* Adapted from Lowry (2023).

Figure 4.2a. At equilibrium (location 3), G, the Gibbs free energy, is at a minimum and $\Delta G_{reaction}$ is zero. This curve shows that at equilibrium, the extent of the reaction favors products. This is also the point on the Gibbs free energy versus the extent of the reaction curve where the slope is zero. The system has reached a state where there is no further net change in the concentrations of reactants and products, meaning that no further spontaneous change can occur unless the equilibrium is perturbed say by the addition of more reactants or products or change in temperature.

The *actual* or *in situ* $\Delta G_{reaction}$ values are also shown in Figure 4.2a as the solid curved line. The *actual* or *in situ* $\Delta G_{reaction}$ curve is at its minimum value at the equilibrium, location #3. $\Delta G_{reaction}$ is negative on the left-hand side of the equilibrium position. It is positive for the reverse reaction shown on the right-hand side of the equilibrium position.

Figure 4.2b shows the relationship between the extent of a reversible reaction and G, $\Delta G_{reacrtion}$, $G^o_{reactants}$, $G^o_{products}$, and $\Delta G^o_{reaction}$ when $\Delta G^o_{reaction} > 0$.

In Figure 4.2b, $\Delta G^o_{reaction} = (G^o_{products} - G^o_{reactants})$ is positive. This means that $G^o_{products} > G^o_{reactants}$ which can be seen from the figure. Figure 4.2b also shows that the reverse reaction is favored, and the reaction equilibrium position is closer to the left-hand side of the extent of reaction axis.

4.8 Gibbs Free Energy and Chemical Potential: Introduction to the Fundamental Equations for Open Systems

As discussed in Section 2.43, energy and matter can flow in and out of an open system. Open systems can be homogeneous, such as solutions, or heterogeneous, like suspensions that contain different phases. For simplicity, this section will only discuss homogeneous systems.

It can be shown that for an open system where matter can be exchanged with the surroundings at constant temperature and pressure $\Delta G = \sum_{i=1}^{k} \mu i \, \Delta ni$. μi is the system's *chemical potential* and *the partial molar Gibbs free energy* of the ith component. Δni is the change in the number to moles of component i of k different components. ΔG is equal to the sum of the product of $\mu i \, \Delta ni$ for all system components. This equation is used to determine ΔG for a system that experiences changes in the number of moles of material at constant temperature and pressure.

Chemical potential plays a key role in important thermodynamic processes such as chemical equilibria, phase equilibria, phase changes, and solubility to name a few. The chemical potential equation, $\Delta G = \sum_{i=1}^{k} \mu i \, \Delta ni$, can be derived from the internal energy state function shown next (Daniels and Alberty, 1975; Long-Qing, 2019).

From Eq. (3.4), the infinitesimal change in internal energy is $dU = \delta q + \delta w + \mu dn$, where δq is the differential heat absorbed by the system, δw is the differential work done on the system, μ is the chemical potential, n is the number of moles of a substance or compound, and μdn is the differential energy provided to the system by the addition of moles of material. The incremental change in internal energy for a reversible open system can be written as:

$$\Delta U = q_{rev} + w + \mu \Delta n \tag{4.45}$$

From Eq. (3.71), $q_{rev} = T\Delta S$, and from Eq. (3.6), $w = -P\Delta V$, and $T\Delta S$ and $-P\Delta V$ are substituted into the respective terms of Eq. (4.45) to give:

$$\Delta U = T\Delta S - P\Delta V + \mu \Delta n \tag{4.46}$$

If there are k different components or substances, the $\mu \Delta n$ term can be written $\sum_{i=1}^{k} \mu i \, dni$ which leads to:

$$\Delta U = T\Delta S - P\Delta V + \sum_{i=1}^{k} \mu i \, \Delta ni \tag{4.47}$$

Equation (4.4) related Gibbs free energy to internal energy by $\Delta G = \Delta U + P\Delta V - T\Delta S$. Substituting ΔU from Eq. (4.47) into Eq. (4.4) and solving for ΔG give:

$$\Delta G = \cancel{T\Delta S} - \cancel{P\Delta V} + \sum_{i=1}^{k} \mu i \ dni + \cancel{P\Delta V} - \cancel{T\Delta S} \tag{4.48}$$

and crossing out common terms gives the system's relationship between Gibbs free energy and its chemical potential.

$$\Delta G = \sum_{i=1}^{k} \mu i \, \Delta ni \tag{4.49}$$

This equation shows that the Gibbs free energy in an open system at constant pressure and temperature is the sum of the product of each component's chemical potential and number of moles or particles of various substances or components of the system.

A more generalized incremental equation for ΔG where $G = f(T, P, n_1, n_2 \ldots)$ combines Eq. (4.7), $\Delta G = V\Delta P - S\Delta T$, and Eq. (4.49):

$$\Delta G = V\Delta P - S\Delta T + \sum_{i=1}^{k} \mu i \, \Delta ni \tag{4.50}$$

The infinitesimal form of Eq. (4.50) is:

$$dG = VdP - SdT + \sum_{i=1}^{k} \mu i \, dni \tag{4.51}$$

As noted previously, at constant temperature and pressure, the Gibbs free energy for an open system is:

$$dG_{T,P} = \sum_{i=1}^{k} \mu i \, dni \tag{4.52}$$

Chemical potential is defined as the change in Gibbs free energy, *G*, when an infinitesimal amount of that component (*dni*) is added, while the temperature (*T*), pressure (*P*), and the number of moles of all other components ($nj \neq i$) are held constant.

$$\mu_i = (\partial G/\partial ni)_{T,P,nj \neq i} \tag{4.53}$$

Chemical potential is also defined by Eq. (4.54)

$$\mu = G/n = G_M \tag{4.54}$$

which shows that the chemical potential, μ, is equal to the Gibbs free energy divided by the number of moles, G/n. G_M is known as the *molar Gibbs free energy or Gibbs molar free energy*. As indicated by Eq. (4.54), the chemical potential equals the molar Gibbs free energy.

It is important to note that chemical potential, μ, is an intensive property. To have consistent unit algebra, μ must have units of *energy per mole,* as G/n and G_M have units of energy per mole. Chemical potential can be thought of as a molecule's escaping tendency.

4.8.1 Chemical Potential Dependence on Temperature and Pressure

As shown in Eq. (4.54), the chemical potential, μ, is equal to the molar Gibbs free energy, G_M. Therefore, for component *i*, the chemical potential is also a function of temperature, pressure, and moles of each substance in the system. That is,

$$\mu_A = \mu_A(T,\ P,\ n_i, \ldots\ldots) \tag{4.55}$$

The differential of Eq. (4.55) is:

$$d\mu_A = (\partial\mu_A/\partial P)_T\ dP + (\partial\mu_A/\partial T)_P\ dT + \sum (\partial\mu_A/\partial n_i\ldots)_{T,P}\ dn_i\ldots \tag{4.56}$$

From Eq. (4.9a), $(\partial G/\partial P)_T = V$; Eq. (4.9b), $(\partial G/\partial T)_P = -S$; and Eq. (4.54), $\mu = G_M$, Eq. (4.56) can be rewritten as:

$$dG_M = d\mu_A = V_M\ dP - S_M\ dT + \sum (\partial\mu_A/\partial n_i\ldots)_{T,P}\ dn_i\ldots \tag{4.57}$$

where dG_M is the infinitesimal change in the molar Gibbs free energy, $d\mu_A$ is the infinitesimal change in the chemical potential of substance A, V_M is the molar volume of substance A, S_M is the molar entropy of compound A, and dn_i is the infinitesimal change in number of moles of substance A. This is the same equation as Eq. (4.51) that is converted to molar quantities.

4.8.2 Gibbs-Duhem Equation for a One-Phase Two-Component System

At constant temperature and pressure, the Gibbs-Duhem equation shows that the partial molar component properties of a mixture are not independent of each component. Thus, if the chemical potential of one of the components in a binary system is known, the chemical potential of the other component can be determined using the Gibbs-Duhem equation.

Equation (4.52), $dG_{T,P} = \sum_{i=1}^{k} \mu i\ dni$, can serve as the starting point for the derivation of the Gibbs-Duhem equation.

Integration of Eq. (4.52) gives:

$$G_{T,P} = \sum\nolimits_{i=1}^{k} \mu i\ ni \tag{4.58}$$

Differentiation of Eq. (4.58) using Eq. (1.88) gives:

$$dG_{T,P} = \sum_{i=1}^{k} \mu i \, dni + \sum_{i=1}^{k} ni \, d\mu i \tag{4.59}$$

Substituting for $dG_{T,P}$ from Eq. (4.52), $dG_{T,P} = \sum_{i=1}^{k} \mu i \, dni$, into Eq. (4.59) leads to:

$$\cancel{\sum_{i=1}^{k} \mu i \, dni} = \cancel{\sum_{i=1}^{k} \mu i \, dni} + \sum_{i=1}^{k} ni \, d\mu i \tag{4.60}$$

where $\sum_{i=1}^{k} ni \, d\mu i$ must equal to zero at constant temperature and pressure for (4.60) to be true. Importantly,

$$\sum_{i=1}^{k} ni \, d\mu i = 0 \text{ is the Gibbs-Duhem equation} \tag{4.61}$$

For a one-phase two-component system at constant temperature and pressure, Eq. (4.61) can be written as:

$$n_1 \, d\mu_1 + n_2 \, d\mu_2 = 0 \tag{4.62}$$

It follows that

$$n_1 d\mu_1 = -n_2 d\mu_2 \tag{4.63}$$

Dividing both sides by n_1 gives:

$$d\mu_1 = -(n_2 d\mu_2)/n_1 \tag{4.64}$$

which relates the chemical potential of component 1 to compound 2.

At constant temperature and pressure, the partial molar component properties of a mixture are also not independent of each component.

1. Graphical Observations of the Partial Molar Volume Interdependence Predicted by the Gibbs-Duhem Equation

Example 4.2

The Gibbs-Duhem equation describes the constraint on the interdependence of the intensive variables

Figure 4.3 illustrates the interdependence of the *partial molar volume*, V_{pm}s, predicted by the Gibbs-Duhem equation. V_{pm} is an intensive thermodynamic variable. Figure 4.3 plots the water and

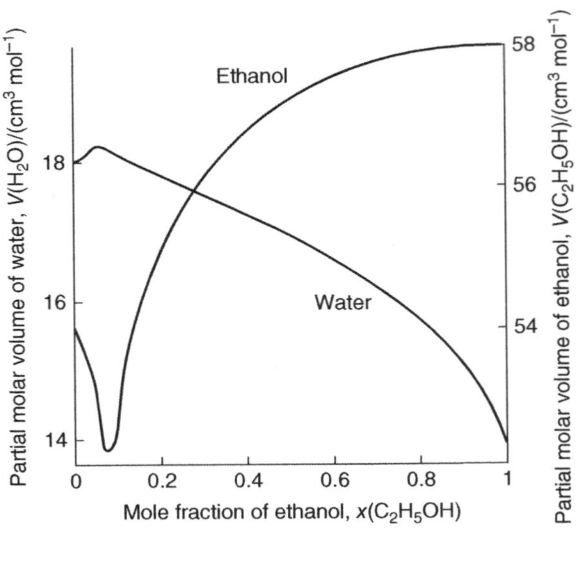

Figure 4.3 Partial molar volume of ethanol and water in ethanol-water mixture. *Source:* Adapted from Sudare et al. (2015).

ethanol V_{pm}s versus their mole fraction. On visual inspection of the water and ethanol V_{pm}s curves, does a trend of any kind stand out? If you are given the Gibbs-Duhem equation in terms of partial molar volumes, $X_{water}\, dV_{pm,\ water} + X_{ethanol}\, dV_{pm,\ ethanol} = 0$, does this equation illicit any visual trends? How about if this equation is rearranged to:

$$X_{water}\, dV_{pm, water} = -X_{ethanol}\, dV_{pm, water}$$

Several observations from Figure 4.3 may now become clearer.

1) The interdependence of partial molar volumes, as described by the Gibbs- Duhem equation, is observed. It is noted that a decrease in the partial molar volume of water with changing composition is associated with an increase in the partial molar volume of ethanol and vice versa.
2) It is also observed that due to the Gibbs-Duhem interdependence and constraints, a positive slope for the water partial molar volume curve is countered with a negative slope for the ethanol partial molar volume curve and vice versa.

2. Calculate the Final Mixture Volume when 100 mL of Ethanol is Added to 100 mL of Water at 20 °C

Given:

Density of ethanol at 20 °C is 0.789 g mL^{-1}
Density of water at 20 °C is 0.998 g mL^{-1}
Ethanol MW 46.07 g mol^{-1}
Water MW 18.02 g mol^{-1}

$$V_{final\ T,P} = n_{ethanol}\ V_{M,ethanol} + n_{water}\ V_{M,water} \tag{4.65}$$

where $V_{final\ T,P}$ is the final volume of the mixture at constant temperature and pressure, $n_{ethanol}$ and n_{water} are the moles of ethanol and water, respectively, and $V_{M,\ ethanol}$ and $V_{M,\ water}$ are the molar volumes of ethanol and water.

Use Figure 4.3 to determine $V_{M,\ ethanol}$ and $V_{M,\ water}$.

Solution

Step 1. Calculate the number of moles of ethanol and water in 100 mL of each.

$$n_{etthanol} = 100\,\cancel{\text{mL}}\,(0.789\,\cancel{\text{g}}/\cancel{\text{mL}})/(46.07\,\cancel{\text{g}}/\text{mol}) = 1.713\ \text{mol} \tag{4.66}$$

$$n_{water} = 100\,\cancel{\text{mL}}\,(0.998\,\cancel{\text{g}}/\cancel{\text{mL}})/(18.02\,\cancel{\text{g}}/\text{mol}) = 5.538\ \text{mol} \tag{4.67}$$

Step 2. Calculate the mole fraction of ethanol and water.

$$\text{Mole Fraction Ethanol} = 1.713/(1.713 + 5.538) = 0.236 \tag{4.68}$$

$$\text{Mole Fraction Water} = 5.538/(1.713 + 5.538) = 0.764 \tag{4.68a}$$

Step 3. Determine the molar volumes for ethanol and water from Figure 4.3.
Ethanol molar volume:

a) From the Eq. (4.68), the ethanol mole fraction is 0.236. From Figure 4.3, use a ruler with 1 mm divisions to estimate the location of 0.236 on the x-axis.
b) At this point, draw a parallel line to the y-axis until it intersects with the ethanol curve.

c) At this point on the ethanol curve, draw a line parallel to the x-axis that intersects the y-axis on the *right* side of the Figure 4.3
d) Estimate the ethanol molar volume.

Ethanol $V_{M,\ ethanol}$ is ~ 55.7 mL mol^{-1}
Water molar volume:

a) From Eq. (4.68a), the water mole fraction is 0.764. The location of 0.764 mole fraction water on Figure 4.3 x-axis is 1 – 0.764 = 0.236 which as determined in "a" above for ethanol mole fraction.
b) At 0.236, draw a parallel line to the y-axis until it intersects with the water curve.
c) At this point on the water curve, draw a line parallel to the x-axis that intersects the y-axis on the *left* side of Figure 4.3.
d) Estimate the water molar volume.

Water $V_{M,\ water}$ is ~ 17.7mL mol^{-1}
Step 4. Substitute the determined molar volume values into Eq. (4.65)

$$V_{final\ T,P} = n_{etthanol}\,V_{M,ethanol} + n_{water}\,V_{M,water}.$$
$$1.713\ \cancel{\text{mol EtOH}}\ (55.7\ \text{mL}/\cancel{\text{mol EtOH}}) + 5.538\ \cancel{\text{mol H}_2\text{O}}$$
$$(17.7\ \text{mL}/\cancel{\text{mol H}_2\text{O}}) =$$
$$= 95.4\ \text{mL} + 98.0\ \text{mL} = \underline{193\ \text{mL mixture (3 significant figures)}}$$

4.9 Molar Gibbs Free Energy and Chemical Potential for an Ideal Gas

The Gibbs free energy change for solids, liquids, and gases at constant temperature is described by Eq. (4.8a), $dG = VdP$. For solids and liquids, the molar volume is generally considered to be independent of pressure. For gases, volume is inversely proportional to pressure according to the gas law and $V = nRT/P$. Substituting for V into Eq. (4.8a) gives:

$$dG = (nRT/P)\,dp \tag{4.69}$$

For an ideal gas, Eq. (4.69) can be integrated and evaluated at its standard state of 1 bar or 1 kPa.

$$\int_G^{\circ} dG = \int_{P^\circ = 1}^{P} (nRT/P)\,dp \tag{4.70}$$

where G° and P° are the standard-state values for Gibbs free energy and pressure. $P^\circ = 1$ kPa. After integration, Eq. (4.70) is:

$$G - G^\circ = nRT\ln\,(P/P^\circ) \tag{4.71}$$

The right-hand side of the Eq. (4.71) is arrived at using separate integration and logarithm operations on Eq. (4.70). First, it is noted that using the indefinite integral Eq. (1.102), $\int (1/x)\,dx = \ln x + c$. In this case, the definite integral is evaluated from P° to P, giving ln (P) – ln (P°). The logarithm operation Eq. (1.47), $\ln x - \ln y = \ln(x/y)$, leads to the ln (P/P°) term in (4.71). Dividing both sides of Eq. (4.71) by the number of moles, n, gives

$$(G - G^\circ)/n = RT\ \ln\,(P/P^\circ) \tag{4.72}$$

and

$$G/n = G^\circ/n + RT \ln(P/P^\circ) \tag{4.73}$$

From Eq. (4.54), $\mu = G/n$, the general form of Eq. (4.73) becomes:

$$\mu = \mu^\circ + RT \ln(P_2/P_1) \tag{4.74}$$

where P_1 has been substituted for P° in Eq. (4.73), the standard-state gas pressure. P_2 has also been substituted for P. This provides a more generalized form of Eq. (4.73). μ° represents the standard-state chemical potential of one mole of pure gas at 1 atm and 25 °C.

4.9.1 Chemical Potential for a Nonideal System

Gases (except the noble gases), liquids, and solids experience noncovalent forces called van der Waals forces such as dispersion or London forces and hydrogen bonds. These intermolecular interaction forces lead to nonideal behavior. The concept of *fugacity, f,* was developed to address the nonideal behavior of gases. In the case of a nonideal gas, fugacity replaces *P,* the gas pressure in Eqs. (4.69) through (4.74). Fugacity can be referred to as the *escaping tendency* of a substance. For mixtures of a gas, the fugacity of an individual component is the *effective or apparent pressure* of the species as opposed to the actual formulated or prepared pressure. A *fugacity coefficient,* φ, is associated with fugacity (*f*) and pressure (P) of a gas by:

$$\phi = f/P \tag{4.75}$$

The fugacity coefficient of a gas is a measure of the magnitude of the nonideality of the system. As φ approaches, unity f and P (actual formulated pressure) will equal each other, and the gas approaches an ideal state. From Eq. (4.75), $\phi P = f$, where $P_1 = 1$ and

$$\mu = \mu^\circ + RT \ln(f) \tag{4.76}$$

In essence, fugacity is a "corrected" pressure that allows use of ideal-like equations to describe the behavior of real systems that exhibit interactions and other nonideal behavior. Eq. 4.76 can be rewritten using the concept of the fugacity coefficient

$$\mu = \mu^\circ + RT \ln(\phi P) = \mu^\circ + RT \ln(\phi) + RT \ln(P) \tag{4.77}$$

which shows that the contribution to nonideality of the chemical potential is $RT \ln(\phi)$, where $\phi = f/P$ is the measure of nonideal behavior.

A similar situation occurs for mixtures of liquids and solids. For liquids and solids, the *activity, a,* replaces fugacity and γ is called the *activity coefficient.* The activity coefficient is a measure of the nonideality of liquids and solids. As with gases, the species activity is a measure of its escaping tendency. For a two-phase system of the same component, the two phases have the same molar Gibbs free energy, chemical potential, and activity.

4.10 Gibbs Free Energy and Non-Pressure-Volume Work

The change in Gibbs free energy is a quantitative measure of the maximum energy available to perform nonPV work at constant temperature and pressure in a closed or open system. The following is a derivation showing that $dG = -dw_{nonPV}$. Here, dw_{nonPV} represents nonPV work done by the system.

The change in the total work, *dw,* can be partitioned into PV work of expansion (PV work) and nonPV work. This can be written as:

$$dw = dw_{PV} + dw_{nonPV} \tag{4.78}$$

A list of five equations that are employed for the derivation are given:

1) Equation (3.4) $dU = dq - dw$ (where the minus sign indicates that the work is done by the system)
2) Equation (3.75) $dq = Tds$ (for reversible process at constant pressure)
3) Rearrangement and derivatization of Eq. (3.30), $H = U + PV$ gives $dU = dH - PdV$
4) Equation (3.6) $dw_{PV} = PdV$ (work of expansion in a closed system when work is done on the system)
5) Equation (4.2) $\Delta G = \Delta H - T\Delta S$ infinitesimal form $dG = dH - TdS$

For this example, to help direct the reader to the equation of interest, the equation of interest will be identified by citing the "listed" number and the associated equation like (1) Eq. (3.4). The derivation begins by substituting Tds from (2) Eq. (3.75) for dq in (1) Eq. (3.4) to give:

$$dU = Tds - dw \tag{4.79}$$

From (3) substitute $dH - PdV$ for dU in Eq. (4.79).

$$dH - PdV = Tds - dw \tag{4.80}$$

From Eq. (4.78) substitute $dw_{PV} + dw_{non\text{-}PV}$ for dw *in* Eq. (4.80).

$$dH - PdV = Tds - dw_{PV} - dw_{nonPV} \tag{4.81}$$

From (4) Eq. (3.6), substitute dw_{PV} for PdV in Eq. (4.81) and cancel common terms.

$$dH - \cancel{dw_{PV}} = Tds - \cancel{dw_{PV}} - dw_{nonPV} \tag{4.82}$$

Rearrange Eq. (4.82) to give:

$$dH - Tds = -dw_{nonPV} \tag{4.83}$$

From (5) Eq. (4.2), substitute dG for $dH - TdS$ in Eq. (4.83).

$$dG = -dw_{nonPV} \tag{4.84}$$

Therefore, $-dG = dw_{nonPV}$. For a reversible process at constant temperature and pressure, the decrease in Gibbs free energy is equal to the maximum *useful work* or *non-pressure-volume expansion work* done by the system. Useful work may be done in the form of electrical work, mechanic work, osmotic work, an increase in surface area, and others. It is also noteworthy to recognize that the concept of "free energy" does not imply that the energy itself is limitless or without origin. It refers to the portion of a system's total energy that is "free" or available to do useful work under constant temperature and pressure conditions, without violating the laws of thermodynamics. The change in free energy is the maximum amount of energy that is available that can be freed to do work beyond PV work (Doan and Lower, 2022).

4.11 Gibbs Free Energy and Phase Changes

Equation (4.15) states that $\Delta G_{system} = \Delta H_{system} - T\Delta S_{system}$. At the phase transition temperature, the phases are in equilibrium. The most common phase transitions encountered by a pharmaceutical scientist are melting, polymorphic, evaporation, and sublimation transformations. At the temperature of the phase transition,

$$\Delta G_{phase\ transtion} = 0 \text{ and } \Delta H_{transition} = T_{trasition}\,\Delta S_{transition} \tag{4.85}$$

For a simple melting process, the solid is heated, and its temperature rises until the melting temperature is reached and melting begins. It is important to note that during the melting process, heat is absorbed by the system (endothermic), but the temperature of the system does not increase until all the solid has melted. Once the solid is completely melted, the temperature of the liquid (melt) will again increase depending on its heat capacity.

At the melting temperature, ΔH_{system} is known as the molar enthalpy of fusion or, simply, the enthalpy of fusion ΔH_{fusion}. ΔH_{fusion} is the energy required to melt one mole of solid. Enthalpies of transition may be positive or negative depending on if the energy is given up by the system (exothermic) or absorbed by the system (endothermic).

The transition enthalpies follow Hess's law. For example, the molar enthalpy of sublimation can be determined by knowing the enthalpy of fusion and enthalpy of vaporization.

$$\Delta H_{sublimation} = \Delta H_{fusion} + \Delta H_{vaporization} \tag{4.86}$$

4.11.1 Determination of the Transition Entropy

From Eq. (4.85), $\Delta H_{transition} = T_{transition}\, \Delta S_{transition}$ can be rearranged to:

$$\Delta S_{transition} = \Delta H_{transition}/T_{transition} \tag{4.87}$$

which shows that the system transition entropy can be calculated knowing the transition enthalpy and transition temperature.

Example 3.6 illustrates how differential scanning calorimetry data can be used to calculate the transition entropy. The difference in the entropies of fusion is used to correlate the crystal lattice packing differences and the fusion entropy of two closely related local anesthetics, lidocaine and prilocaine.

4.12 Application of Gibbs Free Energy and Tertiary Protein Structure and Stability

A globular protein's therapeutic function critically depends on the protein remaining in its most stable *3D tertiary globular form*. A globular protein's function is also strongly dependent on its *secondary* structure made of *helical* and *sheath* structures and the protein's *primary amino acid sequence*. A protein's globular therapeutically active form is called its *native* state. This native state tends to minimize its surface area, forming a compact "globule" or ball. The *core* of the globule generally houses hydrophobic amino acids. The core also experiences van der Waals interactions involving dispersion and London forces. The core of the globular protein is considered hydrophobic in nature. The surface of the globule protein is generally made up of hydrophilic amino acids and is considered hydrophilic in nature.

Denaturation and *unfolding* of a protein can be caused by many environmental factors such as pH, ionic strength, heat, freezing, light, adsorptive surfaces, and oxygen. Mechanical physical stress such as shaking and sheer can also lead to denaturation. Figure 4.4 depicts a native folded globular protein unfolding to become a denatured unfolded nontherapeutically inactive protein. When a native protein unfolds, it exposes many amino acid functions to the aqueous environment, resulting in hydrogen bond formation between water and the hydrophilic secondary and primary amino acids of the protein.

Insulin and albumin are two highly studied globular proteins. Figure 4.5 shows the tertiary structure of insulin.

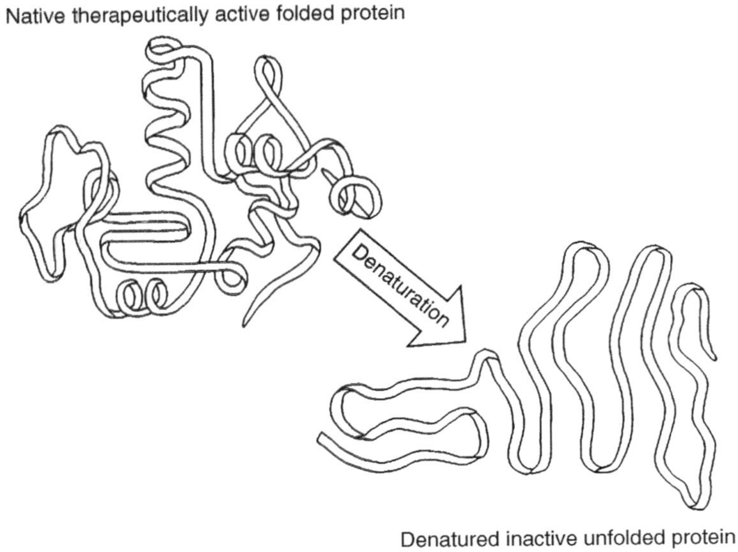

Figure 4.4 Native therapeutically active folded globular protein unfolding to become a denatured unfolded inactive protein.

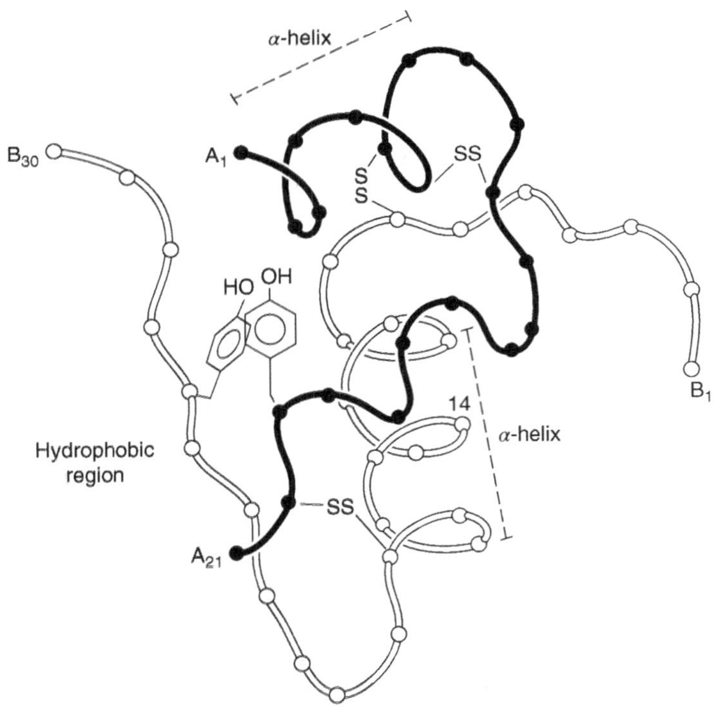

Figure 4.5 Tertiary structure of insulin showing chain A and chain B. Each chain has an alpha helix. A hydrophobic interaction region involving tyrosine on the A and B chains is also shown. The primary amino acid sequence is represented by the open and closed circles of the A and chains. *Source:* Adapted from Wang (1992).

Table 4.4 Comparison of Insulin and Albumin Physical and Chemical Properties.

Property	Insulin[a]	Albumin[b]
Number of amino acids	51	584
Molecular weight (Daltons)	5808	65,000
Stable chemical pH range	7–7.8	2–10; at pH 1-2, albumin elongates reversibly
Stable temperature range	2 to 8° C for 18–24 months; it should not be frozen	2–25° C for 36 months; can be pasteurized at 60° C for 10 hours; can be frozen
Solubility at isoelectric point	Precipitates at pH 5.3–5.5, its isoelectric point	One of the few proteins that is soluble at its isoelectric point
Stability to shear	Precipitates on shaking; mix by rolling gently	Stable to shear

[a] Human insulin https://www.accessdata.fda.gov/drugsatfda_docs/label/2011/018780s120lbl.pdf
[b] Wang and Hanson (1988).

Insulin's physical-chemical stability is comparable to most proteins. On the other hand, albumin has been shown to be a remarkably stable protein. As a result of its stability, albumin has been used as a stabilizer and cryoprotectant in freeze-dried proteins. Table 4.4 compares insulin's and albumin's size and physical-chemical stability when exposed to various environmental conditions.

Application of Gibbs free energy is critical to the thermodynamic understanding of the physical stability of globular proteins. A protein's stable conformation depends on a critical balance of compensating energies. Figure 4.3 and Eq. (4.88) depict a protein denaturation reaction.

$$\text{Native Folded State} \rightarrow \text{Denatured Unfolded State} \tag{4.88}$$

The arrow points to the right to indicate that the denaturation reaction is generally not reversible. The $\Delta G_{reaction}$ for this reaction is shown in Eq. (4.89).

$$\Delta G_{reaction} = G_U - G_N \tag{4.89}$$

where G_U is the Gibbs free energy of the unfolded globular protein and G_N is the Gibbs free energy of the native folded state of the globular protein.

Δ Equation (4.90) follows from Eq. (4.2)

$$\Delta G_{folded} = \Delta H_{folded} - T\Delta S_{folded} \tag{4.90}$$

where the enthalpy, ΔH_{folded}, and entropy, $T\Delta S_{folded}$, contributions to ΔG_{folded} and the stability of a protein's stable folded conformation can be elucidated. The energies favoring the folded and unfolded states are discussed in the next section.

4.12.1 Enthalpic Energy Favoring the Native Folded State

Hydrogen bonds favor the native folded state. Most of the core hydrogen bonds result in the formation of the protein's secondary helix and sheet structures. This hydrogen bond formation occurs on the "inside" of the structures leaving the "outer" portions of the helix and sheet with the hydrophobic portions of the amino acids exposed to the core. These energetically favored configurations lead to a net decrease in enthalpy. That is, heat is released upon hydrogen bond formation giving a $-\Delta H_{folded}$.

van der Waals interactions, including *dispersion and London force* interactions, lead to additional stability of the native state and a $-\Delta H_{folded}$.

4.12.2 Entropic "Hydrophobic Effect" Favoring the Native Folded State

Water's involvement with the *hydrophobic effect* is the primary reason the native protein conformation is a globule. The hydrophobic effect is associated with water's entropic structural change. If water is exposed to the hydrophobic regions of a protein, the hydrophobic residues cause the large water-solvent volume to form a *highly organized cage-like shell* around the protein. This large number of constrained water molecules results in a decrease in water's entropy.

The formation of the folded globule state, with its hydrophobic core, releases the ordered water molecules. As the water is released, the hydrophobic moieties become "buried" in the interior of the protein resulting in its globular form that reduces its surface area. The release of the ordered water molecules increases the water's entropy, which is thermodynamically favorable and drives the folding process.

4.12.3 Entropic Conformational Freedom Favoring Denaturation and Unfolding

When a protein is in its unfolded state, its entropy is large. This conformational freedom is the major driving force for denaturation and unfolding.

4.12.4 Application: Overall Thermodynamic Effects on an Insulin-like Protein's Folded Conformation

Table 4.5 provides the overall energetics involved in maintaining a natively folded protein of 50 amino acids. For perspective, insulin is a 51 amino acid protein.

The positive sign for the hydrogen bonds and van der Waals interactions indicates the energy that contributes to the folded state. That is, it would take a total of 175 kJ mol^{-1} to disrupt the hydrogen and van der Waals bonds that keep the protein in its folded state. Likewise, it would take 210 kJ mol^{-1} to overcome the hydrophobic effects that maintain the protein's folded state.

Table 4.5 Enthalpic and Entropic Contributions to Maintain a 50 Amino Acid Protein in its Native Folded State.

Enthalpic Contributions	Entropic Contributions[a]	Energy[b]
Hydrogen bonds		+50 kJ mol^{-1}
Van der Waals interactions		+125 kJ mol^{-1}
	Hydrophobic effect	+210 kJ mol^{-1}
	Conformational entropy	−365 kJ mol^{-1}
Net Stabilization Energy		+20 kJ mol^{-1}

[a] Entropy values have been converted to Gibbs free energy units using $\Delta G = -T\Delta S$.
[b] + indicates contributions that support a folded state, – indicates contributions that support denaturation and the unfolded state.
Source: from LibreTexts: https://bio.libretexts.org/Under_Construction/OLI/Biochemistry/Unit_2:_Biochemistry/Module_4:_Protein_Structure/Module_4.4:_Tertiary_Structure_and_Protein_Stability

The -365 kJ mol^{-1} associated with the conformational entropy of the unfolded state indicates that 365 kJ mol^{-1} would be lost if a large degree of microstates were given up during protein folding.

The net $+20$ kJ mol^{-1} suggest that unfolding is not a spontaneous process. In other words, $+20$ kJ mol^{-1} contribute to the folded state making the native folded state the more stable protein form.

From Table 4.5, it can be seen that the net energy maintaining the protein in its native folded state is relatively small compared to the contributions favoring folding and those favoring unfolding. It is important to note that the major contribution to protein folding is the entropic hydrophobic effect. The 2.5 times greater positive impact of van der Waals forces on maintaining the folded state compared to the impact hydrogen bonding is also interesting. The magnitude of the negative compensational unfolding conformation entropy is impressive. As mentioned previously, this fine balance favoring folding can be easily disrupted by many environmental factors.

References

Chang, R. (1998). *Chemsitry*, 6e. Boston: McGraw Hill.

Connors, K.A. and Mecozzi, S. (2010). *Thermodynamics of Pharmaceutical Systems: An Introduction to Theory and Applications*, 2e. Hoboken, NJ: John Wiley & Sons.

Daniels, F. and Alberty, R.A. (1975). *Physical Chemistry*, 4e. New York: John Wiley & Sons.

Doan, C. and Lower, S. (2022). https://chem.libretexts.org/Bookshelves/Physical_and_Theoretical_Chemistry_Textbook_Maps/Supplemental_Modules_(Physical_and_Theoretical_Chemistry)/Thermodynamics/Energies_and_Potentials/Free_Energy/Gibbs_(Free)_Energy (accessed March 2022).

LibreTexts. Module 4.4 Tertiary Structure and Protein Stability. https://bio.libretexts.org/Under_Construction/OLI/Biochemistry/Unit_2%3A_Biochemistry/Module_4%3A_Protein_Structure/Module_4.4%3A_Tertiary_Structure_and_Protein_Stability (accessed February 2022).

Human Insulin (2011). https://www.accessdata.fda.gov/drugsatfda_docs/label/2011/018780s120lbl.pdf (accessed March 2025).

Janscó, J. and Fenby, D.V. (1983). Thermodynamics of dilute solutions. *J. Chem. Ed.* 60: 382–386.

Long-Qing, C. (2019). Chemical potential and Gibbs free enegy. *MRS Bull.* 44: 520–523. https://www.cambridge.org/core/journals/mrs-bulletin/article/cemical-potential-and-gibbs-free-energy/2BACF973D95A697C947E741630951F46 (accessed January 2022).

Lower, S. (2023). LibreTexts Chemistry. Chapter 15.4: Free energy and the gibbs function. https://chem.libretexts.org/Bookshelves/General_Chemistry/Book%3A_Chem1_(Lower)/15%3A_Thermodynamics_of_Chemical_Equilibria/15.04%3A_Free_Energy_and_the_Gibbs_Function (accessed April 2023).

Sudare, T., Ueno, T., Wattanaphanit, A., and Saito, N. (2015). Accelerated nanoparticles synthesis in alcohol-water mixture based solution plasma. *Phys. Chem. Chem. Phys.* 17: 30255–30259.

Wang, Y.-C.J. and Hanson, M.A. Technical Report No. 10(1988). Parenteral formulations of proteins and peptides: stability and stabilizers. *J. Parenter. Sci. Technol.* 42: S3–S26. Suppl. No. 2S.

Wang, Y.-C.J. (1992). Parenteral products of peptides and proteins. In: *Pharmaceutical Dosage Forms; Parenteral Medications*, vol. 1 (ed. K.E. Avis, A. Liebermann, and L. Lachman). New York: Marcel Dekker.

Chapter 4 Problem Set

4.1 Fill in the table.

ΔH_{system} sign	ΔS_{system} sign	$-T\Delta S_{system}$ term sign	ΔG_{system} sign	Spontaneity description
– Exothermic	+			
– Exothermic	–			
+ Endothermic	+			
+ Endothermic	–			

4.2 Fill in the figure when $\Delta H_{system} < 0$ and $\Delta S_{system} > 0$.

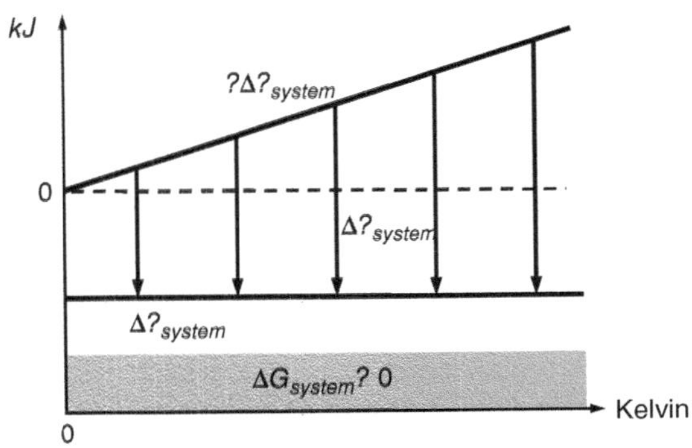

4.3 Explain the above figure.

4.4 Fill in the figure when $\Delta H_{system} > 0$ and $\Delta S_{system} < 0$.

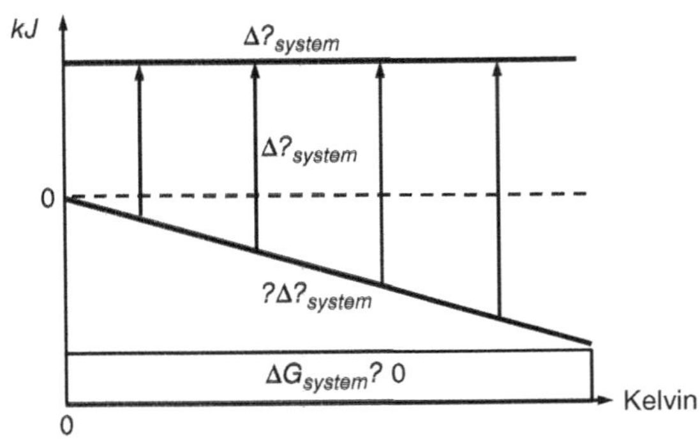

4.5 Explain the above figure.

4.6 Fill in the figure.

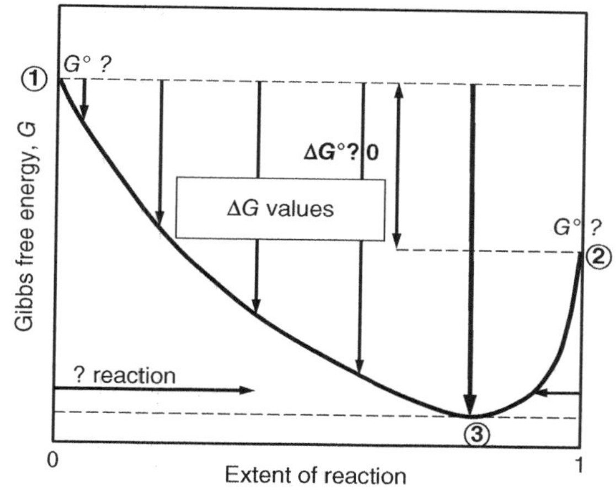

4.7 Explain the above figure.

4.8 Calculate the final mixture volume when 100 mL of 1-propanol is added to 100 mL of water at 20 °C.
Given:

1-propanol density at 20 °C is 0.803 g mL^{-1}
Water density at 20 °C is 0.998 g mL^{-1}
1-propanol MW is 60.09 g mol^{-1}
Water MW 18.02 g mol^{-1}
At a mole fraction of 0.194, 1-propanol, the molar volume of 1-propanol is 72 mL mol^{-1} and water molar volume is 18 mL mol^{-1}

4.9 Show that $dG = -dw_{non\text{-}PV}$

5

Equilibrium

5.1 Introduction

A general definition of *equilibrium* is a state in which opposing forces or influences on a system are balanced. At equilibrium, the internal energy of the system does not change with time and the change in the Gibbs free energy is zero. As stated in section Eq. (4.35), the forward reaction rate equals the reverse reaction rate. Although the opposing influences are balanced and the forward and reverse reaction rates are equal, the equilibrium system is not static. At equilibrium, the system is dynamic with individual molecules or particles in continuous exchange moving in the forward and reverse directions.

There are two types of equilibrium: homogeneous and heterogeneous equilibrium. Homogeneous equilibrium takes place when all the component materials are present in the same phase. A reversible aqueous solution reaction of reactants and products is a homogeneous equilibrium. Melting is an example of a heterogeneous equilibrium where the solid and liquid phases are in equilibrium with each other.

Chemical equilibrium was introduced in several sections of Chapter 4 (see Sections 4.3 and 4.6). This chapter will augment these earlier discussions and expand the equilibrium discussion to include phase equilibrium and phase transitions, ionic equilibrium, and noncovalent binding equilibrium. Solubility and absorption equilibria will be discussed in Chapters 6 and 8.

5.1.1 Brief Review of Chemical Equilibrium

5.1.1.1 Law of Mass Balance

The law of mass balance is an accounting of material and charge of the system. Fundamental to mass balance is the principle of conservation of energy which was discussed in Section 2.4. Basically, matter, charge, and energy can neither be created nor destroyed; rather, they can only be transformed or transferred from one form to another.

This accounting for chemical reactions and equilibrium is done by having the same number of atoms for each element and same total charge for the reactants and products. When this is true, the reaction or equilibrium is said to be *a balanced equation*. That is, mass and charge are balanced on both sides for the equation.

Thermodynamics for Drug Product Design: A Guide for Pharmaceutical Scientists, First Edition. William Craig Stagner.

5.1.1.2 Law of Mass Action

The law of mass action states that the rate of any given chemical reaction is proportional to the product of the activities (or concentrations) of the reactants raised to the power of their stoichiometric amounts. This is illustrated in the simple reaction case shown:

$$aA + bB \rightarrow cC + dD$$
$$\text{Rate} \propto [A]^a[B]^b$$
$$\text{Rate} = k[A]^a[B]^b$$

The proportion sign can be replaced by the "equals sign" and a "constant." k is the rate constant for the reaction; a, b, c, and d are the stoichiometric number of moles for the reaction. A and B are the reactants, and C and D are products of the reaction.

5.1.1.3 Equilibrium State

The equilibrium state occurs when the forward reaction and reverse reaction occur at equal rates. This is shown as:

$$aA + bB \rightleftharpoons cC + dD$$

From the law of mass action:

$$\text{The forward reaction rate} = k_f[A]^a[B]^b$$
$$\text{The reverse reaction rate} = k_r[C]^c[D]^d$$

where k_f and k_r are the forward and reverse rate constants.

As mentioned previously, the equilibrium state occurs when the forward and reverse reaction rates are equal. Setting these rates equal to each other gives:

$$k_f[A]^a[B]^b = k_r[C]^c[D]^d$$

And the equilibrium constant, K, becomes:

$$K_{equilibrium} = k_f/k_r = [\text{Products}]/[\text{Reactants}] = [C]^c[D]^d/[A]^a[B]^b$$

5.1.1.4 Equilibrium State and Gibbs Free Energy

For the reaction $aA + bB \leftrightarrows cC + dD$,

$$\Delta G = \sum G_{products} - \sum G_{reactants} \tag{5.1}$$

Chemical potential, μ, is defined as the change in Gibbs free energy when one mole of substance is added to the system with temperature, pressure, and other system components held constant (Eq. 4.53).

$$\mu = G/n \tag{4.53}$$

and

$$G = n \cdot \mu \tag{5.1a}$$

For a closed system at constant temperature and pressure and substituting Eq. (5.1a) into Eq. (5.1) gives:

$$\Delta G = (c\mu_C + d\mu_D) - (a\mu_A + b\mu_B) \tag{5.1b}$$

where a, b, c, and d are the molar stoichiometric coefficients of the reactants and products.

The chemical potential, μ, is given in Eq. (4.76a),

$$\mu_i = \mu_i^\circ + RT\ln(a_i) \tag{4.76a}$$

The chemical potential from Eq. (4.76) using activity instead of fugacity can be substituted into Eq. (5.1b) to rewrite the equation as:

$$\Delta G = [(c\mu_C^\circ + d\mu_D^\circ) - (a\mu_A^\circ - b\mu_B^\circ)] + \left[(RT\ \ln a_C{}^c + RT\ \ln a_D{}^d) - (RT\ \ln a_A{}^a + RT\ \ln a_B{}^b)\right] \tag{5.1c}$$

Equation (5.1c) can be rearranged to consolidate standard Gibbs free energy and activity terms where $\Delta G^\circ = [(c\mu_C^\circ + d\mu_D^\circ) - (a\mu_A^\circ - b\mu_B^\circ)]$ and using Eq. (1.46) ln x + ln y = ln (xy), and Eq. (1.47), ln x –ln y = ln(x/y) to give:

$$\Delta G = \Delta G^o + (RT)\ \ln\left[a_C{}^c \cdot a_D{}^d)/(a_A{}^a \cdot a_B{}^b)\right] \tag{5.1d}$$

and

$$\Delta G = \Delta G^o + RT\ \ln Q \tag{5.2}$$

From the law of mass action, Eq. (4.34), $[a_C{}^c\ .\ a_D{}^d)/(a_A{}^a\ .\ a_B{}^b)] = Q$, *the reaction quotient*. At equilibrium Q becomes K, the equilibrium constant. At equilibrium, $\Delta G = 0$, and Eq. (5.2) becomes:

$$\Delta G^o = -(RT)\ \ln K \tag{5.3}$$

where Q is replaced by the equilibrium constant, K.

Table 5.1 summarizes the effect of K on ΔG° and Figure 5.1 illustrates the effect of the relationship between the extent of a reversible reaction and $G^\circ{}_{reactants}$, $G^\circ{}_{products}$, and $\Delta G^\circ{}_{reaction}$.

Figure 5.1 shows a Gibbs free energy curve as a function of the extent of a reversible reaction $aA + bB \leftrightarrows cC + dD$. When the extent of the reaction is 0, the reaction mixture is 100% reactants A and B. When the extent of the reaction is 1, the reaction mixture is 100% products C and D. As indicated in Figure 5.1, $G^\circ{}_{products}$ (location 2) < $G^\circ{}_{reactants}$ (location 1). In this case, $\Delta G^\circ{}_{reaction} = (G^\circ{}_{products} - G^\circ{}_{reactants}) < 0$, and the products are favored as shown in Figure 5.1. Depending on the position of the reaction along the extent axis, the reaction will move either toward the reactants or products to achieve equilibrium.

The equilibrium, location 3, is the minimum of the ΔG curve. At this point the largest negative change in the Gibbs free energy is observed. This means that the availability of useful work is maximized when the system is in equilibrium. At the minimum in the ΔG *curve*, the slope of the curve is zero and the concentrations of reactants and product do not change.

Table 5.1 Three Cases Showing the Relationship Between the Equilibrium Constant, K, and ΔG°.

K	Ln K	ΔG°	State of products and reactants at equilibrium
>1	+	–	Extent of reaction favors products
=1	0	0	Product and reactant concentrations are equal
<1	–	+	Extent of reaction favors reactants

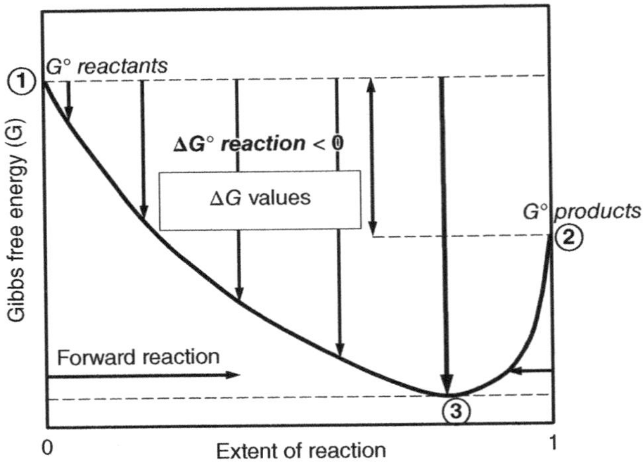

Figure 5.1 Relationship between the extent of a reversible reaction, and $G°_{reactants}$, $G°_{products}$, and $\Delta G_{reaction}$ when $\Delta G°_{reaction} < 0$. *Source:* Adapted from Lower (2023). https://chem.libretexts.org/Bookshelves/General_Chemistry/Book%3A_Chem1_(Lower)/15%3A_Thermodynamics_of_Chemical_Equilibria/15.04%3A_Free_Energy_and_the_Gibbs_Function

5.2 Le Châtelier's Principle

Le Châtelier's principle states that if any *change or stress* is placed on a system that is in *equilibrium*, the system will adjust to a *new* equilibrium *counteracting* the *original change or stress.* Le Châtelier's principle is critical to the understanding, interpretation, and prediction of equilibrium systems and processes. The position of an equilibrium can be changed by a change in concentration, pressure, temperature, and addition of a catalyst. As stated previously, a general definition of *equilibrium* is a state in which opposing forces or influences on a system are balanced. This next section uses a "balance" as an analogy to understand, interpret, and predict reaction and positional equilibrium changes when concentration, pressure, and temperature are changed; and stress the equilibrium system.

5.2.1 Effect of Concentration Change Stress on Equilibrium

The reaction between ferric cation and thiocyanate anion will be used to discuss the effect of concentration change or stress on a system in equilibrium. The mass-charge balanced equilibrium reaction is shown in Eq. (5.4).

$$Fe^{+3}{}_{(aq)} + SCN^{-1}{}_{(aq)}\ \text{(colorless)} \rightleftharpoons FeSCN^{+2}\ \text{(red)} \tag{5.4}$$

There are "four" concentration stresses that can be applied to this equilibrium system. The concentration of reactants can be increased (1) or decreased (2) and the concentration of the products can be increased (3) or decreased (4). Example 5.1 illustrates the effect of adding reactant when the system is in equilibrium. Example 5.2 shows the effect of decreasing the reactant when the system is in equilibrium.

Example 5.1 Effect of Adding Reactant to Eq. (5.4) which is in the Equilibrium State. A Balance is Used as an Analogy

Initial State, Figure 5.2a:

The balance beam is in the "balanced horizontal" position, illustrating that the rate of forward reaction and reverse reaction are equal and the concentrations of reactants and products remain constant at constant conditions.

Added Stress, Figure 5.2b:

The equilibrium is stressed by adding a reactant to the system. The reactant side of the balance shifts downward representing the added stress caused by the increased amount of reactant added.

Counteraction to Stress, Figure 5.2c:

The stress caused by the addition of the reactant is counteracted by the system forming more product. That is, the equilibrium moves toward the right, forming more product. Over time, the product side increases in red color intensity showing increased product formation until a new equilibrium between reactants and products is established.

Rebalanced Equilibrium State, Figure 5.2d:

The balance beam is once again horizontal, indicating that a new equilibrium state has been achieved. The products of the new equilibrium have increased as indicated by the increase in red color intensity. This example follows the Le Châtelier's principle which states that if any *change or stress* is placed on a system that is in *equilibrium*, the system will adjust to a *new* equilibrium *counteracting* the *original change or stress.*

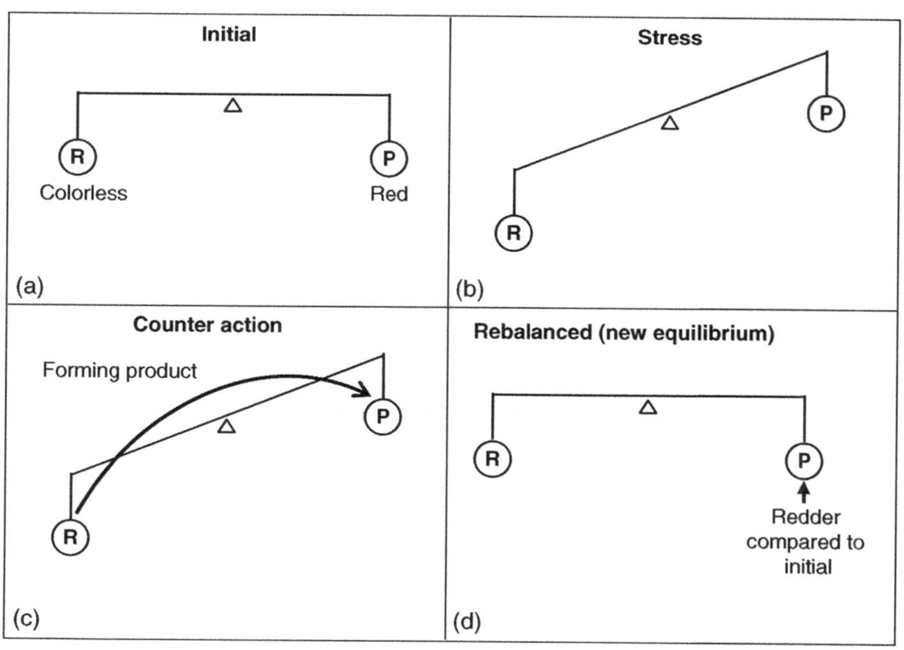

Figure 5.2 (a–d) Effect of adding reactant to a system at equilibrium.

Example 5.2 Effect of Decreasing the Reactant) in Eq. (5.4)
In this example, silver nitrate is added to the solution show in Eq. (5.4). The silver cation reacts with the thiocyanate anion to form a white precipitate. This reaction is shown in Eq. (5.5).

$$Ag^{+1}{}_{(aq)} + SCN^{-1}{}_{(aq)} \rightleftharpoons AgSCN_{(solid)} \tag{5.5}$$

This decreases the amount of $SCN^{-1}{}_{(aq)}$ in solution, and the effect on the equilibrium shown by Eq. (5.4) is discussed.

Initial State, Figure 5.3a:
The balance beam is in the "balanced horizontal" position, illustrating that the rates of forward reaction and reverse reaction for Eq. (5.4) are equal and the concentrations of reactants and products remain constant at constant conditions for the equilibrium state.

Added Stress, Figure 5.3b:
The equilibrium is stressed by decreasing the reactant, $SCN^{-1}{}_{(aq)}$ in Eq. (5.4). This is accomplished by precipitating SCN^{-1} with Ag^{+1} by adding silver nitrate to the equilibrium solution (Eq. 5.4). The left-hand side of the balance beam will move upward as the SCN^{-1} reactant is precipitated and lost from the reactant solution. The right-hand side of the balance beam will move downward because there is more product compared to the original equilibrium condition. Using the balance analogy, the product side becomes "heavier," while the reactant side becomes "lighter."

Counteraction to Stress, Figure 5.3c:
The stress caused by the precipitation and loss of reactant is counteracted by the system forming more reactant. That is, the equilibrium is moved toward the left side of Eq. (5.4) forming more reactant. Over time, the SCN^{-1} reactant is increased until a new equilibrium between reactants and products is established.

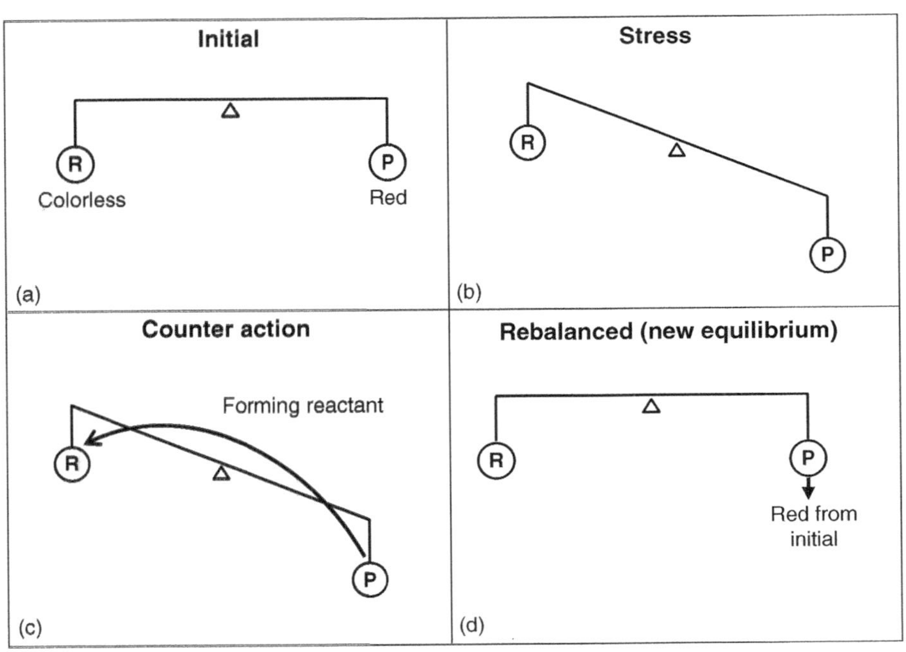

Figure 5.3 (a–d) Effect of decreasing the reactant from a system at equilibrium.

Rebalanced Equilibrium State, Figure 5.3d:

The horizontal balance beam indicates that a new equilibrium state has been achieved. The product of the new equilibrium has decreased as indicated by the decrease in the red color. The new equilibrium has readjusted to the loss of SCN^{-1} caused by precipitating this reactant with silver cation.

5.2.2 Effect of Pressure Change Stress on Equilibrium

In general, pressure changes do not affect solution or solid-state equilibrium systems. Pressure changes affect gas equilibrium systems when the reversible gaseous reactions change the total number of gas molecules as measured by the gas' partial pressure.

Gas pressure is related to the number of gas molecules per volume. More gas molecules in an equal volume result in a higher gas pressure. The pressure of a gaseous system can be changed by changing the volume or adding or decreasing the amount or partial pressure of the mixture components. Addition of gaseous components places stress on the equilibrium. The stress on the equilibrium is reduced by shifting equilibrium to lower pressures and fewer gas molecules.

It is important to note that adding inert gas(es) to a reversible reaction mixture does not affect its equilibrium state. Only addition or reduction of gases involved with the reaction will affect the equilibrium.

There are two pressure stresses that can be placed on a reversible gaseous equilibrium reaction. The equilibrium system can be subjected to (1) an increase in pressure or (2) a decrease in pressure.

Example 5.3 Effect of Adding Pressure by Reducing the Volume to the Gaseous Reaction Mixture Shown in Eq. (5.6) Where Nitrogen Dioxide is Converted to Dinitrogen Tetroxide

$$2NO_{2\,(g)}\ \text{(red-brown)} \rightleftharpoons N_2O_{4\,(g)}\ \text{(colorless)} \tag{5.6}$$

Initial State, Figure 5.4a:

The balance beam is in the "balanced horizontal" position. This means that the opposing forces or influences on a system are balanced. The system is considered in equilibrium. The concentration or partial gas pressure remains constant if the system conditions are not changed.

Added Stress, Figure 5.4b:

The equilibrium is stressed by decreasing the volume of the system which increases the system's pressure. The side of Eq. (5.6) that has the larger number of gas molecules will experience the stress caused by increasing the pressure on the system. The left-hand side of Eq. (5.6) has two molecules of NO_2 compared to one molecule of N_2O_4 on the right-hand side. Therefore, the NO_2 side of the equation "feels" the stress of increased pressure.

The balance analogy illustrates this added stress by showing the "reactant" arm moving downward.

Counteraction to Stress, Figure 5.4c:

To counteract the stress caused by increased pressure, the equilibrium shifts to the right and creates more product, N_2O_4. The equilibrium position shifts to reduce the pressure. This is accomplished by shifting in the reaction direction to fewer gas molecules and lower pressure. Over time, the red-brown color intensity associated with NO_2 will decrease indicating the shift of the equilibrium to N_2O_4.

Rebalanced Equilibrium State, Figure 5.4d:

The horizontal balance beam indicates that a new equilibrium state has been achieved and has been rebalanced. In the new equilibrium, the number of N_2O_4 molecules has increased at the expense of NO_2. The equilibrium solution red-brown color intensity has also decreased.

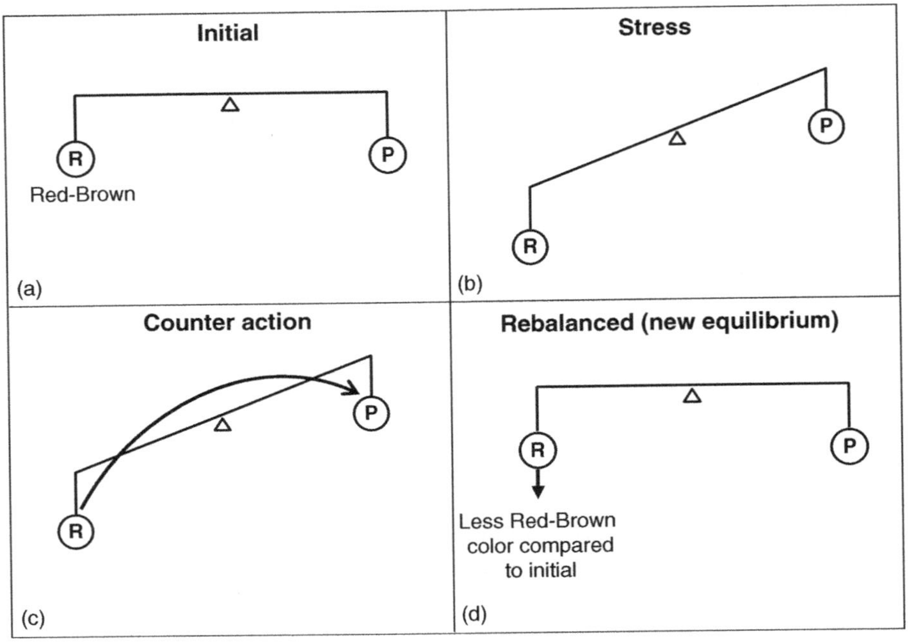

Figure 5.4 (a–d) Effect of adding pressure by reducing the volume to the gaseous reaction mixture.

5.2.3 Effect of Temperature Change Stress on Equilibrium

There are two temperature stresses that can be placed on a reversible equilibrium system. The equilibrium system can be subjected to (1) an increase in temperature or (2) a decrease in temperature. The direction of the counteraction to these changes in temperature depends on whether the equilibrium reversible reaction is endothermic (absorbs heat) or exothermic (gives up heat). Endothermic and exothermic heats were discussed in Section 3.3.2.

For an endothermic equilibrium system, the absorbed heat of the reaction is considered a reactant in the equilibrium equation. An exothermic heat of reaction is a product of the reaction.

Example 5.4 Effect of Increasing the Temperature of an Endothermic Reaction Process
The reaction between hydrogen and iodine is an endothermic reaction having a heat of reaction of $+52\ \text{kJ mol}^{-1}$. The plus sign indicates that the system has absorbed heat. As mentioned previously, endothermic heats are considered a reactant, so it is incorporated on the reactant side of the equation, Eq. (5.7). An increase in the temperature of the equilibrium acts as a stress on the system and causes the equilibrium to shift toward the production of more product.

$$H_{2(g)}\ (\text{colorless}) + I_{2(s)}\ (\text{purple}) + 52\ \text{kJ/mol} \rightleftharpoons 2HI_{(g)}\ (\text{colorless}) \tag{5.7}$$

Initial State, Figure 5.5a:

The balance beam is in the “balanced horizontal” position representing the Eq. (5.7) system to be in equilibrium.

Added Stress, Figure 5.5b:

The temperature of the system is increased. This adds kinetic energy and heat to the reactant side of the equilibrium. Using the balance analogy, the reactant arm moves down, indicating the added heat and increased reactants.

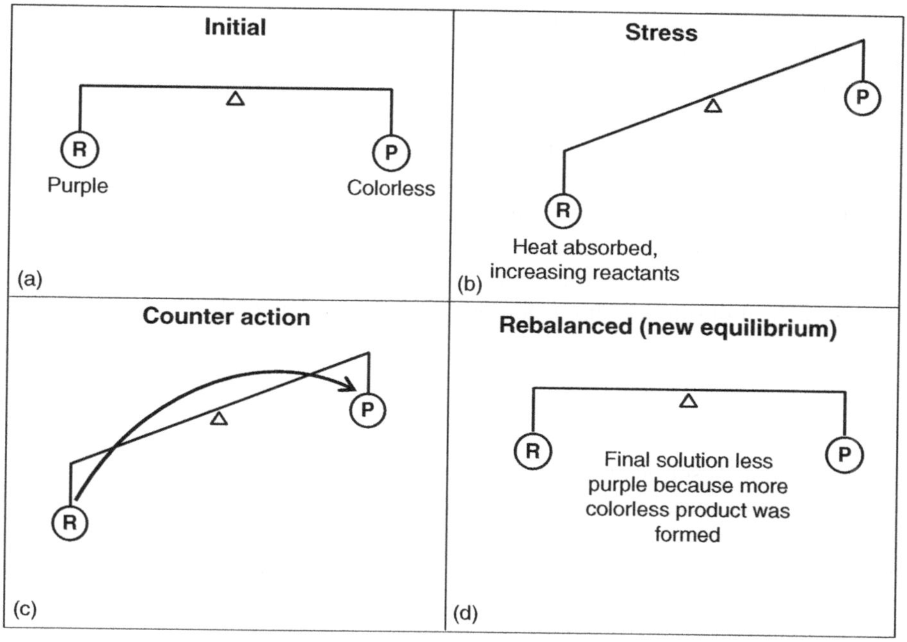

Figure 5.5 (a–d) Effect of increasing the temperature of an endothermic process.

Counteraction to Stress, Figure 5.5c:

The system reactants absorb the added heat. The additional absorbed energy shifts the reaction to the right, creating more product, HI $_{(gas)}$. Over the period of rebalancing, the purple color intensity decreases, as I_2 $_{(s)}$ (purple) reactant is used up to form more colorless HI $_{(gas)}$.

Rebalanced Equilibrium State, Figure 5.5d:

Extending the balance analogy, the unbalanced beam in Figure 5.5c, is rebalanced once a new equilibrium is re-established. This equilibrium has more HI $_{(gas)}$ product and less reactants which is confirmed by a less intense purple-colored equilibrium solution.

5.2.4 Author's Anecdote – Solving an Expensive Production Efficiency Improvement Misadventure by Applying Heat of Solution, Le Châtelier's Principle, and the Nernst and Brunner Equation

One of my students was doing an internship at a major pharmaceutical company. He was assigned to a team that was responsible for delivering meaningful production cost savings by significantly reducing the manufacturing time of a sterile product. After analyzing the various manufacturing unit operations, it was determined that the time required to produce normal saline offered the biggest opportunity to significantly decrease the overall manufacturing time of the sterile product. Attacking this unit operation also seemed to carry the least product quality risks.

Once this decision was made, the next step was to run some trial batches in the pilot plant. The goal, as noted previously, was to reduce the manufacturing time. To achieve a reduction in the manufacturing time, a decrease in the dissolution time was proposed. This required an increase in the dissolution rate of NaCl in water. The manufacturing scientists wrongly assumed that the NaCl dissolution rate could be increased by increasing the temperature of the water. A simple study design evaluated the effect of NaCl dissolution time with incremental increases in sterile water for injection temperature. The results showed that increasing the water temperature had very little effect on decreasing the NaCl dissolution time or increasing its dissolution rate. The scientists decided on an experiment to significantly raise the water temperature as a last-ditch effort to increase the NaCl's dissolution rate. Unfortunately, the desired decrease in dissolution time was not met and there was a significant increase in the amount of "rouging" or corrosion of the expensive stainless steel tanks with the increase in temperature.

My student called me at this point in the project to see whether I could offer any thoughts regarding the inability to increase the dissolution rate of NaCl at higher temperatures. I need to briefly digress to discuss dissolution rate and parameters that affect the dissolution rate. I also need to expand a little upon the heat of solution discussion presented in Section 2.7.1.

The Nernst (1904) and Brunner (1904) equation (Eq. 5.8) describes a drug's or compound's dissolution rate

$$dw/dt = [D(A)/(h)]\,(C_s - C) \tag{5.8}$$

where w is the mass of solute dissolved at time t, D is the diffusion coefficient of the drug, A is the area of the drug's dissolving surface, h is the diffusion layer thickness, C_s is the saturation solubility of the drug, and C is the solution concentration at a given time. From Eq. (5.8), the dissolution rate is directly proportional to the solubility of the drug. This fact helps us to be able to circle back and create a *hypothesis* regarding the inability to increase the dissolution rate of NaCl, to any significant degree, by increasing the solution temperature.

Using the Nernst and Brunner equation, one can hypothesize that the inability to increase the dissolution rate of NaCl with an increase in solution temperature is because *the solubility of NaCl is not affected by temperature to any significant degree.*

Now, I need to provide an additional piece of information regarding NaCl's heat of solution. Figure 5.6 shows the solubility of various salts as a function of temperature. Sodium chloride's heat of solution is +3.9 kJ mol^{-1}. This is an endothermic reaction. So, as shown in Example 5.4, an increase in temperature of an endothermic system results in an increase in product. In the case of a solubility equilibrium, the product is additional amount of solid going into solution. From Figure 5.6, one can see over the temperature range for 0 °C to 100 °C, the solubility increase of NaCl

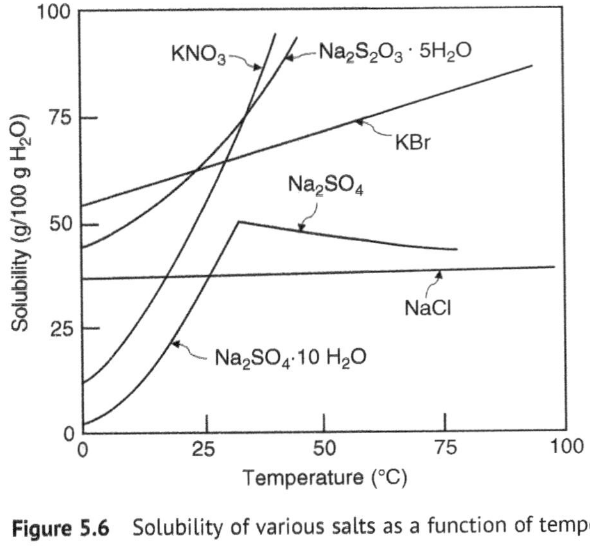

Figure 5.6 Solubility of various salts as a function of temperature.

is very modest. To put the magnitude of the NaCl heat of solution into perspective, KNO_3, also shown in Figure 5.6, has an endothermic heat of solution of +34.9 kJ mol^{-1} almost 9-fold larger than NaCl's. The solubility of KNO_3 increases rapidly from 0 °C to 25 °C.

Had NaCl's heat of solution been negative, as shown for anhydrous Na_2SO_4 at temperatures greater than ~27 °C, NaCl's solubility and dissolution rate would have decreased with temperature increases.

Now, getting back to the original concern explained by my student: "why was the dissolution rate of NaCl not increased as the solution temperature was increased." The hypothesis stated previously: "the inability to increase the dissolution rate of NaCl with an increase in solution temperature is because *the solubility of NaCl is not affected by temperature to any significant degree.*" Knowing that NaCl solubility is only marginally affected by increases in temperature also impacts the results of the Nernst and Brunner dissolution rate equation where the dissolution rate is directly proportional to the NaCl solubility. Since the solubility does not increase to any great extent with increase in temperature, NaCl's dissolution rate does not increase much with increase in temperature.

Had the concepts of the Nernst and Brunner dissolution equation, heat of solution, and Le Châtelier's principle been considered at the outset of this production efficiency project, this costly (time and effort of numerous staff, opportunity cost of tying up the pilot dissolution vessel and room, and refurbishment cost for the rouged dissolution vessel) misadventure would have not occurred.

5.3 Phase Equilibrium of a Pure Substance

Pure substance phases are states of matter that can be characterized by discriminating macroscopic properties. A phase is a homogeneous state of matter. The primary phases discussed in this section are gas, liquid, and solid phases. Amorphous and glassy phases will also be included. Ferromagnetic and superconducting phases which are less likely to be encountered in pharmaceutical product design will not be discussed.

Single (unary)-phase diagrams generally summarize the solid, liquid, and gas phases of a pure substance. There are six-phase transitions that these three phases can undergo: melting, freezing, evaporation, condensation, sublimation, and deposition. These phase transitions are defined in Table 5.2.

Table 5.2 Pharmaceutical Examples of Phase Transitions for the Three States of Matter.

Transition	Phase transition definition	Example
Melting	Solid to liquid	Liquefaction during melt granulation where the granulation agent is melted to form a liquid granulation mass
Freezing	Liquid to solid	Solidification of water into ice during the initial freezing stage of a freeze-drying cycle
Evaporation	Liquid to gas	Fluid-bed drying of water
Condensation	Gas to liquid	Steam sterilization where steam or superheated steam condenses on cool glass container closure systems
Sublimation	Solid to vapor/gas	Vaporization of ice to water vapor during low-temperature low-pressure drying of a freeze-drying process
Deposition	Vapor/gas to solid	Chemical vapor deposition encapsulation of drugs including plasma-enhanced vapor deposition (PE-CVD) and initiated chemical vapor deposition (iCVD)

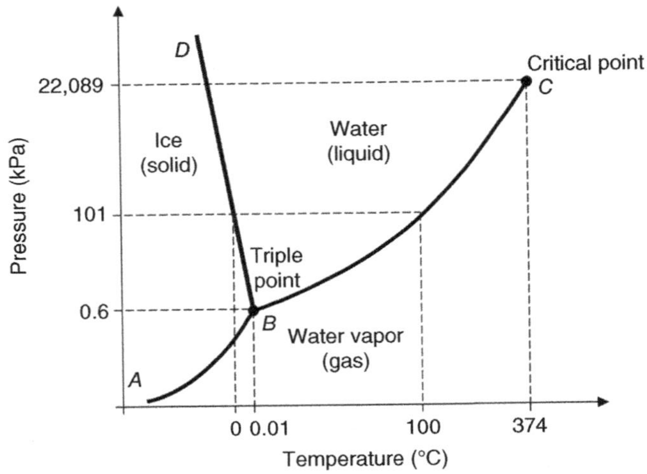

Figure 5.7 Water pressure-temperature phase diagram showing the gas, liquid, and solid phases. The sublimation-deposition, evaporation-condensation, and melting-freezing phase equilibriums are depicted by the broad solid black lines. *Source:* Adapted from Flowers et al. (2019) and Clark (accessed Feb 2022).

The newest advances in drug delivery involve molecular chemical vapor deposition of mostly polymers onto drug surfaces. Acrylic acid, polydimethylsiloxane, and ethylene use PE-CVD technology to coat drugs. Methylacrylic acid, ethylene dimethacrylate, and aminostyrene have also been used to coat particles with iCVD.

The water pressure-temperature phase diagram is shown in Figure 5.7.

Water sublimation-deposition, evaporation-condensation, and melting-freezing phase equilibriums occur along the AB, BC, and BD lines, respectively. The individual phases are described by their respective 2D planes or spaces. Two-phase equilibriums are observed along the 1D lines. The *triple point* describes the location where gas, liquid, and solid phases are all in equilibrium. The triple point of water occurs at 0.01 °C and 0.6 kPa.

The melting point, boiling point, and critical point of water occur at 0 °C and 101 kPa, 100 °C and 101 kPa, and 374 °C and 22,064 kPa, respectively. At the critical point, C, the liquid phase and vapor phase lose their distinct physical identity. The phases coalesce and the densities of liquid and vapor become equal which eliminates the boundary between the two phases. At this point, there is no longer an equilibrium present because the phases are no longer distinct from each other.

5.4 Gibbs Phase Rule

The *Gibbs phase rule* relates the number of phases, components, and intensive variables (see Section 2.5)) that are required to define a particular state of an equilibrium system. The Gibbs phase rule can be applied to closed systems. The phase rule can only be applied for a system in equilibrium.

Consider the closed water pressure-temperature phase diagram (Figure 5.7). As mentioned previously, water sublimation-deposition, evaporation-condensation, and melting-freezing phase equilibriums occur along the AB, BC, and BD lines, respectively. The point "B" represents the triple point where all three phases, vapor/liquid/solid, coexist in equilibrium. The lines ABD, DBC, and

ABC divide the phase diagram into three distinct regions. These areas are known as the solid (ABD area), liquid (DBC area), and vapor or gas (ABC area) regions.

To describe a state or position on the BD line, which represents the melting-freezing equilibrium process, one is *free* to select either a pressure or a temperature on the BD line to maintain the solid-liquid equilibrium phases of interest. Once one variable is selected, the other is defined. Therefore, in this case, one has the freedom to select one variable to define the state or position on the BD solid-liquid line. This is referred to as *one degree of freedom*. Stated slightly differently, degrees of freedom are the number of variables that one is free to choose while keeping the phase or phases of interest in equilibrium.

Anywhere in the solid, liquid, or vapor phase region of the diagram, one is free to choose a pressure *and* temperature to remain in the desired phase region. In this case, there are two degrees of freedom.

At point B, one has no freedom to choose either pressure or temperature, they are defined. The state or position B is invariant, and the *degrees of freedom are zero.*

The Gibbs phase rule equation formalizes the calculation of *degrees of freedom:*

$$F = C - P + 2 \tag{5.9}$$

where F is the number of degrees of freedom, C is the number of independent components in the system, P is the number of phases, and 2 accounts for the intensive variables: pressure, and temperatures.

In general, the *number of degrees of freedom* is defined as the number of intensive variables required to describe the system completely. An intrinsic physical variable that is independent of the size of the system is called an *intensive* property or variable, such as temperature, pressure, surface tension, density, melting point, viscosity, concentration, and others which were discussed in Section 2.5. In most cases, temperature, pressure, and concentration are sufficient to describe a system.

Example 5.5 Calculating the Degrees of Freedom on the Solid-liquid Equilibrium BD line from Figure 5.7

Using Eq. (5.9) and the BD line of Figure 5.7, $C = 1$ where water is a pure substance. $P = 2$ since the BD line represents solid-liquid equilibrium.

Substituting into Eq. (5.9) $F = C - P + 2$, the degrees of freedom, F can be calculated.

$$F = 1 - 2 + 2 = 1$$

This means that if we select a particular temperature on the BD line (one free choice), then the pressure is constrained by the equilibrium to a specific value.

The Gibbs phase rule provides the theoretical foundation, based on thermodynamics, for characterizing the chemical or physical state of a system. It can be used to predict equilibrium relations of the phases present as a function of physical conditions such as pressure and temperature. Later in this chapter, we will do just that. The Gibbs phase rule will be used to determine the number of phases present at the eutectic point of a eutectic mixture.

5.5 Temperature-Composition Phase Equilibrum Diagrams of a Two-Component System

Phase diagrams provide knowledge about phase changes that can occur in a particular system. This section will only discuss two component systems. A single-phase, two-component system can be completely described by specifying three independent variables: temperature, pressure, and composition. Since pressure is assumed to be constant for experimentation, phase equilibria diagrams can be fully described by temperature and composition.

It is beyond the scope of this book to discuss all the possible phase equilibria for a two-component system. Avner (1974) devoted a chapter to discussing four main categories of two-component phase changes and seven types of phase diagrams including solid solutions (Type I), insoluble solid eutectic reaction (Type II), partial solubility in solid state, (Type III), formation of congruent-melting intermediate phase (Type IV), peritectic reaction (Type V), monotectic reaction (Type VI), and components insoluble in the melted state and insoluble in the solid state (Type VII). Other phase reactions included reversible polymorphic transitions, order-disorder phase changes, eutectoid transitions, and peritectoid transformations.

5.5.1 Classic Two-Component Eutectic Phase Diagram

Pharmaceutical eutectic mixtures are gaining importance as drug products and drug delivery systems. Research efforts to characterize and understand the molecular and microstructure of eutectic mixtures has lagged the characterization of other multicomponent solids like salts, solid solutions, molecular complexes, and cocrystals.

A eutectic mixture is defined as a binary mixture of two solids whose melting point is lower compared to the individual components. Raoult's law asserts that the *melting/freezing point* of a pure compound will be lowered by the presence of another compound if both compounds are soluble in each other in the molten liquid state and insoluble in the solidified state.

Cherukuvada and Nangia (2014) suggested that a eutectic is a conglomerate or mixture of solid lattice structures whose domains are held together in the liquid melt by weak interphase boundary interactions. These conglomerates often adopt domains of a lamellar arrangement of a liquid-solution component and microcrystalline solid solutions. These domains form a unique *eutectic microcrystalline structure.*

This unique microstructure results in higher thermodynamic energy than found for either individual component. Eutectics are high energetic systems associated with higher Gibbs free energy, enthalpy, and entropy compared to its crystalline counterparts (Cherukuvada and Nangia, 2014).

A classic eutectic phase diagram is shown in Figure 5.8.

This classic eutectic phase diagram has a number of key features. The x-axis is the weight percent of component B, and the y-axis is temperature. The individual melting points of A and B are at M and N, respectively. The *solidus* line is MFGN. A solid phase exists at any point on the solidus line. The two *liquidus* lines are ME and NE. The liquidus line represents an equilibrium condition between the molten homogeneous solution liquid A–B and solid A or solid B.

The *eutectic melting temperature* is line FG. The *eutectic melting point* is E and is located where the liquidus lines of ME and NE intersect the eutectic melting temperature line, FEG. The eutectic point marks the position of the *eutectic mixture composition,* which in this case is 40%A–60%B. The eutectic point is an invariant point ($F = 0$) and is described completely by the eutectic composition, eutectic melting point, and the experimental pressure condition. The FEG line is the tie-line for the eutectic point and eutectic composition. The FEG tie-line bounds solid A and solid B. This means that the eutectic mixture is a three-phase system of solid A, solid B, and the liquid melt at the specific eutectic composition.

The Gibbs phase rule can also be used to confirm that the eutectic mixture is a three-phase system. As noted previously, the eutectic point is invariant. The Gibbs phase rule is $F = C - P + 1$, where pressure is constant. F equals 0 and $C = 2$. Solving for P using the Gibbs phase rule gives $P = -F + C + 1$. Substituting 0 for F and 2 for C gives $P = 0 + 2 + 1 = 3$. The Gibbs phase rule and the interpretation of the phase diagram are consistent.

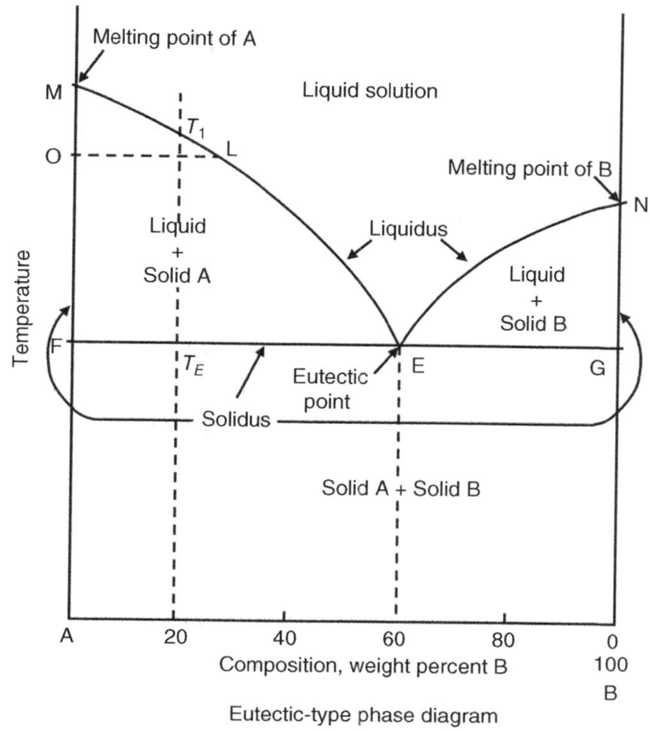

Figure 5.8 Classic eutectic phase diagram of a binary system of components A and B. *Source:* Adapted from Avner (1974).

There are four phase regions in Figure 5.8: (1) below the eutectic temperature (FEG line), there is a mixture of crystalline solid A and B, (2) the region FEM is a mixture of solid A and the liquid A–B solution, (3) the MEN region is a homogeneous liquid solution of the melts of A and B, and (4) the GEN region is a mixture of solid B and liquid A–B solution. The horizontal line OL in the FEM region is known as a *tie-line*. The tie-line connects the vertical solidus line MF and the liquidus line ME. The ME line represents an equilibrium between the molten homogeneous liquid A–B and solid A. The two-phase regions are easier to identify if the single-phase area is designated first.

In Section 5.5.3, the tie-line will be used to determine phase composition and percent of each phase of a *point* on the tie-line.

5.5.2 Application: Commercialized Eutectic Product and Its Eutectic Phase Diagram

Eutectic Mixture of Local Anesthetics (EMLA®) has been successfully commercialized as topical local anesthetic. The product provides improved penetration and faster anesthesia than the individual components of lidocaine and prilocaine found in other commercial products. The phase diagram of the lidocaine-prilocaine eutectic mixture is shown in Figure 5.9.

Points A and D are the melting points of lidocaine and prilocaine, respectively. The solidus line for a classical eutectic phase diagram would be lines AB for lidocaine (B is inadvertently missing) and DE for prilocaine. The experimental solidus line shows a slight deviation from the y-axis. The authors were unable to show that a lidocaine or prilocaine solid phase existed above the eutectic temperature of $18 \pm 1\,°C$ at high lidocaine (90–100) and prilocaine (85–100) mol%, respectively. It is

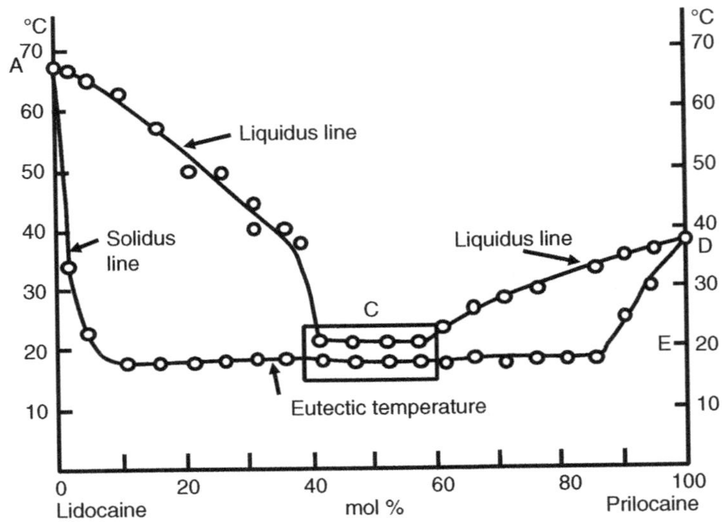

Figure 5.9 Phase diagram of a lidocaine-prilocaine eutectic mixture exhibiting near-classic behavior. *Source:* Adapted from Brodin et al. (1984).

also noted that this phase diagram does not reveal the classic nonvariant single eutectic mixture point at the eutectic melting temperature. The eutectic composition in this case covers the range of 40–60 mol% prilocaine. Curiously, at region C, the liquidus lines do not intersect the eutectic melting point.

After extensive investigations using hot-stage microscopy, differential thermal analysis, X-ray diffraction, infrared spectroscopy, three-week equilibration period, and solubility measurements to try to explain the "complications" of the phase diagram, the author's concluded that the lidocaine-prilocaine was a classic eutectic containing "no intermediate compounds." They claimed the eutectic composition was close to 1 : 1 mol%. The eutectic temperature was reported to be 18 ± 1 °C. As mentioned previously, no solid solubility was detected.

5.5.3 Application: Complex Two-Component Equilibrium Phase Diagram – NaCl and Water

A very important two-component pharmaceutical system is sodium chloride and water. A 0.9% w/v solution of sodium chloride is isotonic with blood and tissues, which means that it has the same tonicity as blood and tissues. A 0.9% w/v sodium chloride solution is also known as normal saline by the medical and pharmaceutical professions. It is widely used as an intravenous dosage form to dilute inline intravenous drugs and to replace fluid. Normal saline solution is also used as a vehicle to delivery nasal sprays, respiratory aerosols, irrigation fluids, eye drops, or other dosage forms. Sodium chloride (NaCl) is also used in freeze-dried parenterals for reconstitution as a tonicity-adjusting agent. A constant-pressure NaCl-water phase diagram is shown in Figure 5.10. The temperature range of this phase diagram is –40 °C to 40 °C.

This common two-component system is relatively complex and has seven-phase regions shown: (a) ice, solidus line AL; (b) ice plus brine (solution of NaCl in water), ABC; (c) brine solution, ACDE; (d) brine plus solid NaCl dihydrate, CDFG; (e) brine and anhydrous NaCl solid, EDFJK; (f) solid mixture of ice plus NaCl dihydrate solid, BLHG; (g) and solid NaCl dihydrate plus solid anhydrous NaCl, FGHIJ.

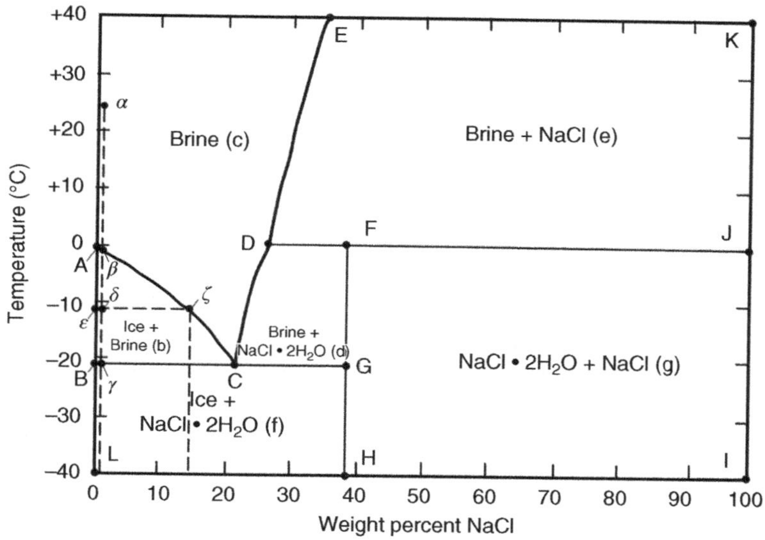

Figure 5.10 A constant-pressure NaCl-water phase diagram over the temperature range of −40 °C to 40 °C. *Source:* Adapted from Badawy (2015).

Note that the *x*-axis is given as weight percent NaCl, not percent w/v, which is fine for our discussion. The *y*-axis is temperature in centigrade. For purposes of this discussion, melting point and freezing/crystallization point will mean that the same temperature even though melting is a thermodynamic equilibrium process and freezing or crystallization is a rate process. At point A, the freezing point of pure water is zero when the NaCl concentration is zero.

A sterile freeze-dried parenteral product for reconstitution may be formulated at room temperature, 25 °C to contain 0.9% w/w NaCl. See point α. The first step in freeze-drying is to freeze all the contents before drying can occur. In this case, the system composition is 0.9% w/w NaCl. As the solution is cooled along the dashed line, it will reach line AC where water begins to freeze and separate from the solution. Point β on the dashed line is on the AC line. The AC line represents the equilibrium between region (b) which has two phases, ice and brine. Brine is a solution of NaCl and water. The brine phase is region (c).

As the temperature is lowered further along the dashed line, the proportion of ice increases which leaves less liquid water in the system. This results in an increase in NaCl concentration in the remaining liquid water until the temperature of the system reaches −21.1 °C, at point γ. The *eutectic temperature* is −21.1 °C. At this temperature, the NaCl concentration has increased to ~23% w/w, the *eutectic composition*. This is an ~25.6-fold increase in NaCl concentration!

As the temperature is continually decreased along the dashed line below −21.1 °C, phase region (f) is reached. This phase is composed of ice and solid NaCl dihydrate crystals.

How many phases are there at the eutectic point if equilibrium is maintained? The answer is three phases. At the eutectic point, point C, there are four-phase regions that "touch" each other and are in equilibrium: region (b), ice plus brine; region (c), brine; region (d), brine plus NaCl dihydrate solid; and region (f), ice plus NaCl dihydrate solid. These four regions have three phases that consist of ice (solid), NaCl dihydrate (solid), and brine (NaCl-water solution). Microscopic examination of the eutectic mixture reveals that the two solid phases, ice and NaCl $2H_2O$, and the brine phase can form a *lamellar matrix* solid dispersion of microcrystals.

Example 5.6 Determination of the Phase Composition and Percent of Each Phase of a 0.9% w/w NaCl-Water Solution Prepared at 25 °C at 1 atm Pressure and Then Cooled to −11 °C

The phase composition and percent of each phase can be determined from the phase diagram by using *tie-lines* that "tie" the equilibrium boundaries together. A tie-line can be drawn through point δ, located on the dashed vertical line at −11 °C. A tie-line is drawn from ε to ξ or the $\varepsilon\xi$ tie-line. The $\varepsilon\xi$ tie-line connects ε located at NaCl w/w % = 0 or 100% w/w ice and ξ = 15% w/w NaCl (brine).

The phase composition in region (b) at point δ is a liquid phase containing 15% w/w NaCl in water (brine) and a solid phase of ice.

To determine the percent of each phase in region (b), the *lever rule* concept needs to be introduced. The lever rule can be explained by using an analogy of balancing a teeter-totter for a heavy and lighter child by using a fulcrum and adjusting the distance between the two children. The lever rule for determining the proportion of each phase of a two-phase system is depicted in Figure 5.11.

Lever rule

$$\text{Portion phase 1 (ice phase)} = (C_2 - C)/(C_2 - C_1) \tag{5.10}$$

$$\text{Portion phase 2 (brine solution)} = (C - C_1)/(C_2 - C_1) \tag{5.11}$$

Using Eqs. (5.10) and (5.11), the proportions of the ice and NaCl solution phase can be calculated.

From Eq. (5.10) Ice phase = (15 − 0.90)/(15 − 0) = 14.1/15 = 0.94 or 94 % ice

From Eq. (5.11) NaCl solution phase = (0.90 − 0)/(15 − 0) = 0.060 or 6.0 % brine solution

Therefore, at delta (δ), the proportion of ice and brine is 94% w/w ice, while 6.0% w/w is made up of the brine solution that is 15% w/w NaCl. It is important to note that as the water becomes ice, the remainder of the unfrozen H_2O becomes concentrated with NaCl.

This NaCl concentration effect that occurs during freezing needs to be kept in mind because proteins are generally unstable at hypertonic conditions.

The remainder of the NaCl-water phase diagram will be briefly discussed. Region (d) is two phases, brine plus solid NaCl dihydrate (hydrohalite). Line CD is the solubility line for $NaCl{\cdot}2H_2O$. A *peritectic transition* occurs at 0.1 °C (point D), and anhydrous NaCl is the thermodynamic stable species at this temperature and above. A peritectic transition involves a liquid and solid phase where the solid changes composition. In this case, the peritectic transition is the dehydration of $NaCl{\cdot}2H_2O$ to the anhydrous form.

Region (e) is also a two-phase region of brine and solid anhydrous NaCl. The solubility of anhydrous NaCl in water occurs above the perieutectic temperature and along the DE line.

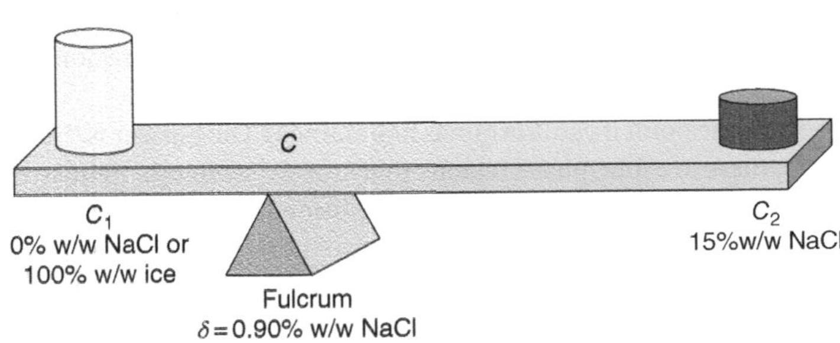

Figure 5.11 Depiction of lever rule for a 0.9% NaCl solution at −11 °C.

Phase region (g) is composed of solid anhydrous NaCl and solid $NaCl{\cdot}2H_2O$ at a composition greater than approximately 38% w/w NaCl and at and below the perieutectic melting point.

5.6 Phase Equilibrium of a Three-Component System

Around the 1960s, self-emulsifying drug delivery systems (SEDDSs) were conceived to enhance the solubility of poorly soluble drugs. It has also been shown that the maximum absorbable dose is directly proportional to drug solubility. The SEDDS concept was based on creating a thermodynamically stable isotropic anhydrous system of drug, lipid/oil, surfactant, and cosolvent. This isotropic drug containing liquid would spontaneously form an emulsion when diluted with water. This SEDDS can carry a higher payload of drug than conventional liquids.

Over recent years, self-nanoemulsifying drug delivery systems (SNEDDSs) have been shown to increase drug solubility, dissolution rate, and oral bioavailability of poorly water-soluble drugs to an even greater extent than SEDDS. SNEDDSs are isotropic liquid drug mixtures that spontaneously form oil-in-water nanoemulsions when diluted with water in the body. Due to high surfactant/co-surfactant to oil ratios, SNEDDSs have more drug-loading capacity. SNEDDSs are generally formulated in a liquid state and encapsulated into soft gelatin capsules. SNEDDSs also provide long-term stability due to the absence of water.

Figure 5.12 is a *ternary phase diagram* of a SNEEDS nonaqueous system.

The dark region in Figure 5.12 represents the isotropic ternary mixture of surfactant, lipid, and cosolvent SNEDDS that spontaneously forms stable nanoemulsions when diluted with water. Drug solubility varies within the black bounded space. More than one phase exits outside the black area.

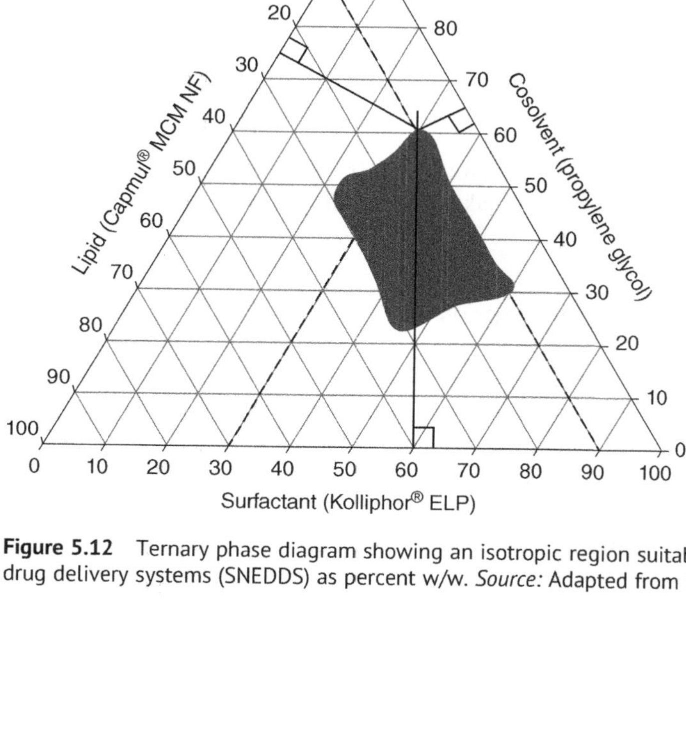

Figure 5.12 Ternary phase diagram showing an isotropic region suitable for self-nanoemulsifying drug delivery systems (SNEDDS) as percent w/w. *Source:* Adapted from Patel et al. (2023).

Example 5.7 Determination of the Phase Composition Indicated by the Arrow in the Ternary Phase Diagram (Figure 5.12)

The composition of the surfactant, lipid, and cosolvent system can be defined by determining the percentage of two components. The third component can be determined by knowing 100% − % component 1 − % component 2 = % of component 3. In other words, a ternary system has two degrees of freedom. One can choose the percentage of two components, and the third component can be immediately calculated. A ternary system's composition can be depicted by 2D equilateral triangle as shown in Figure 5.12 based on the fact that a ternary system has two degrees of freedom.

The three sides of the equilateral triangle represent the zero percentage of each component. For example, the "bottom" side in Figure 5.12 is labeled Surfactant (Kolliphor® ELP) and represents 0% w/w surfactant. The peak opposite this side of the triangle represents 100% w/w of that component (Kolliphor® ELP). Please note that there are evenly spaced parallel lines drawn from the surfactant baseline to the peak opposite the surfactant side. These parallel lines are labeled in 10% increments starting at 10% and going to 90% just before the peak which is 100%.

The same is true for the other two sides. The left-hand side of the triangle labeled lipid represents 0% w/w lipid. Lines parallel to this side increases incrementally from 10% to 90% lipid until at the peak opposite the lipid side is 100%. The right-hand side of Figure 5.12 triangle has similar features.

To determine the phase composition at the arrow requires the following steps:

Step 1. At the bottom of the triangle (surfactant side), draw a perpendicular line from the bottom side to the point indicated by the arrow.

Step 2. Following this perpendicular line, determine the distance from the surfactant side. In this case, the distance equals 60% w/w surfactant. This distance can be found by finding the parallel lines that move from the bottom of the triangle (labeled 0%) to the peak opposite the bottom labeled 100%.

Step 3. From the left-hand (lipid) side of the triangle, draw a perpendicular line from the lipid side of the triangle to the point designated by the arrow.

Step 4. See Step 2. Follow the perpendicular line, to find the distance from the lipid side to the point marked by the arrow. This distance is 30% w/w lipid.

Step 5. From the right-hand (cosolvent) side of the triangle, draw a perpendicular line from the cosolvent side of the triangle to the point designated by the arrow.

Step 6. See Step 2. From the right-hand (cosolvent) side of the triangle, draw a perpendicular line from the cosolvent side of the triangle to the point designated by the arrow. This distance is 10% w/w.

The ternary composition at the point indicated in Figure 5.12 is 60% w/w surfactant, 30% w/w lipid, and 10% w/w cosolvent. This shows that surfactant makes up most of the isotropic ternary mixture.

5.7 Phase Transitions of Pure Substances

Table 5.2 gave pharmaceutical examples of the six phase transitions for the three states of matter. The enthalpy of fusion has been discussed in some detail in Section 3.3.4. All six phase transitions have heats associated with their transition. These are called *enthalpies of transition*. An *enthalpy of transition is the amount of heat absorbed or released by one mole of substance as it undergoes a phase change at its transition temperature.*

Melting/freezing, vaporization/condensation, and sublimation/deposition conjugate phase pairs have respective endothermic/exothermic conjugate enthalpies of transition. Melting, evaporation, and sublimation phase transitions require energy. As such, substances undergoing these transitions

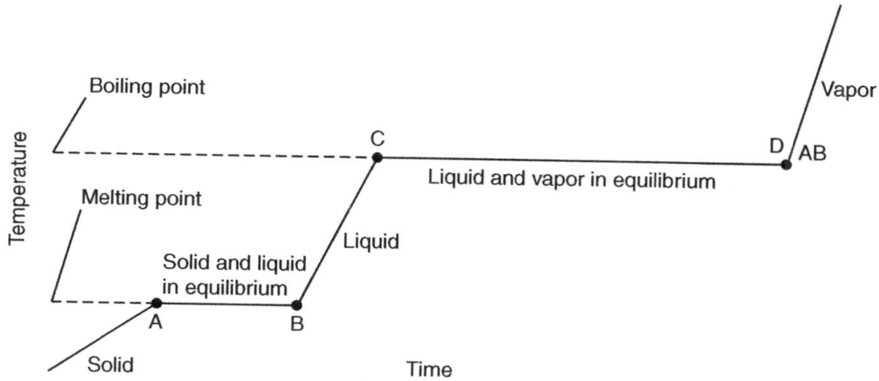

Figure 5.13 A temperature-time melting, boiling, and vaporization profile. *Source:* Adapted from Chang (1998).

absorb heat, an endothermic process. Freezing, condensation, and deposition phase transitions release energy during these transitions, which is an exothermic process.

Phase transitions are equilibrium processes. For example, *melting point is defined as the temperature where the solid and liquid phases coexist and are in equilibrium*. The same is true for all three conjugate phase transition pairs.

Figure 5.13 is a temperature-time profile of a solid material that is heated through its melting and vaporization transitions.

As the solid is heated its temperature increases over time until point A, the material's melting point. At the melting point, the material absorbs heat which is required to provide energy to overcome the interaction forces that maintain the crystal structure. During this period, line AB, the temperature remains constant. This heat is called the *molar enthalpy of fusion*. This heat is also known as the *molar latent heat of fusion* because the absorbed heat does not result in an increase in the kinetic energy or temperature of the material. If there is solid material present, the system is in thermal equilibrium between the solid and liquid melt. When all the solid is melted at point B, the absorbed heat increases the kinetic energy of the liquid melt, which results in an increase in the temperature of the liquid along line BC. Point C is the *boiling point* of the liquid melt. The liquid starts to vaporize, and the liquid and vapor are now in thermal equilibrium shown by line CD. As long as there is a liquid phase, the temperature remains constant, and the heat absorbed is known as the *molar heat of vaporization* or *molar latent heat of vaporization*. At point D, all the liquid has been transformed into vapor and the vapor temperature increases due to its increase in kinetic energy.

Two additional important points need to be made. First, note that the slopes of solid, liquid, and vapor lines are different. The slopes of these lines are determined by the specific heat capacity of each phase. The slope of the vapor line > liquid line > solid line. This makes sense since the heat capacity of a material is related to the substance's *molecule mobility* (see Section 3.3.7). The molecular mobility of the gas is greater than its liquid phase, which is greater than its solid phase.

5.7.1 The Clapeyron Equation: Vapor Equilibrium Boundary

The Clapeyron equation is an exact equation that characterizes the *vapor phase* equilibrium between a solid and vapor, and a liquid and vapor. The two common forms of the Clapeyron equation are:

$$dP/dT = \Delta S_M/\Delta V_M \tag{5.12}$$

and

$$dP/dT = \Delta H_M / T\Delta V_M \tag{5.13}$$

where P is the saturated vapor pressure along the solid-vapor or liquid-vapor equilibrium line. ΔS_M and ΔH_M are molar entropy and enthalpy of sublimation/deposition or vaporization/condensation. ΔV_M is the molar volume change between the two conjugate phases. Equations (5.12) and (5.13) are known as the *Clapeyron equation*. Equations (5.12) and (5.13) indicate that the slope of the equilibrium solid-vapor or liquid-vapor line in the pressure versus temperature plot is equal to $\Delta S_M/\Delta V_M$ or $\Delta H_M/T\Delta V_M$ where $\Delta S_M = \Delta H_M/T$ (see Section 3.4.4).

The derivation of the *Clapeyron equation* is provided knowing that at equilibrium, the Gibbs free energies of the two phases in equilibrium are equal. The liquid-vapor ΔG at equilibrium is equal to zero, Eq. (5.14).

$$\text{Liquid} \rightleftharpoons \text{Vapor}$$
$$\Delta G = G_l - G_v = 0 \tag{5.14}$$

and

$$G_l = G_v \text{ and } dGl = dG_v \tag{5.15}$$

From Eq. (4.7), $dG = VdP - SdT$. $VdP - SdT$ is substituted for dG in Eq. (5.15) to give:

$$V_l\, dP - S_l\, dT = V_v\, dP - S_v\, dT \tag{5.16}$$

And rearranging Eq. (5.16) gives:

$$(V_l - V_v)\, dP = (S_l - S_v)\, dT \tag{5.17}$$

As stated previously, the slope of the equilibrium line is dP/dT. Solving Eq. (5.17) for dP/dT *is*

$$dP/dT = (S_l - S_v)/(V_l - V_v) = \Delta S_{M,\text{phase tansition}}/\Delta V_{M,\text{phase tranition}} \tag{5.17a}$$

From Eq. (3.77), $\Delta S = \Delta H/T$ which can be substituted into Eq. (5.12) to give Eq. (5.13):

$$dP/dT = \Delta H_{M,\text{phase transition}}/T\Delta V_{M,\text{phase transition}} \tag{5.17b}$$

See Mander (2013) for more discussion regarding the Clapeyron equation.

5.7.2 Derivation of the Clausius-Clapeyron Equation: Vapor Equilibrium Boundary

Several assumptions are made in deriving the Clausius-Clapeyron equation, so it is not as exact as is the Clapeyron Equation. However, it is often more useful because the $\Delta V_{M,\text{ phase transition}}$ is not required to use the equation. The derivation of the Clausius-Clapeyron equation starts with the Clapeyron equation, $dP/dT = \Delta H_{M,\text{ phase transition}}/T\Delta V_{M,\text{ phase transition}}$. Since the volume change for sublimation and vaporization is large, $\Delta V_{M,\text{ phase transition}}$ can be considered the molar vapor volume, $V_{M,\text{ vapor}}$. If the vapor is assumed to approximate an ideal gas, $V_{M,\text{ vapor}} = RT/P$. Note the RT/P term does not include n, the number of moles, because the volume term is molar volume where V is divided by n. Applying the approximations and substituting $V_{M,\text{ vapor}} = RT/P$ for $\Delta V_{M,\text{ phase transition}}$ into Eq. (5.13) gives:

$$dP/dT = P\Delta H_{M,\text{phase transition}}/RT^2 \tag{5.18}$$

The Clausius-Clapeyron equation gives the approximate slope of solid-vapor or liquid-vapor equilibrium line of Figure 5.7.

Equation (5.18) can be rearranged as:

$$dP/P = \left[\Delta H_{M,phase\ transition}/RT^2\right] dT \tag{5.19}$$

where $dP/P = d \ln P$ from Eq. (1.85). Substituting $dP/P = d \ln P$ into Eq. (5.19) leads to:

$$d \ln P = \left[\Delta H_{M,phase\ transition}/RT^2\right] dT \tag{5.20}$$

Equation (5.20) can be integrated (see integral Eq. 1.100):

$$\int_{P_1}^{P_2} d\ \ln P = \left[\Delta H_{M,phase\ transition}/R\right] \int_{T_1}^{T_2} dT/T^2 \tag{5.21}$$

to give:

$$\ln P_2/P_1 = -\Delta H_{M,phase\ transition}/R\ [1/T_2 - 1/T_1] \tag{5.22}$$

Using $T_2\ T_1$ as a common denominator for the right-hand side of the Eq. (5.22) gives:

$$\ln P_2/P_1 = -\Delta H_{M,phase\ transition}/R\ [(T_1 - T_2)/\ T_2\ T_1] \tag{5.23}$$

If both sides of Eq. (5.23) are multiplied by −1, it becomes:

$$\ln P_1/P_2 = \Delta H_{M,phase\ transition}/R\ [(T_2 - T_1)/\ T_2\ T_1] \tag{5.24}$$

which is the primary Clausius-Clapeyron operating equation.

5.7.3 Application of the Clausius-Clapeyron Equation

The Clausius-Clapeyron equation can be used to estimate the enthalpy of vaporization, assess the effect of pressure change on the boiling temperature, and approximate the effect of temperature change on the vapor pressure.

Example 5.8 Determine the Saturation Water Vapor Pressure at −20 °C during a Freeze-Drying Process

Note: To achieve sublimation in the freeze-dryer, the freeze-dryer chamber pressure must be lower than the determined saturation vapor pressure at −20 °C.

Given:

Ice enthalpy of fusion at 0 °C is 6.01 kJ mol^{-1}
Water enthalpy of vaporization at 0 °C is 45.06 kJ mol^{-1}
Water vapor pressure at 0 °C is 611.148 Pa (1 atm = 101,325 Pa)
Assume that the enthalpy of vaporization is not affected by temperature change

First, to determine the water vapor pressure at −20 °C, we need to know the molar enthalpy of sublimation which we were not given. Since we are given the water's enthalpy of vaporization and ice's heat of fusion, we can apply Hess's law (3.3.3) to find the enthalpy of sublimation.

Application of Hess's law to determine the molar enthalpy of sublimation of ice.

Step 1: Melting

$H_2O_{(s)} \rightarrow$ ~~$H_2O_{(l)}$~~

Step 2: Vaporization

~~$H_2O_{(l)}$~~ $\rightarrow H_2O_{(v)}$

Step 3: Application of Hess's law. The same compounds that appear as both reactants and products are cancelled, giving the final reaction. The sublimation process is:

$$H_2O_{(s)} \rightarrow H_2O_{(v)}$$

The enthalpy of sublimation is the sum of Step 1 and Step 2.
The ice enthalpy of sublimation at 0 °C is:

$$6.01\ \text{kJ/mol} + 45.06\ \text{kJ/mol} = \underline{51.07\ \text{kJ/mol}} \tag{5.25}$$

Now, we can employ the Clausius-Clapeyron Eq. (5.24) to determine the vapor pressure of ice at −20 °C.

$$\ln P_1/P_2 = \Delta H_{M,phase\ transition}/R\ [(T_2 - T_1)/\ T_2\ T_1] \tag{5.26}$$

Using Eq. (1.47),
$\ln 611.148 - \ln P_2 = (51.07\ \cancel{\text{kJ mol}}^{-1})/8.314\text{E-3}\ \cancel{\text{kJ K}}^{-1}\cdot\cancel{\text{mol}}[(253.15\ \cancel{\text{K}} - 273.15\ \cancel{\text{K}})/253.15\ \cancel{\text{K}}\cdot 273.15\ \cancel{\text{K}}]$
$\ln 611.148 - \ln P_2 = (6142.65)[(-20)/69147.92] = 1.7767$
$6.4153 - \ln P_2 = 1.7767$
$\ln P_2 = 6.4153 - 1.7767 = 4.6386$, take the antilogarithm of both sides

$$\underline{P_2 = 103.4\ \text{Pa}} \tag{5.27}$$

The literature value for the saturated water vapor pressure over ice at −20 °C is 103.2 Pa (Barley, 2022). This is a ~4.9-fold decrease in saturated water vapor pressure; [(611.148 Pa −103.2)/103.2 Pa)! A typical primary freeze-drying chamber pressure range is between 13.3 and 40 Pa. For sublimation to occur, the partial pressure of water vapor in the freeze-dryer chamber must be lower than the vapor pressure of the ice. Since the chamber pressure (13.3–40 Pa) is well below the saturation vapor pressure of ice at −20 °C (103.2 Pa), a sufficient vapor pressure gradient exists to drive the sublimation process, allowing water to be removed from the frozen product as vapor.

5.7.4 Ehrenfest Classification of Phase Transition Order

In 1933, Paul Ehrenfest developed a classification system for thermodynamic equilibrium phase transitions. The classification system was based on the lowest derivative of the Gibbs equation that resulted in a discontinuity in a thermodynamic property during a phase transition or *the crossing of the coexistence curve of the two phases*. In general, only first- and second-order phase transitions are important.
First-order phase transitions have the following common thermodynamic properties:

1) Example of first-order transitions are melting, vaporization, and sublimation.
2) The change in the Gibbs free energy is continuous for a change in pressure and temperature, although there is a distinct change in the slope at the phase transition (see Figure 5.7 and Eq. (4.7)).

$$dG = VdP - SdT \tag{4.7}$$

3) Entropy and volume change are discontinuous and exhibit an abrupt *step change* at the phase transition. Intuitively, one can visualize the step changes in material volume and entropy with especially vaporization and sublimation. The partial derivatives of the Gibbs free energy Eq. (4.7) with respect to pressure and temperature are shown by Eqs. (5.28) and (5.29) (see Section 1.13) for a discussion of partial differential equations)

$$-(\partial G)/\partial T)_P = S \tag{5.28}$$

$$(\partial G)/\partial P)_T = V \tag{5.29}$$

4) The system exhibits a *latent heat* that is not associated with an increase or decrease in temperature. That is, during the phase transition, the temperature remains constant. The two phases that coexist at the phase transition temperature, $T_{transition}$, have different entropies, S_1 and S_2 and volumes, V_1 and V_2. Therefore, the system must absorb or release heat. This *latent heat* or *latent enthalpy of transition* or simply *heat* or *energy of transition* is due to the difference in entropies of the two phases (Eq. 5.30).

$$\Delta H_{transiton} = T_{transition}(S_1 - S_2) \text{ or } T_{transition}\Delta S \tag{5.30}$$

5) Most drug polymorphic phase transitions are first-order. Drug polymorphic transitions often lead to discontinuous changes in volume, density, and entropy. The typical polymorphic endothermic peak has a latent enthalpic nature where the material temperature remains constant until the phase change is complete.

According to Ehrenfest, the second derivative of the Gibbs free energy is discontinuous for thermodynamic properties like heat capacity and thermal expansion coefficient. A Curie point phase transition is considered a classic second-order phase transition. While glass and amorphous material are characterized by glass and enthalpic recovery transitions and corresponding transition temperatures, the phase changes are time-dependent. These transitions also exhibit discontinuities in heat capacity which are observed on DSC (differential scanning calorimetry) and MDSC (modulated differential calorimetry). However, they do not meet the Ehrenfest assumption requiring an equilibrium phase transition. Because of this distinction, glass and amorphous transitions are referred to as second-order. They are also often loosely called second-order phase transitions.

Second-order phase transitions have the following common thermodynamic properties:

1) The change in the Gibbs free energy is continuous, and there is no distinct change in the slope at the phase transition.
2) The entropy and volume changes are continuous and exhibit a distinct change in slope at the phase transition as the material changes from a relatively ordered state to a more disordered state.
3) There is no latent heat associated with the second-order phase transitions.
4) The system's heat capacity is discontinuous. This discontinuous thermodynamic property is shown as a step change on differential scanning calorimetry. The second-order partial derivative of Gibbs free energy is heat capacity and given by Eq. (5.31):

$$-T(\partial^2 G)/\partial T^2)_P = T(\partial S)/\partial T)_P = C_P \tag{5.31}$$

5) Other properties are also discontinuous. The thermal expansion coefficient and thermal compressibility are discontinuous.

For a more extensive theoretical discussion of the Ehrenfest transition order classification, see phase transitions: https://itp.uni-frankfurt.de/~gros/Vorlesungen/TD/6_Phase_transitions.pdf.

5.8 Pharmaceutical Application of Chemical Equilibrium

Example 5.9 First-Order Epimerization Equilibrium Reaction between Tetracycline and Epitetracycline (Thorstein, 2014)

Over time, the majority of drug degradation reactions go to completion yielding mostly drug degradants. A few drugs undergo reversible degradation where the drug reaches an equilibrium with one or more drug related degradants. Tetracycline degradation is an example of a reversible degradation mechanism. Tetracycline epimerization has been shown to occur in the pH range of 2–6.

Epitetracycline is a degradation product of tetracycline. It is not as potent an antibiotic as tetracycline, and it is more toxic. The equilibrium for this reversible reaction can be written:

$$\text{Tetracycline } (Tet) \rightleftharpoons \text{Epitetracycline } (ET) \tag{5.32}$$

A detailed discussion of chemical kinetics and stability is beyond the scope of this text; however, a brief overview is available by Al-Achi et al. (2023). The purpose of this example is to show that some drugs degrade to an equilibrium state. Another purpose is to show that forward and reverse rate constants can be determined from an equilibrium process. For a pharmaceutical product design scientist, these rate constants are extremely valuable. Two important uses are to predict drug stability at different pHs and temperature and to determine the shelf-life of the product. Let us begin the journey of determining the forward and reverse rate constants for tetracycline epimerization from equilibrium data.

The rate constant for the *forward reaction* (tetracycline to epitetracycline) is given the symbol k_f. The *reverse reaction rate constant* (epitetracycline to tetracycline) is k_r. These rate processes can be described by Eq. (5.33).

$$d(Tet)/dt = -k_f(\text{Tet}) + k_r(ET) \tag{5.33}$$

where $d(Tet)/dt$ is the change in the concentration of *Tet* over time, *Tet* is the tetracycline concentration at a given time, and *ET* is the epitetracycline concentration at given time. Equation (5.33) shows that the *net rate decrease, d(Tet)/dt,* of *Tet* is given by the rate at which *Tet* decreases in forward step, plus the rate of *Tet* formation in the reverse step. The mass balance for ET is:

$$ET = Tet_0 - Tet \tag{5.34}$$

where Tet_0 is the tetracycline concentration at time zero.

ET from Eq. (5.34) can be substituted into Eq. (5.33) to give:

$$d(Tet)/dt = -k_f\ (Tet) + k_r\ (Tet_0 - Tet) \tag{5.35}$$

At equilibrium, integration of (5.35) yields:

$$(k_f + k_r)t = \ln\ (Tet_0 - Tet_{eq})/\ (Tet - Tet_{eq}) \tag{5.36}$$

and

$$t = 1/(k_f + k_r) \cdot \ln\ (Tet_0 - Tet_{eq})/\ (Tet - Tet_{eq}) \tag{5.37}$$

A detailed step-by-step derivation of this integration and the required algebraic manipulations are beyond the scope of this discussion.

The slope of Eq. (5.37) is *S (for slope)*

$$S = 1/(k_f + k_r) \tag{5.38}$$

when time is plotted on the y-axis and $\ln\ (Tet_0 - Tet_{eq})/\ (Tet - Tet_{eq})$ is plotted on the *x*-axis.

From Figure 5.14a, the equilibrium tetracycline concentration was 34% of its initial concentration. Epitetracycline reached an equilibrium concentration of 66%, which is almost twice the concentration of tetracycline.

S is determined from a plot of time versus $\ln\ (Tet_0 - Tet_{eq})/(Tet - Tet_{eq})$ based on Eq. (5.37). From Figure 5.14b, Slope $= S = 1/(k_f + k_f) = 40$ minutes,and

$$\underline{S = (k_f + k_f) = 1/40 \text{ min} = 2.5\text{E-}2 \text{ min}^{-1}} \tag{5.39}$$

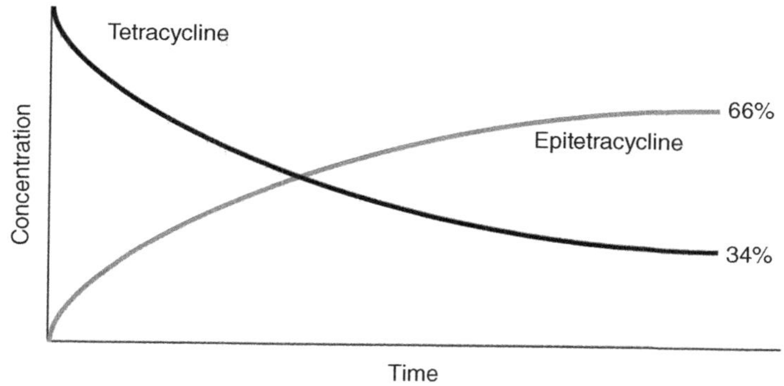

Figure 5.14a Equilibrium between tetracycline and epitetracycline. *Source:* Adapted from Loftsson (2014).

Figure 5.14b Plot of Time versus ln $(Tet_0 - Tet_{eq})/(Tet - Tet_{eq})$. *Source:* Adapted from Loftsson (2014).

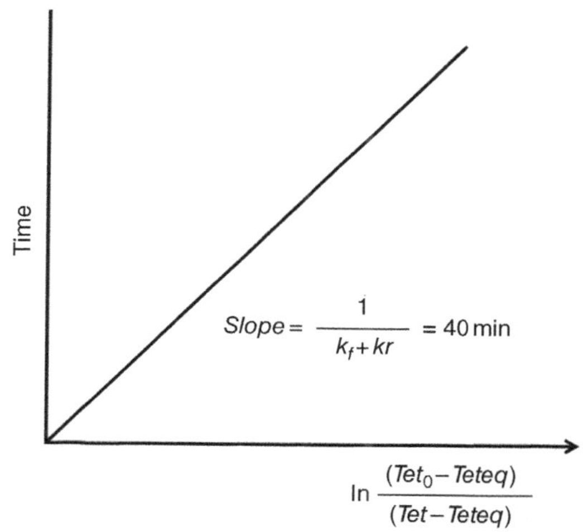

At equilibrium, Eq. (5.33) becomes:

$$d(Tet)/dt = -k_f(Tet_{eq}) + k_r(ET_{eq}) = 0 \tag{5.40}$$

At equilibrium,

$$k_r(ET_{eq}) = k_f(Tet_{eq}) \tag{5.41}$$

and the equilibrium constant is equal to the expressions provided in Eq. (5.42) with concentrations given in percentages.

$$K = ET_{eq}\,\% \,/\, Tet_{eq}\,\% \text{ and } K = k_f/k_r \tag{5.42}$$

Therefore, from Figure 5.14a and Eq. (5.42)

$$\underline{K = 66/34 = 1.9} = k_f/k_r \tag{5.43}$$

The K tells us the same thing as Figure 5.14a. Epitetracycline at equilibrium is about twice the concentration compared to tetracycline.

Knowing the equilibrium constant, k_f and k_r can be calculated using $k_f / k_r = 1.9$ (Eq. 5.43). From Eq. (5.43), k_f is

$$k_f = 1.9\, k_r. \tag{5.44}$$

k_r is determined by substituting for k_f from Eq. (5.44) for k_f in Eq. (5.39)

$$S = k_f + k_r = 2.5\text{E-2 min}^{-1} \tag{5.39}$$

giving

$$1.9\, k_r + k_r = 2.5\text{E-2 min}^{-1} \text{ and } \underline{k_r \text{ equals } 8.6\text{E-3 min}^{-1}} \tag{5.45}$$

Now, we can substitute 8.6E-3 min^{-1} for k_r in Eq. (5.39) and solve for k_f.

$$\underline{k_f = 1.6\text{E-2 min}^{-1}} \tag{5.46}$$

This shows that the forward rate constant, 1.6E-2 min^{-1}, is about 1.9 times larger than the reverse rate constant, 8.6E-3 min^{-3}. This is consistent with the equilibrium value, *K*, and the ET_{eq} % / Tet_{eq} % ratio given in Eq. (5.43).

5.9 Ionic Equilibrium

Pharmaceutical liquid formulations such as emulsions, suspensions, and solutions for oral, topical, inhalation, ophthalmic, and injectable formulations use buffers to optimize solubility, stability, and diminish tissue irritation or damage. Knowledge of acid-base ionic buffer equilibrium is a key component to product design. Several aspects of weak acid and weak base ionic equilibria and acid-base buffer systems are discussed in this section.

5.9.1 Ionization Equilibrium of Weak Acids and Bases

Brønsted–Lowry defined an *acid as a proton (H^+) donor* and a *base as a proton acceptor*. When a Brønsted–Lowry acid loses a proton, a *conjugate* base is formed. Likewise, when a Brønsted–Lowry base gains a proton, a *conjugate* acid is formed. These conjugate reactions are shown later. *Conjugate* means coupled or connected, and in the case of weak acids and bases, we refer to them as *conjugate acid-base pairs*. Weak acids undergo partial dissociation in water which results in relatively low *hydronium or hydrogen ion* concentrations. Typical weak acids like acetic acid, benzoic acid, and water have pH values that range from 4 to 7. Weak bases like aniline and trimethylamine have pHs that generally range from 7 to 10. pH is a function of the (1) ionization constant and (2) the concentration of the weak acid or base and its conjugate base or conjugate acid.

On the other hand, strong acids like HCl and strong bases like NaOH completely dissociate in water. As such, they become totally ionized. In the case of HCl and NaOH, the concentration of the hydrogen ions or hydroxide ions is equal to the concentration of HCl and NaOH added to the water phase.

The general equation for the ionization of a weak acid shows the weak acid donating a hydrogen ion to water to form a hydronium ion:

$$HA_{(aq,\,acid\ 1)} + H_2O_{(l,\,base\ 1)} \rightleftharpoons H_3O^+{}_{(aq,\,acid\ 2)} + A^-{}_{(aq,\,base\ 2)} \tag{5.47}$$

where $HA_{(aq,\ acid\ 1)}$ is a weak acid, $H_3O^+{}_{(l,\ acid\ 2)}$ is the *hydronium ion*, and $A^-{}_{(aq,\ base\ 2)}$ is the conjugate base of $HA_{(aq,\ acid\ 1)}$. In the remainder of this section, $H_3O^+{}_{(aq)}$ with be referred to as *hydrogen ion* and represented by H^+. The equilibrium constant, K, for this dissociation ionization process is (see Section 5.1.1.3):

$$K = [\text{Products}]/[\text{Reactants}] = [H_3O^+][A^-]/[HA][H_2O] \tag{5.48}$$

where $[H_3O^+]$, $[A^-]$, $[HA]$, and $[H_2O]$ are the molar concentrations of each of the species. From Eq. (5.48) and this specific case, the molar units cancel resulting in K, the equilibrium constant, being a unitless constant.

The concentration of water is approximately 55 M in aqueous systems and can be considered constant compared to the smaller concentrations of the other products and reactants in this ionization process. For example, at pH 1, the hydrogen ion concentration is 1E-1 M. Therefore, K can be rewritten as a new constant called the *weak acid ionization constant, K_a.*

$$K_a = K \cdot [H_2O] = [H_3O^+][A^-]/[HA] \tag{5.49}$$

From Eq. (5.49), dimensional analysis (see Section 1.2) gives molar units for K_a.

The general equation of a weak base is:

$$B_{(aq, base\ 1)} + H_2O_{(aq, acid\ 1)} \rightleftharpoons BH^+{}_{(aq, acid\ 2)} + OH^-{}_{(aq, base\ 2)} \tag{5.50}$$

where $B_{(aq, base\ 1)}$ is a weak base, $BH^+{}_{(aq, acid\ 2)}$ is the conjugate acid of $B_{(aq, base\ 1)}$, and $OH^-{}_{(aq, base\ 2)}$ is the hydroxide ion solution concentration. Here, the weak base $B_{(aq, base\ 1)}$ accepts a hydrogen ion from water. The equilibrium constant, K, for this dissociation ionization process is (see Section 5.1.1.3):

$$K = [\text{Products}]/[\text{Reactants}] = [BH^+][OH^-]/[B][H_2O] \tag{5.51}$$

where $[BH^+]$, $[OH^-]$, $[B]$, and $[H_2O]$ are the molar concentrations of these species. Considering the water molar concentration being constant, the weak base ionization constant becomes:

$$K_b = K \cdot [H_2O] = [BH^+][OH^-]/[B] \tag{5.52}$$

From Eq. (5.47), $A^-{}_{(aq)}$ is the conjugate base of $HA_{(aq)}$. $A^-{}_{(aq)}$ can react with water which is shown in Eq. (5.53):

$$A^-{}_{(aq)} + H_2O_{(l)} \rightleftharpoons HA_{(aq)} + OH^-{}_{(aq)} \tag{5.53}$$

and the

$$K_b = [HA][OH^-]/[A^-] \tag{5.54}$$

K_b has units of molar.

The *autoionization* of water is shown in Eq. (5.55).

$$H_2O_{(l, acid\ 1)} + H_2O_{(l, base\ 1)} = H_3O^+{}_{(aq, acid\ 2)} + OH^-{}_{(aq, base\ 2)} \tag{5.55}$$

and

$$K = [H_3O^+][OH^-]/[H_2O][H_2O] \tag{5.56}$$

where the autoionization constant for water or *ion product of water* is:

$$K_w = K \cdot [H_2O] \cdot [H_2O] = [H_3O^+][OH^-] \tag{5.57}$$

In this case, the dimensional units of K_w is molar squared, $[M]^2$

It can now be shown that $K_w = K_a \cdot K_b$ by using Eqs. (5.49) and (5.54).

$$K_w = K_a \cdot K_b = [H_3O^+][\cancel{A^-}]/[\cancel{HA}] \cdot [\cancel{HA}][OH^-]/[\cancel{A^-}] = \underline{[H_3O^+][OH^-]} \tag{5.58}$$

K_w is 1E-14 M^2 at 25 °C. At pH 7, the hydrogen ion and hydroxide ion concentrations are equal.

5.9.2 pH and pK_a

pH is a measure of the *acidity* or *alkalinity* of a solution. Aqueous solutions at 25 °C with a pH less than seven are considered acidic, while those with a pH greater than seven are considered basic or alkaline. pH 7.00 is considered neutral at 25 °C, because at this pH, the concentration of H^+ equals the concentration of OH^-. pH is formally dependent upon the activity of hydrogen ions (H^+), but for very pure dilute solutions, molarity may be used as a substitute with modest sacrifice of accuracy.

pH is shorthand for a mathematical transformation. In chemistry, a small "p" is used in place of writing $-\log_{10}$. In the case of pH, the H represents $[H^+]$, the molar concentration of hydrogen ions. The symbol "p" is a mathematical operator that turns a quantity into its negative base 10 logarithm ($-\log_{10}$ or $-\log$).

K_a and pK_a describe the strength of the acid. The larger the K_a, the stronger the acid. That is, more H^+ is produced as product (see Eq. 5.49) compared to the reactants. It is also true that the lower the pK_a, the stronger the acid, since p$K_a = -\log K_a$ or $\log 1/K_a$ (see Section 1.11). It is important to be understand this inverse relationship between pK_a and K_a. This inverse relationship also exists for the relationship between pH and $[H^+]$ where pH $= -\log [H^+] = \log [H^+]^{-1} = \log (1/[H^+])$.

Over the years, the literature has slowly abandoned the use of pK_b to describe the strength of the base. More commonly, the K_a or the pK_a of the base's conjugate acid is used to describe the strength of the base which gives a common equilibrium constant to compare the strengths of acids and bases. In the instance of bases, the smaller the K_a, the stronger the base, and the larger the pK_a, the stronger the base. For example, Eq. (5.50) can be written in terms of the base's conjugate acid and the equilibrium constant in this case is K_a.

$$BH^+{}_{(aq)} + H_2O_{(l)} \rightleftharpoons B_{(aq)} + H_3O^+{}_{(l)} \tag{5.59}$$

$$K_a = \left[B_{(aq)}\right]\left[H_3O^+{}_{(l)}\right]/\left[BH^+{}_{(aq)}\right] \tag{5.60}$$

5.9.3 Henderson-Hasselbalch Equation

The *Henderson-Hasselbalch* equation describes the condition of an acid or base equilibrium system. The *generic* form of the *Henderson-Hasselbalch* equation is provided in Eq. (5.61).

$$\text{pH} = \text{p}K_a + \log\,[\text{conjugate base}]/[\text{conjugate acid}] \tag{5.61}$$

This equation uses Bronsted and Lowry acid-base terminology of conjugate acid and conjugate base. It is *generic* because the same equation can be used for a weak acid or weak base. The Henderson-Hasselbalch equation derivation for a weak acid and base is shown.

Using Eq. (5.47) as the general weak acid ionization equilibrium:

$$HA_{(aq)} + H_2O_{(l)} \rightleftharpoons H_3O^+{}_{(aq)} + A^-{}_{(aq)} \tag{5.47}$$

and the K_a from Eq. (5.49):

$$K_a = [H_3O^+][A^-]/[HA] \tag{5.49}$$

and the logarithm operation rules from Section 1.11, the base 10 logarithm is taken of both sides of Eq. (5.49) to give Eq. (5.62).

$$\log K_a = \log\,[H_3O^+] + \log[A^-]/[HA] \tag{5.62}$$

Now, log K_a is moved to the right side of the equation and log [H_3O^+] is moved to the left side of Eq. (5.62) to result in:

$$-\log [H_3O^+] = -\log K_a + \log [A^-]/[HA] \quad (5.63)$$

The "p" operation that was introduced in Section 5.92 is employed where $p = -\log$ and is substituted into Eq. (5.63). In addition, [A^-] is the conjugate base and [HA] is the conjugate acid in Eq. (5.63). Equation (5.64) is:

$$pH = pK_a + \log [\text{conjugate base}]/[\text{conjugate acid}] \quad (5.64)$$

which is consistent with the Henderson-Hasselbalch generic form shown by Eq. (5.61), pH = pK_a + log [conjugate base]/[conjugate acid].

To derive the Henderson-Hasselbalch equation for a weak base in the form of the generic Henderson-Hasselbalch, Eq. (5.61) needs to reverse the equilibrium reaction and use the conjugate acid as the reactant. This establishes the equilibrium for the weak base's pK_a. This derivation begins with conjugate acid-conjugate base equilibrium, Eq. (5.59)

$$BH^+_{(aq)} + H_2O_{(l)} \rightleftharpoons B_{(aq)} + H_3O^+_{(aq)} \quad (5.65)$$

The K_a for Eq. (5.65) is Eq. (5.66):

$$K_a = [B_{(aq)}][H_3O^+_{(aq)}]/[BH^+_{(aq)}] \quad (5.66)$$

The Henderson-Hasselbalch derivation now proceeds in a similar fashion as for the weak acid. The log of is taken of both sides of the equation.

$$\log K_a = \log [H_3O^+_{(aq)}] + \log [B_{(aq)}]/[BH^+_{(aq)}] \quad (5.67)$$

Log K_a is moved to the right side of Eq. (5.67) and log [$H_3O^+_{(aq)}$] is moved to the left side.

$$-\log [H_3O^+_{(aq)}] = -\log K_a + \log [B_{(aq)}]/[BH^+_{(aq)}] \quad (5.68)$$

Using the "p" operator, Eq. (5.68) becomes:

$$pH = pK_a + \log [B_{(aq)}]/[BH^+_{(aq)}] \quad (5.69)$$

[$B_{(aq)}$]/[$BH^+_{(aq)}$] is in the generic form of [conjugate base]/[conjugate acid], so the Eq.(5.69) has the generic Henderson-Hasselbalch form of Eq. (5.61).

5.9.4 Le Châtelier's Principle and Common Ion Effect on pH

The equilibrium dissociation ionization for a weak acid is:

$$HA_{(aq)} + H_2O_{(l)} \leftrightarrows H_3O^+_{(aq)} + A^-_{(aq)} \quad (5.47)$$

For acetic acid (AA), the equilibrium is:

$$CH_3COOH + H_2O \rightleftharpoons H_3O^+ + CH_3COO^- \quad (5.70)$$

This equation is valid and independent of where the species come from. The common ion effect occurs when a second source of a common ion is introduced into an equilibrium system. If a Na^+ $Acetate^-$ salt (Na^+A^-) which readily dissociates into Na^+ and A^- ions is added to the Eq. (5.70) equilibrium, more A^- (CH_3COO^-) ions are now present in the system. Le Châtelier's principle

(Section 5.2) predicts how this change or stress placed on the equilibrium system will create a new equilibrium by counteracting the original change or stress.

As a general example, let us say that Na^+A^- salt is added to the Eq. (5.70) equilibrium. The Na^+A^- adds a second source of A^- ions. The additional A^- (CH_3COO^-) ions stress the system on the product side of the equilibrium. To counteract this stress, the equilibrium is pushed to the left, creating more CH_3COOH and decreasing H_3O^+ or H^+. The common ion *suppresses* the ionization of the acid, HA or in this case AA (CH_3COOH). The increased HA (AA) concentration and decrease in hydrogen ion concentration result in an increase in the pH of the system.

Example 5.10a Determine the pH of a 0.2 Molar Acetic Acid Solution

Given:

25 °C
Acetic acid (AA), 0.20 M
Acetic acid pK_a, 4.74 and $K_a = 1.8\text{E-}5$ M
$K_a = [H^+][A^-]/[HA]$
Generic Henderson-Hasselbalch equation,
pH = pK_a + log [conjugate base]/[conjugate acid]

At equilibrium, the concentrations of AA, acetic acid anion, and hydrogen ion achieve a constant concentration that is defined by AA's $K_{a.}$ which is $K_a = [H^+][CH_3COO^-]/[CH_3COOH]$.

Initially, AA's concentration is 0.2 M. At equilibrium, x M of AA will dissociate to give x M of H^+ and x M CH_3COO^-. This change is given by:

$$K_a = [x][x]/[0.20 - x] = 1.8\text{E-}5\text{ M} \tag{5.71}$$

which becomes:

$$x^2 = -1.8\text{E-}5x + 3.6\text{E-}6\text{ M} \tag{5.72}$$

and moving the right-hand term of Eq. (5.72) to the left side of the equation gives:

$$x^2 + 1.8\text{E-}5x - 3.6\text{E-}6 = 0 \tag{5.73}$$

Equation (5.73) can be solved by the quadratic equation, where $ax^2 + bx + c = 0$. a and b are the coefficients, x is the variable, and c is the constant term. The solution of this quadratic equation is:

$$x = [-b \pm \sqrt{(b^2 - 4ac)}]/(2a) \tag{5.74}$$

where x is the concentration of H^+ and CH_3COO^-. Substituting for the terms in Eq. (5.74) leads to:

$$x = [-1.8\text{E-}5 + / - \sqrt{(1.8\text{E-}5)^2 - 4(1)(-3.6\text{E-}6)}]/2(1) \tag{5.75}$$

$$x = [-1.8\text{E-}5 + / - \sqrt{(3.24\text{E-}10 + 1.44\text{E-}5}]/2(1) \tag{5.76}$$

$$\underline{x = 1.88\text{E-}3\text{ M}} \text{ and } -1.92\text{E-}3 \text{ where the positive root is } [H^+]. \tag{5.77}$$

<u>The pH of the 0.2 M AA solution is:</u>

$$\underline{\text{pH}} = -\log 1.88\text{E-}3\text{ M} = -(-2.73) = \underline{2.73}. \tag{5.78}$$

Example 5.10b What Is the pH of a Solution Containing 0.2 M Acetic Acid (AA) and 0.3 M Sodium Acetate?

Given:

25 °C
Acetic acid, 0.20 M
Sodium acetate, 0.30 M
Acetic acid pK_a, 4.74 and $K_a = 1.8E\text{-}5$ M
$K_a = [H^+][A^-]/[HA]$
Generic Henderson-Hasselbalch equation,
$pH = pK_a + \log$ [conjugate base]/[conjugate acid]

Sodium acetate is a strong electrolyte, and it completely dissociates into sodium ion and acetate anion. Acetate ion is the conjugate base of AA, and as a strong electrolyte, its concentration is 0.30 M. The concentration of acetate contributed by AA was shown above to be 1.9E-3 M. This concentration is negligible compared to the 0.30 M acetate from Na Acetate. To determine the pH of this example solution, the concentrations of acetate anion and AA can be directly substituted into the generic Henderson-Hasselbalch equation. The pK_a for AA is also substituted into the generic Henderson-Hasselbalch equation.

$$pH = 4.74 + \log[0.30]/[0.20] \tag{5.79}$$

$$\underline{pH} = 4.74 + \log 1.5 = 4.74 + 0.18 = \underline{4.92} \tag{5.80}$$

Adding 0.30 M acetate to the AA solution increased the pH from 2.73 to 4.92. Addition of the common acetate ion raised the pH 2.19 pH units. Le Châtelier's principle explains this increase in pH with the addition of a common ion. Adding sodium acetate to the AA solution adds stress to the equilibrium by increasing the acetate anion concentration on the right-hand side of Eq. (5.70).

$$CH_3COOH + H_2O \leftrightarrows H_3O^+ + CH_3COO^- \tag{5.70}$$

To counteract this stress, the equilibrium moves to the left. This rebalancing results in an increase in AA and a decrease in hydrogen ion. This decrease in hydrogen ion is equivalent to increasing the pH. Simply put, the addition of the acetate common ion *suppresses* the ionization of acetic acid which decreases the formation of hydrogen ions.

5.10 Effect of Temperature on the Equilibrium Constant: The van't Hoff Equation

From Eqs. (4.2) $\Delta G = \Delta H - T\Delta S$ and (5.3), $\Delta G^\circ = -(RT) \ln K$, ΔG° can be expressed in terms of enthalpy and entropy giving Eq. (5.81).

$$-(RT) \ln K = \Delta H^\circ - T\Delta S^\circ \tag{5.81}$$

Dividing both sides of Eq. (5.81) by $-(RT)$ leads to Eq. (5.82).

$$\ln K = -\Delta H^\circ/RT + \Delta S^\circ/R \tag{5.82}$$

This equation is known at the *van't Hoff* equation. The generalize form is:

$$\ln K = -\Delta H/RT + \Delta S/R \tag{5.83}$$

ΔH is the enthalpy of any equilibrium process such as heat of melting, vaporization, and solution. Likewise, ΔS is the entropy of any equilibrium process. It provides the relationships among the equilibrium constant, enthalpy, entropy, and temperature. A plot of ln K values at a specific temperature in 1/K results in a slope equal to $-\Delta H/R$ and an intercept is $\Delta S/R$. Therefore, the enthalpy and entropy of an equilibrium process can be determined by employing the van't Hoff plot.

The change in the equilibrium constant, K, resulting from a change in temperature can be determined in the following manner where $\ln K_1 = -\Delta H/RT_1 + \Delta S/R$ and $\ln K_2 = -\Delta H/RT_2 + \Delta S/R$.

$$\ln K_1 - \ln K_2 = [-\Delta H/RT_1 + \cancel{\Delta S/R}] - [-\Delta H/RT_2 + \cancel{\Delta S/R}] \tag{5.84}$$

Using Eq. (1.47) for the logarithm operation,

$$\ln (K_1/K_2) = -\Delta H/R\,(1/T_1 - 1/\ T_2) \tag{5.85}$$

and introducing a common denominator for Eq. (5.85) gives Eq.(5.86):

$$\ln (K_1/K_2) = -\Delta H/R\,[(T_2 - T_1)/T_2\,T_1] \tag{5.86}$$

K_2 can be determined by knowing K_1, ΔH, T_1, and T_2. See Connors and Mecozzi (2010) for further information.

Example 5.11 What is a Protein's Unfolding Equilibrium Constant, $K_{unfolding}$, at 37 °C (Glasfeld, 2015)

Given:

N $\leftrightarrows$ U where N is the protein in native state and U is the protein in the unfolded state

$K_{unfolding}$ = [U]/[N] where [N] and [U] are the molar concentrations of N and U

At 22 °C, $K_{unfolding}$ is 0.25,

$\Delta H_{unfolding} = +25\ \text{kcal mol}^{-1}$

K_2 is the unfolding equilibrium constant at body temperature, 37 °C.

Using Eq. (5.86),

$$\ln (0.25/K_2) = -[25\ \text{kcal}/\cancel{\text{mol}}/1.987\text{E-}3\ \cancel{\text{Kcal mol}^{-1}}\cdot\cancel{\text{K}}][(310.15\ \cancel{\text{K}} - 295.15\ \cancel{\text{K}})/[310.15\ \cancel{\text{K}}\cdot 295.15\ \cancel{\text{K}}] \tag{5.87}$$

Using logarithm operation Eq. (1.47) on the left-hand side of the equation and applying simply mathematical operations on the right-hand side gives:

$$-1.386 - \ln K_2 = -1.258\text{E}4\,(15/310.15.15) = -1.258\text{E}4\cdot 1.638\text{E-}4 = -2.060 \tag{5.88}$$

$$-\ln K_2 = -2.060 + 1.386 = -0.674 \tag{5.89}$$

$$\ln K_2 = 0.674 \tag{5.90}$$

Taking the antilog of both sides of Eq. (5.90) yields:

$$\underline{K_2 = 1.962} \text{ and using significant figure rule 1.24, } \underline{K_2 = 2.0} \tag{5.91}$$

The protein's $K_{unfolding}$ increased from 0.25 at 22 °C to 2.0 at body temperature. The percent of folded protein can be calculated for each temperature.

Example 5.12 What is the Percent of Folded Protein at 22 °C and 37 °C?
Given:

$$K_{unfolding,}\ 22\ °C = 0.25$$

$$\text{Let } [U] = x$$
$$\text{Let } [N] = y$$

$$x/y = 0.25 \tag{5.92}$$

$$x + y = 1 \text{ or } 100\% \tag{5.93}$$

Solve for x and y
From Eq. (5.92),

$$x = 0.25y \tag{5.94}$$

Substitute x from Eq. (5.94) into Eq. (5.93) to give:

$$0.25y + y = 1 \tag{5.95}$$

Solving for y from Eq. (5.96)

$$1.25y = 1 \text{ and } y = \underline{0.80 \text{ or } 80\% \text{ folded protein}} \tag{5.96}$$

Solve for x using Eq. (5.94)

$$x = 0.25\ (80\%) = \underline{20\% \text{ unfolded protein}} \tag{5.97}$$

At 22 °C, the protein is 80% folded protein
Using similar calculations, it can be shown that the protein was only 33% folded at body temperature.

5.10.1 The Effect of Temperature on the Ionic Product of Water (K_w)

It is interesting to note that the pK_w decreased by almost 2 pH units over the temperature range of 0 °C to 60 °C.

References

Al-Achi, A., Gupta, M.R., and Stagner, W.C. (2023). *Integrated Pharmaceutics: Applied Preformulation, Product Design, and Regulatory Science*, 2e, 78–107. Hoboken, NJ: John Wiley & Sons.

Avner, S. (1974). Phase diagrams. In: *Introduction to Physical Metallurgy*, 2e. New York: McGraw-Hill.

Badawy, S. (2015). Laboratory freezing desalination of seawater. *Desalin. Water Treat.* 57 (24): 11040–11047. https://doi.org/10.1080/19443994.2015.1041163.

Barley, J. (2022). Basic principles of freeze drying. SP Scientific Freeze Drying/Lyophilization Information: Basic Principles. https://rechneronline.de/air/saturation-vapor-pressure.php (accessed March 2022).

Brodin, A., Nyqvist-Mayer, A., Wadsten, T. et al. (1984). Phase diagram and aqueous solubility of the lidocaine-prilocaine binary system. *J. Pharm. Sci.* 73: 481–484.

Brunner, E. (1904). Reaktionsgeschwindigkeit in heterogenen Systemen. *Z. Phys. Chem.* 47U: 56–102.

Chang, R. (1998). *Chemistry*, 6e. Boston: McGraw-Hill.

Cherukuvada, S. and Nangia, A. (2014). Eutectics as improved pharmaceutical materials: design, properties, and characterization. *Chem. Commun.* 50: 906–923.

Clark, J. https://chem.libretexts.org/Bookshelves/Physical_and_Theoretical_Chemistry_Textbook_Maps/Supplemental_Modules_(Physical_and_Theoretical_Chemistry)/Equilibria/Physical_Equilibria/Phase_Diagrams_for_Pure_Substances. (accessed February 2022).

Connors, K.A. and Mecozzi, S. (2010). *Thermodynamics of Pharmaceutical Systems: An Introduction to Theory and Applications*, 2nde. Hoboken, NJ: John Wiley & Sons.

Flowers, P., Theopold, K., Langley, R., and Robinson, W. (2019). Liquids and solids. In: *Chemistry*, 2e, 493. openstax™. https://openstax.org/details/books/chemistry-2e (accessed March 2023).

Glasfeld, A. (2015). http://people.reed.edu/~glasfeld/Chem391/notes/ChStability_2015.pdf.

Harned, H.S. and Owen, B.B. (1958). Properties of weak electrolytes. In: *The Physical Chemistry of Electrolytic Solutions*, 3rde. New York: Reinhold Publishing Corporation.

Loftsson, T. (2014). *Drug Stability for Pharmaceutical Scientists*. Oxford, UK: Academic Press.

Lower, S. (2023). LibreTexts Chemistry. Chapter 15.4: Free Energy and the Gibbs Function. https://chem.libretexts.org/Bookshelves/General_Chemistry/Book%3A_Chem1_(Lower)/15%3A_Thermodynamics_of_Chemical_Equilibria/15.04%3A_Free_Energy_and_the_Gibbs_Function (accessed April 2023).

Mander, J. (2013). Carnotcycle: the classic blog on thermodynamics. https://carnotcycle.wordpress.com/2013/04/01/the-clausius-clapeyron-equation/ (accessed March 2022).

Nernst, W. (1904). Theorie der Reaktionsgeschwindigkeit in heterogenen Systemen. *Z. Phys. Chem.* 1904 (47U): 52–55.

Patel, V.D., Rathod, B., Haware, R.V., and Stagner, W.C. (2023). Optimized L-SNEEDS and spray-dried S-SNEDDS using a QbD-DM^3 rational design for model compound ketoprofen. *Int. J. Pharm.* https://doi.org/10.1016/j.ijpharm.2022.122494.

Phase Transitions. https://itp.uni-frankfurt.de/~gros/Vorlesungen/TD/6_Phase_transitions.pdf (accessed August 2022).

Thorstein, L. (2014). *Drug Stability for Pharmaceutical Scientists*. Oxford, UK: Academic Press.

Chapter 5 Problem Set

5.1 You routinely need to prepare a phosphate buffer for dissolution studies. Preparing large quantities takes a long time. You decide to heat the phosphate to increase its solubility and so it will dissolve faster. The company chemical store has two sodium phosphates to choose from: disodium phosphate and disodium phosphate dodecahydrate.
- a) Does it matter which disodium phosphate you choose?
- b) If it matters, what is your rationale?

5.2 a) Explain the sodium sulfate curve shown in Figure 5.6.
- b) From 0 °C to ~30 °C, what sodium sulfate species would you find as the excess solid? At ~30 °C and higher, what sodium sulfate species would you find as the excess solid?

5.3 The vapor pressure of water changed 27.17 mmHg from 99 °C to 100 °C. The corresponding volume change from water to vapor over this 1 °C was 3.0114E-2 $m^3\ mol^{-1}$. Calculate the water enthalpy of vaporization (adapted from https://carnotcycle.wordpress.com/2013/04/01/the-clausius-clapeyron-equation/).

5.4 Using the table provided (data determined from Table 5.3), calculate water's autoionization-dissociation enthalpy. What does the enthalpy of autoionization-dissociation mean/tell us?

Table 5.3 The Ionic Product of Water (K_w), pK_w, and Water's Neutral pH at Different Temperatures[a]

°C	K_w	pK_w	Water's neutral pH
0	1.14E-15	14.944	7.472
10	2.92E-15	14.535	7.268
20	4.51E-15	14.346	7.173
25	1.01E-14	13.997	6.998
30	1.47E-14	13.833	6.917
35	2.09E-14	13.680	6.840
40	2.92E-14	13.535	6.768
50	5.47E-14	13.262	6.631
60	9.61E-14	13.038	6.519

[a] Harned and Owen (1958).

1/*T*	ln K_w
0.00353	−33.467
0.00341	−33.032
0.0033	−31.851
0.00319	−31.165
0.00309	−30.593
0.00300	−29.972

5.5 Using the table, calculate the enthalpy and entropy of protein unfolding. The equilibrium constant for each temperature is: $K =$ *concentration unfolded/concentration folded*. Discuss the meaning of the enthalpy and entropy terms. Data obtained from Glasfeld (2015).

1/*T*	ln *K*
0.00345	−2.12
0.00339	−1.38
0.00333	−0.67
0.00328	0.017
0.00323	0.68
0.00317	1.33
0.00313	1.95

6

Drug Solubility Equilibrium

6.1 Introduction

Solubility in general, and drug solubility specifically, is an equilibrium process and system. *Solvation* is the interaction of a solvent with the dissolved solute. *Dissolution* is the process of dissolving a solute in a solvent and is a kinetic process. The equilibrium concepts discussed in Chapter 5 apply to drug solubility. A drug's solubility is one of the most important physical-chemical properties that impact drug product design.

A *solution* is a homogeneous mixture of two or more substances. There are seven types of homogeneous mixtures of solutes and solvents. This chapter focuses on the solubility of solids in liquids. Solid-liquid solubility is the most frequent and important pharmaceutical system of the seven types of homogeneous mixtures. A *binary solution* has only two *components* or *constituents*: a *solvent* and a *solute*. By definition, the solvent makes up the greater quantity of the solution and serves as the "carrier" or "vehicle" for solution. The solute makes up the lesser quantity of the binary system. By convention, since the solvent is present in excess, it is designated *component 1,* and as the lesser component, the solute is identified as *component 2.*

Solubility or *saturation solubility* is defined as the maximum amount of solute that can dissolve in a given quantity of solvent at a specified set of conditions. The most common conditions that affect solubility at atmospheric pressure are temperature, pH, and solvent dielectric constant. *Solubility* can also be defined as the concentration of solute in a *saturated solution* at a specified set of conditions. Both solubility definitions depend on the concept of a saturated solution and are equivalent to the term, "*equilibrium solubility*." For a two-component system of solute and solvent which has two phases, drug solution and excess solid at constant temperature and pressure, the system is invariant with zero degrees of freedom. The Gibbs phase rule, Eq. (5.9), $F = C - P + 2$, can be used to explain these various solubility terms. F is the number of degrees of freedom, C is the number of independent components in the system, P is the number of phases, and 2 accounts for the intensive variables, pressure, and temperature. A drug's solubility, saturated solubility, and equilibrium solubility are invariant at constant temperature and pressure. Using the Gibbs phase rule at constant temperature and pressure, $F = 0 = 2\ (C) - 2\ (P)$. Since solubility is measured at constant temperature and pressure, the general Gibbs phase rule becomes $F =$ 2 (components) − 2 (phases) + 2 (intrinsic variables; temperature and pressure) − 2 (intrinsic variables; controlled temperature and pressure).

Thermodynamics for Drug Product Design: A Guide for Pharmaceutical Scientists, First Edition. William Craig Stagner.

A *saturated solution* exists when the solute is in equilibrium with the solute's undissolved solid phase. It is important that one completely comprehends the importance of this general premise and the nuance of this definition. As discussed in Sections 4.9 and 4.9.1, the chemical potential of a substance is associated with substance's "escaping tendency." In the case of solubility, drugs with higher escaping tendencies, larger chemical potentials, will have greater amounts of drug "escape" from its undissolved solid-state phase into the solvent. The solubility equilibrium between the solute drug substance and solid-state drug phase is driven by the drug's chemical potential. This important relationship between the amount of solute in solution and the identity of the conjugate undissolved solid-state phase is often overlooked or neglected during solubility studies.

There are two other types of solutions: *unsaturated* and *supersaturated* solutions. An unsaturated solution contains less solute in solution than the solvent has capacity to dissolve. A supersaturated solution accommodates more solute in solution than the solvent is capable of dissolving with respect to the thermodynamically stable solid-state drug. Supersaturated solutions are thermodynamically unstable, and the excess solute will precipitate over time in order to reach its thermodynamically stable saturated solubility. The induction time for precipitation to occur is a function of supersaturation ratio (Liu, 2000).

6.1.1 Author's Anecdote – Too Often Solubility Studies Emphasize Analyzing the Concentration of Solute in Solution While Neglecting the Critical Importance of the Identity of the Undissolved Solid Phase Whose Chemical Potential is Responsible for the Observed Solubility

The solubility equilibrium can be given by Eq. (6.1):

$$\text{Crystalline solute} + \text{Pure liquid solvent} \rightleftharpoons \text{Saturated solute solution} \tag{6.1}$$

From personal experience, the author has seen too many cases where erroneous solubilities have been reported because the undissolved solid phase's physical state was not characterized. For example, when water is the solvent, hydrates are usually more insoluble than their anhydrate counterparts. One needs to be careful when reporting the solubility of an anhydrate to make sure the excess solid is the anhydrate and not the converted less soluble hydrate form which commonly happens during a solubility study. Care must also be taken to not use common ions of a drug salt in selecting buffer components or when using salts to adjust the ionic strength of a solution intended for a solubility study. Additionally, the salts of buffers may form low or high solubility salts of the drug.

It is much rarer for the solid to undergo some chemical reaction that changes the drug's bonding arrangement. Here, the chemical structure of the solid and solute will be different.

Before reporting solubility data, it is necessary to characterize the undissolved excess solid phase material to confirm it has not changed physically or chemically over the course of the solubility study. X-ray diffraction, infrared spectroscopy, thermal gravimetric analysis, and differential scanning calorimetry can be used to characterize the initial solid form used in the solubility study and the excess solid form at the completion of the study. The results are then compared to confirm the two solids are the same. *Any solubility value is suspect if the solid-state form of the excess solid is not characterized at the end of the solubility study. If the solid-state excess is different from the input material (initial solid), the measured solubility reflects that of the excess solid.*

6.2 Brief Review of Solubility Concentration Scales Used in this Chapter

For thermodynamic discussions, it is important to be able to account for all the species in solution whether they are solute or solvent species. Mole fraction, percent weight by weight, and molality provide this information. These concentration units are also temperature-independent. It should be pointed out however, that the thermodynamic standard states can be based on mole fraction, molal, weight percent, or molarity concentration scales. As mentioned in the Introduction, component 1 refers to the solvent and component 2 is the solute as the lesser component. The solubility definitions use the term "drug" for the more generic term "solute" to emphasize the focus of this textbook on drug product design. In this case, component D and component 2 are equivalent and make reference to the solute component.

Mole fraction (X). Many theoretical thermodynamic discussions use *mole fraction* concentration to describe the system. Mole fraction is the ratio of the number of moles of one substance to the total moles of all the components in the solution or system. Equation (6.2) shows that the drug mole fraction is a unitless number.

$$X_D = n_D/n_T \tag{6.2}$$

X_D is the mole fraction of drug or solute, n_D is the moles of drug, and n_T is the total moles in solution. A more generic mole fraction definition for a binary system of solvent and solute is as follows: X_1 represents the mole fraction of the solvent and X_2 is the mole fraction solute since the solvent is larger fraction of material.

Percent weight by weight (% w/w). Percent weight by weight is the ratio of weight of drug, D, to the total weight of solution or system multiplied by 100 (Eq. 6.3).

$$\%\mathrm{w/w} = w_D/w_T \cdot 100\% \tag{6.3}$$

% w/w is percent weight by weight, w_D is the weight of drug, and w_T is the weight of the solution or system.

Molality (m). Molality is independent of the solution volume, and it is also independent of temperature since weight does not vary with temperature. Molality is a measure of the number of moles of solute per 1 kg of solvent. See Eq. (6.4)

$$m = n_D/w_{solvent} \tag{6.4}$$

where m is molality, n_D is moles of drug, and $w_{solvent}$ is 1 kg of solvent.

Molarity (M). Molarity is the number of moles of drug (solute) in 1 L of solution. This concentration scale does not account directly for all the species in solution. This concentration scale is also dependent on the temperature as the volume is a function of temperature. Molarity is shown in Eq. (6.5)

$$M = n_D/V_T \tag{6.5}$$

where M is the solution molarity, n_D is the number of moles of drug (solute), and V_T is 1 L of solution.

Weight per volume. Common units associated with a weight per volume concentration scale is mg solute per mL of solution ($mg_{solute}/mL_{solution}$).

The concentration scales are not proportional to one another. However, since mole fraction, percent weight by weight, and molality provide the quantity of all the species in solution, it is possible to convert from one scale to the other knowing the molecular weight of each species and the definition of each scale. On the other hand, molarity and weight per volume do not account for the quantities of each species and therefore cannot be converted to any other concentration scale.

In highly dilute solutions, mole fraction and molality and mole fraction and molarity can be considered proportional. This proportionality is revealed in Eqs. (6.6) and (6.7).

$$X = (m \cdot \text{molar mass}_{solvent})/(1000) \tag{6.6}$$

where X is the mole fraction of the solute that is significantly smaller than 1, m is the solution molality, and $MW_{solvent}$ is the molecular weight of the solvent.

$$X = (M \cdot \text{molar mass}_{solvent})/(1000 \cdot p_{solvent}) \tag{6.7}$$

where X is the mole fraction of the solute that is significantly smaller than 1, M is the solution molarity, $MW_{solvent}$ is the molecular weight of the solvent, and $p_{solvent}$ is the density of the solvent.

6.3 Solubility-Related Intermolecular Interactions

6.3.1 Solvent and Solute Interactions

The solubility of a drug is the result of the balance or tension between three intermolecular interactions. This balance or tension involves *solvent-solvent, solute-solute*, and *solvent-solute* interactions. Solvent-solvent intermolecular interactions occur in all solutions and are especially important in water. Solute-solute intermolecular interactions involve the pure drug solute. These interactions become more significant in concentrated solutions. Solvent-solute is the third interaction that plays an important role in all solutions.

6.3.2 Types of Intermolecular Attraction Forces

An ionic bond arises from electrostatic forces of attraction between two oppositely charged ions. Ionic bonds are formed between metals and nonmetals.

A covalent bond, also called a valence bond, occurs as the result of sharing electrons. Covalent bonds join nonmetals to nonmetals.

There are several other general types of interactive attractive forces. These include hydrogen bonds, van der Waals forces, salt electrostatic interactions, and the hydrophobic effect.

Hydrogen bonds generally form strong electrostatic dipole-dipole interactions that involve hydrogen. *Dipole-dipole* forces are attractive forces between the positive end of one polar molecule and the negative end of another polar molecule. As denoted by its name, hydrogen bonds involve specific electron donor and electron acceptor paired interactions between a hydrogen atom and high electronegative atoms like nitrogen, oxygen, and fluorine. An example of a hydrogen bond is the interaction between methanol and water.

van der Waals forces are a group of relatively weak attraction forces. These forces include *Keesom* forces (weak dipole-dipole interactions), *Debye* forces (dipole-induced dipole), and *London* or *dispersion* forces (induced dipole-induced dipole). A dipolar substance can be thought of as having a pair of equal and oppositely charged areas or *poles* that are separated by a distance. A substance's *polarity* or *polarity strength* is measured by its dipole moment, μ. The dipole moment is the distance between a pair of opposite charges times the charge magnitude. The defined direction is toward the positive charge. The separation of charge and differences in electronegativity between the poles create a dipole moment. The larger the separation distance and electronegativity difference leads to a larger dipole moment. In comparison, a nonpolar molecule is considered to have no charged poles with charge distribution evenly spread across the molecular surface. However, the charge distribution is not static and moves/oscillates which can cause a temporary dipole.

Keesom forces can only occur between polar molecules which exhibit permanent dipole moments. The interaction is electrostatic in nature and based on Coulomb's law. The larger the dipole moment, the stronger the attractive force. An example of a Keesom permanent dipole-dipole interaction is methanol and chloroform.

Debye forces occur from a permanent dipole-induced dipole interaction. The permanent dipole induces a polarization of a second molecule that leads to mutual attraction of the induced dipolar molecule. An example of a dipole-induced dipole is an acetone-hexane attraction interaction.

London or dispersive forces are temporary dipole-induced dipole interaction. London forces are caused by random fluctuations of electron density. London forces exist for all substances including nonpolar, polar, or ionic materials. Dispersion forces increase with an increase in the number of electrons. In some cases, dispersion forces are similar or greater than dipole-dipole interactions. An example of London forces is liquid mineral oil which is nonpolar yet a condensed liquid. A second example is the attractive dispersion forces between octane and hexane.

Salts and ionic interactions can form *ion-ion, ion-dipole,* and *ion-dative* or *coordination* bonds. *Ion-ion* bonds typically occur when one atom gives up an electron and another atom gains an electron forming a mutually electrostatic attractive bond. This cationic-anionic attractive force is noncovalent; that is, there is minimal sharing of electrons. Typically, salts are formed by neutralization reactions between a base and an acid. Solution ionic intermolecular interactions are sometimes called *ion pairing* or *salt bridges* An *ion-dipole* interaction usually occurs in polar solvents like water. This interaction force occurs between a cation or anion and a polar molecule. An example of an *ion-dipole* interaction is sodium chloride in water. An *ion-dative* or *coordination* involves a two-center, two-electron covalent bond in which two electrons are shared from the same atom. An example of an *ion-dative* complex is the formation of hexaaquaaluminum, $Al(H_2O)_6^{3+}$, when aluminum chloride dissolves in water.

Hydrophobic interactions have been discussed in Section 4.12.

Table 6.1 shows the relative strengths of the various forces. It is interesting to note that London dispersion forces range from 1–15 kcal mol^{-1} which covers the range of hydrogen bonds. Relatively large London forces are seen with large molecules having a large number of electrons.

Table 6.1 Dissociation Energies for Various Types of Intermolecular Attractive Forces.

Bond type	Dissociation energy (kcal mol^{-1})
Ionic lattice (primary valence bonds)[a]	100–200
Covalent (primary valence bonds)[a]	50–150
Hydrogen bonds[b]	1–12
Keeson dipole-dipole[b]	0.5–2
Debye dipole-induced dipole[c]	Up to 0.5
London dispersion force[b]	<1–15

[a] Martin et al. (1983).
[b] https://en.wikipedia.org/wiki/Intermolecular_force, 2022.
[c] https://williams.chemistry.gatech.edu/structure/molecular_interactions/mol_int.html

6.3.3 Solvent Classification

Solvents can be classified as nonpolar and polar. The polar solvents can be further classified as polar aprotic and polar protic. Table 6.2 gives several solvent properties for some common solvents. Polar aprotic solvents have a dipole moment, but they do not donate hydrogen ions. Polar protic solvents have a dipole moment and can form a hydrogen bond or donate a hydrogen ion. The dipole moment was discussed in Section 6.3.2. Table 6.2 also provides the solvent's *dielectric constant*, ε. The dielectric constant is a measure of a substance's ability to store electrical energy in an electric field. It is also a measure of a substance's ability to insulate and stabilize charges from each other. Simplistically, it is the ability to reduce the solute's internal charge.

It is generally accepted that solvents that have similar polarity and dielectric properties with the solute are more likely to form a solution with the solute. That is, *likes dissolve likes*. Polar solvents dissolve polar solutes, water dissolves sugar and salt. Likewise, nonpolar solvents dissolve nonpolar solutes. Hexane dissolves oils. These generalizations are complicated by the solute-solute interactions related to the solid state such as heat of fusion and melting point.

Table 6.2 Solvent Properties.

Solvent	Boiling point (C)	Density (g mL^{-1})	Dipole moment[a]	Dielectric constant
Nonpolar				
Diethyl ether	35	0.713	1.25	4.3
Chloroform	61	1.498	1.04	4.8
1,4-Dioxane	101	1.033	0.45	2.3
Toluene	111	0.867	0.36	2.4
Cyclohexane	81	0.779	0.00	2.0
Hexane	69	0.655	0.00	1.9
Polar aprotic				
Dimethyl sulfoxide	189	1.092	3.96	47
Acetonitrile	82	0.786	3.92	38
Dimethylformamide	153	0.944	3.82	38
Acetone	56	0.786	2.88	21
Ethyl acetate	77	0.894	1.78	6
Tetrahydrofuran	66	0.886	1.75	8
Dichloromethane	40	1.327	1.60	9
Polar protic				
Water	100	1.000	1.85	80
Acetic acid	118	1.049	1.74	6.2
Methanol	65	0.791	1.70	33
Ethanol	79	0.789	1.69	30
Isopropanol	82	0.785	1.66	18
Formic acid	101	1.21	1.41	58

[a] Winnike, R. (2023).

6.4 Nonelectrolyte Enthalpy of Solution and Hess's Law

A *nonelectrolyte* is an uncharged substance that is not expected to give rise to ions. Nonelectrolytes may be nonpolar or polar and may participate in van der Waal's attractive forces. A commonly accepted solution formation model involves three enthalpic steps. Hess's law can be used to determine the overall enthalpy change that occurs in the solution formation involving these three steps. We were introduced to Hess's law in Section 3.3.3. Hess's law states that the total enthalpy change of a reaction is independent whether the reaction takes place in one step or any number of steps because enthalpy is a state function. Therefore, the enthalpy change depends only on the initial and final state. The solution formation model can be described by the three equilibrium steps shown in Eqs. (6.8)–(6.10). The overall equilibrium process is given by Eq. (6.11) where the common reactants and products from the different steps have been canceled (Connors and Mecozzi, 2010).

Step 1

Crystalline solute $\rightleftharpoons$ ~~Supercooled liquid solute~~ ΔH_{fusion} (6.8)

Step 2

Pure liquid solvent $\rightleftharpoons$ ~~Solvent containing cavity~~ $\Delta H_{cavityformation}$ (6.9)

Step 3

~~Supercooled liquid solute~~ + ~~Solvent containing cavity~~
$\rightleftharpoons$ Saturated solute solution $\Delta H_{solute\text{-}solvent}$ (6.10)

Applying Hess's law, the sum of these equilibrium steps is Eq. (6.11):

$$\text{Crystalline solute} + \text{Pure liquid solvent} \rightleftharpoons \text{Saturated solute solution} \tag{6.11}$$

which is the same as Eq. (6.1). Employing Hess's law, the enthalpy of solution is the sum of enthalpy processes from Step 1 to Step 3 (see Eq. 6.12).

$$\Delta H_{solution} = \Delta H_{fusion} + \Delta H_{cavity\ formation} + \Delta H_{solute\text{-}solvent\ interaction} \tag{6.12}$$

where ΔH_{fusion} is enthalpy of fusion of the crystalline solute, $\Delta H_{cavity\ formation}$ is the energy required to form a cavity in the solvent, and $\Delta H_{solute\text{-}solvent\ interaction}$ is the solute-solvent interaction. ΔH_{fusion} and $\Delta H_{cavity\ formation}$ are always endothermic reactions because breaking up solute and solvent attractive forces requires absorption of heat energy. $\Delta H_{solute\text{-}solvent\ interaction}$ is also known as the solvation enthalpy ($\Delta H_{solvation}$). If water is the solvent, $\Delta H_{solvation}$ is the hydration enthalpy. As seen in Eqs. (6.9) and (6.10), the forming of a "solvent containing cavity" is cancelled according to Hess's summation law where cavity formation is shown as product in Eq. (6.9) and as a reactant in Eq. (6.10). That is, the steps involving solvent cavity formation are part of the model, but the enthalpy of the process is canceled in the overall solution formation enthalpy change. Therefore, Eq. (6.12) becomes:

$$\Delta H_{solution} = \Delta H_{fusion} + \Delta H_{solvation} \tag{6.13}$$

The $\Delta H_{solvation}$ is zero for an ideal solution because the solvent-solvent interactions and solvent-solute interactions are equal. This means that a solute molecule can replace a solvent molecule without energy gain or loss. Therefore, for an ideal solution, the enthalpy of solution equals the enthalpy required to break up the crystal solute lattice structure, that is as a first approximation, the heat of fusion. This is shown by Eq. (6.14)

$$\Delta H_{solution} = \Delta H_{fusion} \tag{6.14}$$

For nonideal solutions, the enthalpy of solution may be an exothermic ($-\Delta H_{solution}$) or endothermic ($+\Delta H_{solution}$) process. depending on the magnitude of the enthalpy of solvation. This is depicted in Figure 6.1a,b.

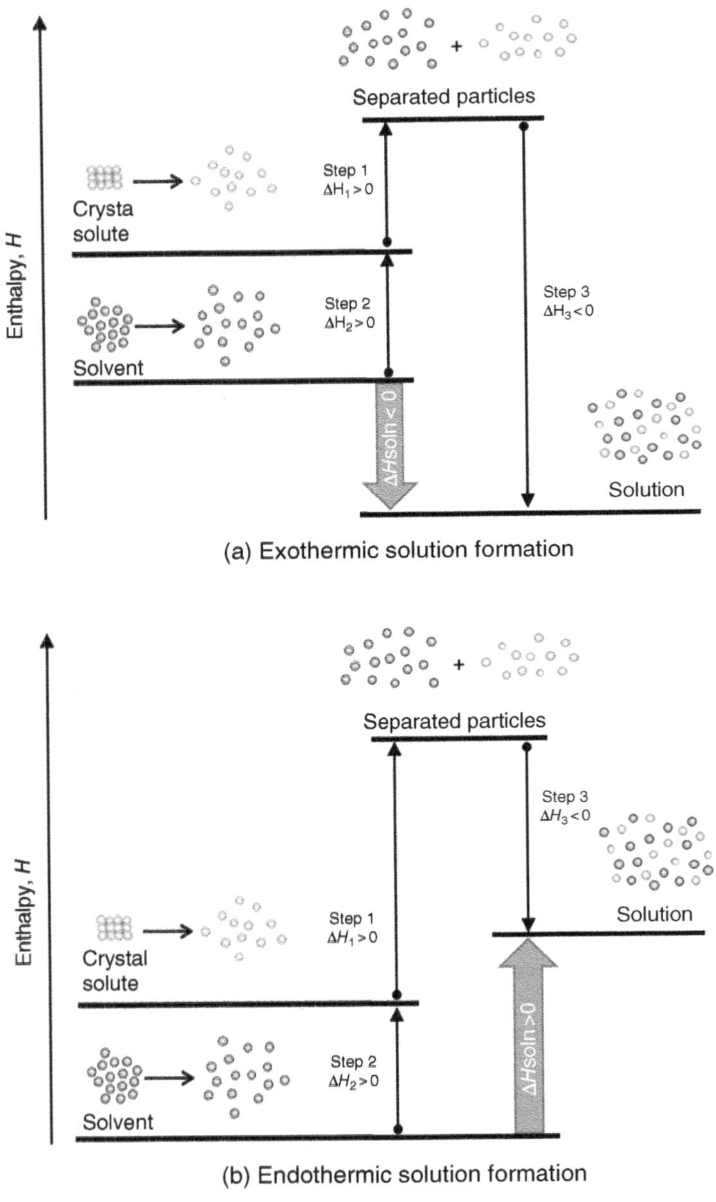

Figure 6.1 (a, b) Enthalpies associated with the enthalpy of solution for a nonideal solution. *Source:* LibreTexts. Anonymous (2022)/The LibreTexts libraries.

Figure 6.1 illustrates the solution formation model that was described by Eqs. (6.8)–(6.11). Figure 6.1a shows the enthalpy changes resulting in an exothermic enthalpy of solution. Here, Step 1 involves the melting of the crystal solute and forming a supercooled liquid. The heat of fusion is endothermic since heat/energy needs to be absorbed to overcome the crystal lattice structure. Step 2 is another endothermic reaction that requires energy to break up the solvent-solvent attraction forces so a cavity can be formed in the solvent. Step 3, the enthalpy of solvation is negative ($-\Delta H_{solvation}$), resulting from intermolecular solute-solvent attraction forces. For an exothermic heat of solution ($-\Delta H_{solution}$) to occur, the negative enthalpy of solvation must be greater than the sum of the endothermic reactions in steps 1 and 2.

Figure 6.1b explains the enthalpy changes that lead to an endothermic heat of solution. In this context, steps 1 and 2 are endothermic in nature like the exothermic solution formation example (Figure 6.1a). Similarly, the enthalpy of Step 3 of Figure 6.1b is negative as noted for Step 3 of Figure 1.6a. The difference in this case is that the negative exothermic enthalpy of solvation is smaller than the sum of the positive endothermic contributions associated with steps 1 and 2. This results in an overall endothermic enthalpy of solution ($+\Delta H_{solution}$).

It is important to note an endothermic heat of solution does not mean that a solution is not formed. This dilemma was discussed in Section 3.4.3 and in the following sections.

6.5 Nonelectrolyte Entropy of Solution

For ideal solutions, the ΔS_{mixing} mixing for a binary solution is:

$$\Delta S_{mixing} = -R(X_1 \ln X_1 + X_2 \ln X_2) \tag{6.15}$$

where R is the gas, X_1 is the mole fraction of the solvent, and X_2 is the mole fraction of the solute. This equation was derived for the mixing of two ideal gases in Section 3.4.5 and Eq. (3.98). For ideal mixing to occur, the size of the solute and solvent should be the same and the interactions should be athermal and to the same degree as the solute-solute and solvent-solvent interactions.

The solution entropy of nonideal solutions is depicted in Scheme (6.16).

$$\Delta S_{solution} = \Delta S_{solid \rightarrow solution} + \Delta S_{solvent\ reorganization} + \Delta S_{mixing} \tag{6.16}$$

where $\Delta S_{solid \rightarrow solution}$ accounts for an increase in microstates and entropy when the solid escapes the crystal lattice and is dispersed into the solvent. The $\Delta S_{solvent\ reorganization}$ term describes the ordering of solvent molecules associated with the solute. This typically leads to a decrease in entropy. ΔS_{mixing} characterizes the comixing of the solute and solvent to provide a homogeneous solution. This always leads to an increase in entropy.

6.5.1 Hydrophobic Effect: The Self-Association of Water in the Presence of Nonpolar Solutes and the Subsequent Self-Association of the Nonpolar Solutes

The hydrophobic effect was also discussed in Section 4.12.2. In the case of solution formation between a nonpolar solute and water, water is not able to interact with the nonpolar material to form dipole-dipole or hydrogen bonds. Instead, water molecules rearrange themselves to maximize their hydrogen bonding ability. This self-association process leads to water forming a *cage-like shell* around the nonpolar drug. The dominant factor limiting the solubility of a nonpolar solute is the unfavorable entropy change associated with the ordering of water molecules around the nonpolar solute. The hydrophobic effect is based on the nonpolar solute molecules self-associating. This decreases the surface area of the nonpolar surface exposed to the cage-structured water. This decrease in nonpolar surface area allows a portion of the structured water return to the bulk water. The return to the bulk water leads to a significant entropy increase as shown for a protein in Section 4.12.4. This increase in water entropy is the primary driving force for the hydrophobic effect.

6.6 Nonelectrolyte Gibbs Free Energy of Solution

The Gibbs free energy of solution is:

$$\Delta G^{\circ}{}_{solution} = \Delta H^{\circ}{}_{solution} - T\Delta S^{\circ}{}_{solution} \tag{6.17}$$

where $\Delta G^{\circ}{}_{solution}$, $\Delta H^{\circ}{}_{solution}$, and $\Delta S^{\circ}{}_{solution}$ are the standard states of the respective state functions.

6.6.1 Application: Using Gibbs Free Energy of Solution to Calculate the Solubility Advantage of Amorphous Compounds

The amorphous solubility advantage for ritonavir (RTV) has been reported (Rathod et al., 2020) using modulated differential calorimetry and Hancock and Parks prediction methodology. It is well known that amorphous solids are higher energetic forms than their crystalline counterparts. These higher energetic forms exhibit higher solubility and dissolution rates than crystalline material. The predicted solubility advantage for amorphous drugs is 12–1652 times its crystalline form (Hancock and Parks, 2000).

Example 6.1 Determination of Amorphous Ritonavir (RTV) Solubility Advantage at 37 ° C Using Modulated Differential Scanning Calorimetry (MDSC) and Thermodynamic Considerations (Adapted from Hancock and Parks, 2000)
Section 4.4 and Section 5.1.1.4 discussed the equation $\Delta G^{\circ} = -RT \ln K$. Mole fraction solute at saturation equilibrium solubility can replace the equilibrium constant, K, in this equation. The equation can be written in terms of the enthalpy and entropy state functions from Eq. (6.17).

$$\Delta G^{\circ}{}_{solution} = -RT \ln X_2 = \Delta H^{\circ}{}_{solution} - T\Delta S^{\circ}{}_{solution} \tag{6.18}$$

The solubility ratio of the *amorphous/crystalline* RTV is $\sigma_T{}^a/\sigma_T{}^c$, where $\sigma_T{}^a$ is the solubility of amorphous RTV and $\sigma_T{}^c$ is the solubility of crystalline RTV at a given temperature T. This ratio can also be considered the amorphous material's *solubility advantage*.

The difference between amorphous and crystalline RTV solubility is:

$$\Delta G^{\circ}{}_a - \Delta G^{\circ}{}_c = -RT \ln \sigma_T{}^a - (-RT \ln \sigma_T{}^c) \tag{6.19}$$

where $\sigma_T{}^c$ is the solubility of the crystalline RTV and $\sigma_T{}^a$ is the solubility of amorphous RTV at 37 °C. To simply the terms, the standard-state symbol, °, will not be explicitly shown in the following terms, but is implied.

First, we note that $\Delta G^{\circ}{}_a - \Delta G^{\circ}{}_c$ can be represented by:

$$\Delta G_T{}^{a,c} = -RT \ln \sigma_T{}^a - (-RT \ln \sigma_T{}^c) \tag{6.20}$$

where T is 37 °C, $\sigma_T{}^a$ is amorphous solubility, and $\sigma_T{}^c$ is crystalline RTV solubility. The RT terms and negative sign on the right-hand side of Eq. (6.20) can be collected to give Eq. (6.21).

$$\Delta G_T{}^{a,c} = -RT\,[\ln\,\sigma_T{}^a - \ln\,\sigma_T{}^c)] \tag{6.21}$$

Now, divide both sides by $-RT$ and use logarithm operator Eq. (1.47) to collect terms on the right-hand side Eq. (6.21).

$$-\Delta G_T{}^{a,c}/\ RT = \ln\,(\sigma_T{}^a/\sigma_T{}^c) \tag{6.22}$$

where the antilog of ln $(\sigma_T{}^a/\sigma_T{}^c)$ gives the RTV solubility advantage of the amorphous material (a; amorphous) over its crystalline (c; crystalline) counterpart at 37 °C. $\Delta G^{a,c}$ can be calculated from MDSC data. MDSC was discussed in Section 3.3.8. The enthalpy and entropy can be experimentally determined using MDSC. $\Delta G^{a,c}$ can be calculated by applying the Gibbs free energy equation:

$$\Delta G_T{}^{a,c} = \Delta H_T{}^{a,c} - T\Delta S_T{}^{a,c} \tag{6.23}$$

where $\Delta H_T{}^{a,c}$ and $\Delta S_T{}^{a,c}$ are the enthalpy and entropy differences between amorphous RTV and its crystalline form.

Equation (6.24) shows the enthalpy difference between the amorphous RTV and the crystalline material. The subscript T has been left out to simplify the terms' identifiers.

$$\Delta H^{a,c} = \Delta H^a - \Delta H_{fusion}{}^c \tag{6.24}$$

where $\Delta H^{a,c}$ is the difference in the amorphous and crystalline RTV enthalpies. $\Delta H_{fusion}{}^c$ is the enthalpy of fusion. The heat capacity of ΔH^a is measured by MDSC using Eq. (3.58) that states $\Delta H = C_p\,(T_{final} - T_{initial})$. In this case,

$$\Delta H_T{}^a = C_p{}^a\ \left(T_{fusion}{}^c - T\right) - C_p{}^c\left(T_{fusion}{}^c - T\right) \tag{6.25}$$

Collecting the temperature terms on the right-hand side of Eq. (6.25) gives:

$$\Delta H_T{}^a = \left(C_p{}^a - C_p{}^c\right)\left(T_{fusion}{}^c - T\right) \tag{6.26}$$

Substituting for $\Delta H_T{}^a$ from Eq. (6.26) into Eq. (6.24) indicates:

$$\Delta H_T{}^{a,c} = \left(C_p{}^a - C_p{}^c\right)\left(T_{fusion} - T\right) - \Delta H_{fusion}{}^c \tag{6.27}$$

$\Delta S_T{}^{a,c}$ can be calculated by knowing

$$S_{final} - S_{initial} = \Delta S = C_p\ \ln\ T_{final}/T_{intial}\ (\text{Eq.}3.85) \tag{6.28}$$

and

$$\Delta S_T{}^{a,c} = \Delta S_T{}^a - \Delta S_{fusion}{}^c \tag{6.29}$$

where $\Delta S_{fusion}{}^c$ is the entropy of fusion. $\Delta S_{fusion}{}^c = \Delta H_{fusion}/T_{fusion}{}^c$ from Eq. (3.77).

$$\Delta S_T{}^a = C_p{}^a\left(\ln\ T_{fusion}{}^c/T\right) - Cp^c\ \left(\ln\ T_{fusion}{}^c/T\right) \tag{6.30}$$

The common terms are collected in Eq. 6.30 to give Eq. 6.31.

$$\Delta S_T{}^a = \left(C_p{}^a - C_p{}^c\right)\left(\ln\ T_{fusion}{}^c/T\right) \tag{6.31}$$

The right-hand side of Eq. (6.31) can be substituted in Eq. (6.29) for $\Delta S_T{}^a$. The right-hand side of $\Delta S_{fusion}{}^c = \Delta H_{fusion}/T_{fusion}{}^c$ can also be substituted into Eq. (6.29) for $\Delta S_{fusion}{}^c$ to give Eq. (6.32).

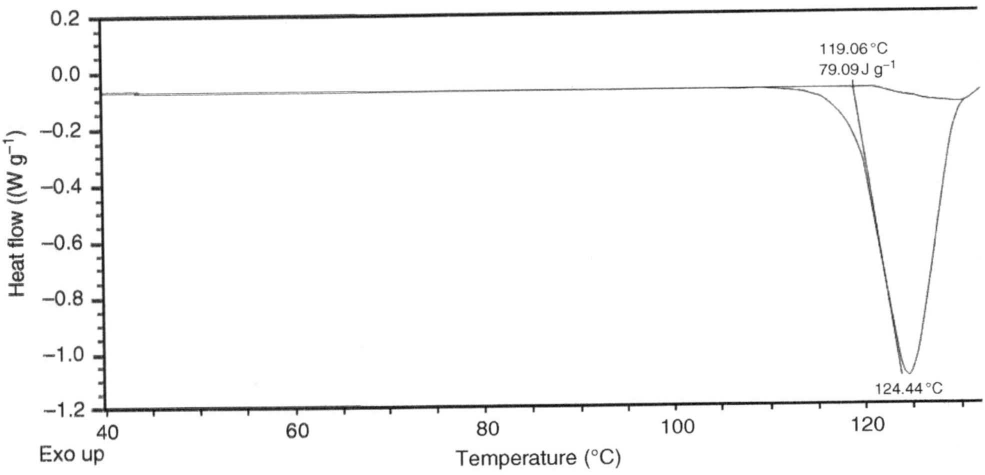

Figure 6.2 Differential scanning calorimetry melting profile for RTV.

$$\Delta S_T{}^{a,c} = \left(C_p{}^a - C_p{}^c\right)\left(\ln\ T_{fusion}{}^c/T\right) - \Delta H_{fusion}/T_{fusion}{}^c \tag{6.32}$$

Now, $\Delta H^{a,c}$ and $\Delta S_{T'}{}^{a,c}$ are expressed in terms that can be determined using MDSC.

Figure 6.2 shows the differential scanning calorimetry profile for the melting of crystalline RTV. The peak melting point, $T_{fusion}{}^c$, is 124.44 °C, and the enthalpy of fusion, $\Delta H_T{}^c$, is 79.09 J g^{-1}.

Figure 6.3 shows the MDSC heat capacity profiles for crystalline and amorphous RTV. The profiles show that the amorphous $C_p{}^a$ is 0.6103 J g^{-1} · K and $C_p{}^c$ is 0.1159 J g^{-1} · K. This makes sense since C_p is a measure of a material's molecular mobility. Amorphous RTV solid should be more mobile than crystalline RTV.

We are now ready to calculate the RTV amorphous solubility advantage.

RTV experimental data and givens:

Crystal melting point, $T_{fusion}{}^c$: 124.44 °C or 397.59 K
Crystal enthalpy of fusion, $\Delta H_T{}^c$: −79.09 J g^{-1}
Crystal entropy of fusion, $\Delta S_{fusion}{}^c$: 0.1989 J g^{-1} · K = (79.09 J g^{-1})/397.59 K

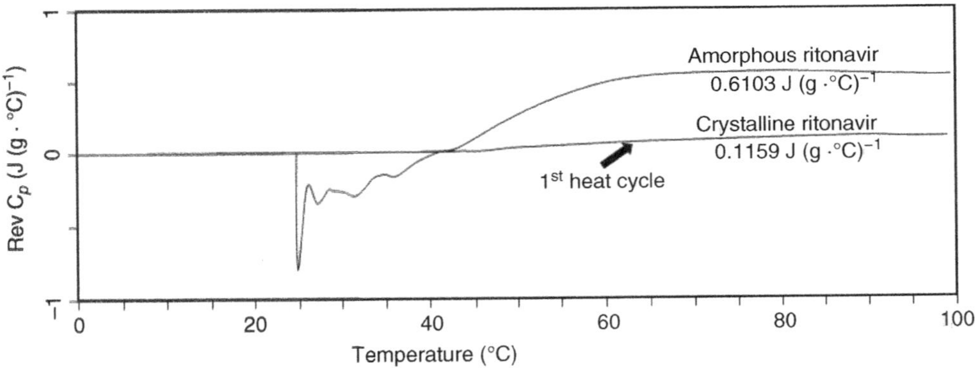

Figure 6.3 Modulated differential scanning calorimetry profile for RTV crystalline and amorphous forms.

Crystal heat capacity, $C_p{}^c$: 0.1159 J g^{-1} · K
Amorphous heat capacity, $C_p{}^a$: 0.6103 J g · K
Temperature, T: 37 °C or 310.15
Gas constant, R: 8.314 J (mol · K)$^{-1}$
RTV molecular weight: 720.9

Calculations by substituting values for the parameters:

1) $$\Delta H_T{}^{a,c} = \left(C_p{}^a - Cp^c\right)\left(T_{fusion}{}^c - T\right) - \Delta H_{fusion}{}^c \tag{6.27}$$

$$= \left(0.6103\,\mathrm{J\,g^{-1}\cdot K} - 0.1159\,\mathrm{J\,g^{-1}\cdot K}\right)(397.59\,\mathrm{K} - 310.15\,\mathrm{K})] - 79.09\,\mathrm{J\,g^{-1}}$$

$$= \left[0.4944\,\mathrm{J}\,(\mathrm{g}\cdot\cancel{\mathrm{K}})^{-1}\right](87.44\,\cancel{\mathrm{K}}) - 79.09\,\mathrm{J\,g^{-1}} \tag{6.33}$$

$$\Delta H_T{}^{a,c} = 43.23\,\mathrm{J\,g^{-1}} - 79.09\mathrm{J\,g^{-1}} = -\underline{35.86\,\mathrm{J\,g^{-1}}}$$

2) $$\Delta S_T{}^{a,c} = \left(C_p{}^a - Cp^c\right)\left(\ln\, T_{fusion}{}^c/T\right) - \Delta H_{fusion}/T_{fusion}{}^c \tag{6.32}$$

Substituting the MDSC values into Eq. (6.32) establishes:

$$= \left(0.6103\,\mathrm{J\,g^{-1}\cdot K} - 0.1159\,\mathrm{J\,g^{-1}\cdot K}\right)(\ln\, 397.59\,\cancel{\mathrm{K}}/310.15\,\cancel{\mathrm{K}}) - 0.1989\,\mathrm{J\,g^{-1}\cdot K}$$

$$= \left(0.4944\,\mathrm{J\,g^{-1}\cdot K}\right)\,\ln\, 1.2819 - 0.1989\,\left(\mathrm{J\,g^{-1}}\right)\cdot\mathrm{K}$$

$$= \left(0.4944\,\mathrm{J\,g^{-1}\cdot K}\right)(0.2483) - 0.1989\,\left(\mathrm{J\,g^{-1}}\right)\cdot\mathrm{K}$$

$$= 0.1228\,\mathrm{J\,g^{-1}\cdot K} - 0.1989\,\mathrm{J\,g^{-1}\cdot K}$$

$$\Delta S_T{}^{a,c} = -\underline{0.0761\,\mathrm{J\,g^{-1}\cdot K}} \tag{6.34}$$

3) $$\Delta G_T{}^{a,c} = \Delta H_T{}^{a,c} - T\Delta S_T{}^{a,c} \tag{6.23}$$

$$= -35.86\,\mathrm{J\,g^{-1}} - \left(310.15\,\cancel{\mathrm{K}}\cdot -\ \ 0.0761\,\mathrm{J\,g^{-1}}\cdot\cancel{\mathrm{K}}\right)$$

$$= -35.86\,\mathrm{J\,g^{-1}} + 23.57\,\mathrm{J\,g^{-1}}$$

$$= -12.29\,\mathrm{J\,g^{-1}} \tag{6.35}$$

$$\Delta G_T{}^{a,c} = -12.29\,\mathrm{J\,g^{-1}}\left(720.9\,\cancel{\mathrm{g}}\,\mathrm{mol^{-1}}\right)$$

$$= -\underline{8.860\mathrm{E}3\,\mathrm{J\,mol^{-1}}}$$

4) From Eq. (6.22),

$$\ln\,(\sigma_T{}^a/\sigma_T{}^c) = -\Delta G^{a,c}/\ RT \tag{6.22}$$

$$= 8.860\mathrm{E}3\,\cancel{\mathrm{J\,mol^{-1}}}/(8.314\,\cancel{\mathrm{J}(\mathrm{mol}}\cdot\cancel{\mathrm{K}})^{\cancel{-1}}\cdot 310.15\cancel{\mathrm{K}})$$

$$\ln\,(\sigma_T{}^a/\ \sigma_T{}^c) = 3.436$$

$$(\sigma_T{}^a/\ \sigma_T{}^c) = \underline{31.1} \tag{6.36}$$

The solubility advantage of amorphous RTV is 31.1. This means that amorphous RTV is 31.1 times more soluble than its crystalline counterpart.

6.7 Nonelectrolyte: Determining the Solution Enthalpy and Entropy Using the van't Hoff Equation

Section 5.10 and Example 5.11 discussed the van't Hoff equation which is brought forward here as Eq. (6.37).

$$\ln K = -\Delta H^{\circ}_{solution}/RT + \Delta S^{\circ}_{solution}/R \tag{6.37}$$

For a material exhibiting equilibrium solubility, mole fraction solute can replace K in this equation to give Eq. (6.37a).

$$\ln X_2 = -\Delta H^{\circ}_{solution}/RT + \Delta S^{\circ}_{solution}/R \tag{6.37a}$$

$\Delta H^{\circ}_{solution}$ and $\Delta S^{\circ}_{solution}$ can be determined by plotting $\ln X_2$ versus $1/T$. This plot gives a slope of $-\Delta H^{\circ}_{solution}/RT$ and an intercept of $\Delta S^{\circ}_{solution}/R$.

6.8 Nonelectrolyte Ideal Solution – Effect of Enthalpy of Fusion and Melting Temperature

The equation that describes the relationship between a nonelectrolyte's ideal solubility and the enthalpy of fusion and melting point is $\ln X_{2,i} = -(\Delta H_{fusion}/R)\,[(T_m - T)/T \cdot T_m]$ where $X_{2,i}$ indicates the ideal mole fraction solubility, ΔH_{fusion} is the heat of fusion of the nonelectrolyte, T_m is the melting point of the compound, and T is the solution temperature.

This remaining section derives the ideal solubility equation. The derivation is meant to demystify the origin of the ideal solubility equation. Starting with Eq. (6.18) $-RT \ln X_2 = \Delta H^{\circ}_{solution} - T\Delta S^{\circ}_{solution}$ and dividing both sides by $-RT$ *gives* $\ln X_2 = -\Delta H^{\circ}_{solution}/RT + \Delta S^{\circ}_{solution}/R$ which is the same as Eq. 6.37a, the van't Hoff equation. Now, we take the derivative of this equation. The general operational Eq. (1.84) $du^n/dx = nu^{n-1}\,du/dx$ and Eq. (1.81) d (constant)/$dx = 0$ are applied to the right-hand side of the equation to give:

$$(d \ln X_2) = -(-\Delta H^{\circ}_{solution})/RT^2) + 0 \tag{6.38}$$

where $\Delta S^{\circ}_{solution}/R$ is treated as a constant. After simplification, this equation becomes:

$$(d \ln X_2) = +\Delta H^{\circ}_{solution})/RT^2 \tag{6.39}$$

Equation (6.39) can be integrated from the solution temperature, T, to the compound's melting point, T_m.

$$\int_{X2}^{Xmelt} d \ln \cdot X_2 = +(\Delta H^{\circ}_{solution}/R) \int_{T}^{Tm} T^{-2} dT \tag{6.40}$$

Employing the integration operation Eq. (1.100), $\int x^n\,dx = (x^{(n+1)})/(n+1)$, to the right-hand side of Eq. (6.40) leads to:

$$\ln X_{melt} - \ln X_2 = -(\Delta H^{\circ}_{solution}/R)\,[1/T_m - 1/T] \tag{6.40a}$$

Multiply the negative sign through the term on the right-hand side of Eq. (6.40a) to give:

$$\ln X_{melt} - \ln X_2 = (\Delta H^{\circ}{}_{solution}/R)\,[1/T - 1/T_m] \tag{6.40b}$$

where the mole fraction of a liquid melt of a pure substance is 1 and ln 1 = 0. Equation (6.40b) becomes (6.41).

$$0 - \ln X_2 = (\Delta H^{\circ}{}_{solution}/R)\,[1/T - 1/T_m] \tag{6.41}$$

Multiplying both sides by −1 gives:

$$\ln X_2 = -(\Delta H^{\circ}{}_{solution}/R)\,[1/T - 1/T_m] \tag{6.42}$$

Using a common denominator of $T \cdot T_m$, Eq. (6.42) becomes:

$$\ln X_2 = (-\,\Delta H^{\circ}{}_{solution}/R)\,[(\cancel{T} \cdot T_m/\cancel{T} - T \cdot \cancel{T_m}/\cancel{T_m})/T \cdot T_m] \tag{6.43}$$

and

$$\ln X_2 = (-\,\Delta H^{\circ}{}_{solution}/R)\,[(T_m - T)/T \cdot T_m] \tag{6.44}$$

Equation (6.14) states $\Delta H_{solution} = \Delta H_{fusion}$ for an ideal solution. ΔH_{fusion} can be substituted into Eq. (6.44) for $\Delta H_{solution}$ to yield Eq. (6.45), the *ideal solubility equation*:

$$\ln X_{2,i} = -\left(\Delta H_{fusion}/R\right)\,[(T_m - T)/T \cdot T_m] \tag{6.45}$$

where $\ln X_{2,i}$ is the nonelectrolyte solute ideal solubility, ΔH_{fusion} is the solute heat of fusion, R is the gas constant, T_m is the melting point of the drug in kelvin, and T is the solution temperature in kelvin.

If the heat of fusion and melting temperature are known, Eq. (6.45) can be used to calculate a drug's ideal solubility. The ΔH_{fusion} and melting point can be determined by differential scanning calorimetry. The values can then be used to estimate mole fraction solubility without performing more time-consuming solubility studies. Estimating rank order solubility of potential lead research compounds can be important in the selection of development candidates. An example calculation of lidocaine's mole fraction solubility using Eq. (6.45) was provided in Section 3.3.4.

Rearranged forms of 6.45 can be used to gain insight into the relationships between mole fraction solubility specifically or solubility in general and ΔH_{fusion} and T_m. For example, ΔH_{fusion} is multiplied through the terms in the right-hand side of Eq. (6.45) to give:

$$\ln X_{2,i} = \left(-\,\Delta H_{fusion}/RT\right) + \left(\Delta H_{fusion}/R \cdot T_m\right] \tag{6.46}$$

where $\Delta H_{fusion}/T_m = \Delta S_{fusion}$. Equation (6.46) shows that the higher the heat of fusion, the lower the drug solubility. Equation (6.46) also indicates that the ratio of the heat of fusion and the melting point impacts drug solubility. The larger the ratio or ΔS_{fusion}, the more positive the impact on drug solubility. Equation (6.46) shows the compensation effect of the enthalpy and entropy of fusion terms on drug solubility.

Using Eq. (6.14) that states $\Delta H_{solution} = \Delta H_{fusion}$, the heat of solution can be substituted into Eq. (6.46) to give:

$$\ln X_{2,i} = (-\Delta H_{solution}/RT) + (\Delta H_{solution}/R \cdot T_m] \tag{6.46a}$$

The heat of solution can be measured by solution calorimetry. Equation (6.46a) points out that drug solubility is inversely proportional to the drug's melting point. That is, the higher the melting point, the lower the solubility.

Equation (6.45) can also be rearranged to show the relationship between drug solubility and ΔH_{fusion}.

$$\ln X_{2,i} = -\left(\Delta H_{fusion}/R\right) [(T_m - T)/T \cdot T_m] \tag{6.45}$$

Multiplying both sides of Eq. (6.45) by −1 gives:

$$-\ln X_{2,i} = \left(\Delta H_{fusion}/R\right) [(T_m - T)/T \cdot T_m]$$

and using logarithm operating Eq. (1.48) on this equation results in:

$$\ln X_{2i}{}^{-1} = \left(\Delta H_{fusion}/R\right) [(T_m - T)/T \cdot T_m]$$

and using exponent operating, Eq. (1.34) gives:

$$\ln 1/X_{2,i} = \left(\Delta H_{fusion}/R\right) [(T_m - T)/T \cdot T_m] \tag{6.47}$$

Equation (6.47) shows that ideal solubility (X_{2i}) is inversely proportional to the enthalpy of fusion, ΔH_{fusion}. This makes sense since the enthalpy of fusion is the amount of energy required to break up the lattice structure of a compound. The larger the ΔH_{fusion}, the stronger the lattice attractive forces. It follows that the stronger the lattice attractive forces, the lower the "escaping" tendency of the lattice molecules. This lower molecular escaping tendency results in lower solubility.

6.9 Nonelectrolyte Ideal Solution Properties and Henry's Law

6.9.1 Ideal Solution Formation of a Nonelectrolyte: Entropy Driven Process

Solution formation is described by Eq. (6.17) which is restated here as Eq. (6.37).

$$\Delta G^{\circ}{}_{solution} = \Delta H^{\circ}{}_{solution} - T\Delta S^{\circ}{}_{solution} \tag{6.37}$$

The enthalpy of solution is described by Eq. (6.13).

$$\Delta H_{solution} = \Delta H_{fusion} + \Delta H_{solvation} \tag{6.13}$$

For an ideal solution, $\Delta H_{solution} = \Delta H_{fusion}$ (6.14). ΔH_{fusion} is always positive since heat energy needs to be absorbed to break the crystal lattice structure.

The ideal solution formation entropy is given by the same Eqs. (3.98) and (6.15).

$$\Delta S_{mixing} = -R(X_1 \ln X_1 + X_2 \ln X_2)$$

The ideal solution entropy of mixing is always positive since the number of microstates increases and the homogeneous solution mixture system is more disordered than the individual solute and solvent systems. This is because there are no intermolecular solute-solvents interactions.

Equation (6.37) can now be rewritten:

$$\Delta G^{\circ}{}_{solution} = \Delta H^{\circ}{}_{fusion} - T\Delta S^{\circ}{}_{mixing} \tag{6.48}$$

The ideal solution process is entropy-driven. Since ΔH_{fusion} is always positive and ΔS_{mixing} is always positive, the entropy term needs to be larger than the enthalpy term for $\Delta G_{solution}$ to be negative and solution formation to be spontaneous.

6.9.2 Determining the Enthalpy of Fusion and Melting Point of a Crystalline Solute Using the van't Hoff Equation

For ideal solutions, Eq. (6.45) which is a rearrangement of common ideal solubility Eq. (6.44) can be used to determine the crystalline solute heat of fusion and melting point.

$$\ln X_{2,i} = \left(-\Delta H_{fusion}/RT\right) + \left(\Delta H_{fusion}/R \cdot T_m\right] \tag{6.45}$$

A plot of $\ln X_2$ versus $1/T$ (a van't Hoff plot) gives a slope of $-\Delta H_{fusion}/R$ and an intercept of $\Delta H_{fusion}/R \cdot T_m$. The enthalpy of fusion can be determined from the slope. Once ΔH_{fusion} is calculated, the melting point can be computed from the intercept.

6.9.3 Henry's Law

Henry's law describes the ideal behavior of nonideal solutions at very low dilution. Henry's law is a gas law which states that the amount of a gas dissolved by a liquid is proportional to the pressure of the gas over the liquid. Henry's equation in terms of mole fraction gas is:

$$P_2 = k_2 X_2 \tag{6.49}$$

where P_2 is the gas pressure over the liquid, k_2 is Henry's law constant, and X_2 is the mole fraction gas solubility in the liquid solvent. For nonideal solutions, this proportionality occurs only at very low solute concentrations where the solution also behaves as an ideal solution. In this case, the chemical potential for the solute is given by:

$$\mu_2 = \mu^{0}{}_{k,2} + RT\ \ln\ X_2 \tag{6.50}$$

where μ_2 is the chemical potential of the solute, $\mu^{0}{}_{k,2}$ is the standard state of the system which incorporates Henry's law constant for the solute, and $RT \ln X_2$ have the meanings established in the above discussions. Henry's law can also be extended to very dilute solutions of crystalline solutes.

The highly dilute concentrations at which nonideal solutions behave as an ideal solution depend on the intermolecular forces of the solvent, solute, and solute-solvent. From examples, solutions having a mole fraction less than 0.02 and molality or molarity less than 10^{-4} have been shown to act as ideal solutions and can be used as approximate concentrations exhibiting ideal behavior.

6.10 Nonideal Nonelectrolyte Solutions

Equation (6.11) describes the equilibrium solution formation process.

$$\text{Crystalline solute} + \text{Pure liquid solvent} \rightleftharpoons \text{Saturated solute solution} \tag{6.11}$$

where

$$\text{Undissolved solute}, u \rightleftharpoons \text{Dissolved solute}, d \tag{6.51}$$

At equilibrium, the chemical potential of the excess undissolved solute, μ_u, and dissolved solute, μ_d, is equal.

$$\mu_u = \mu_d \tag{6.51a}$$

The chemical potential is the Gibbs molar free energy where $\mu = G/n = G_M$ (Eq. 4.54). Likewise, at equilibrium, the thermodynamic activity of the undissolved solute, a_u, and dissolved solute, a_d, is equal, $a_u = a_d$. The thermodynamic activity is a measure of the *effective concentration* of the solute.

By convention, and for convenience, nonideal solution equations take on the same form as ideal solutions. This is accomplished by substituting activity, a, into the ideal solubility chemical potential equation (Eq. 4.74). The relationship between chemical potential and activity is given in Eq. (6.52).

$$\mu = \mu^\circ + RT \ln (a) \tag{6.52}$$

where μ is chemical potential, μ° is the *standard chemical potential* which is a function of temperature and pressure, and a is the activity or *effective concentration* of the solute which depends on temperature, pressure, and composition. Activity can be given in various concentration scales like X, M, and m. Note that the logarithm function is defined for dimensionless numbers. That is pure numbers without any physical units. Activity and effective concentration can be converted to a dimensionless ratio or pure number by using a standard reference quantity like 1 mol L^{-1} for molar concentration and 1 mol kg^{-1} for molal concentration.

Activity or effective concentration is equal to the *product of the prepared or measured analytical concentration times the activity coefficient.* The activity coefficient is a unitless quantity. To maintain dimensional consistency, the actual concentration is converted to a dimensionless pure number in a similar fashion as activity. Although the precision of this language is important to ensure the mathematical integrity when performing logarithmic operations and to maintain dimension consistency, it is very cumbersome. For the remainder of this book, it will be understood that the use of the terms activity, effective concentration, activity coefficient, and actual concentration are dimensionless numbers or can be made dimensionless numbers by using standard reference quantities.

For nonelectrolyte equilibrium saturated solubility measurements,

$$a_{nonelectrolyte} = X_{measured\ solubility} \times \gamma \tag{6.53}$$

where $a_{nonelectrolyte}$ is the activity of the dissolved solute or the solutes effective concentration in the saturated solution, $X_{measured\ solubility}$ is the actual concentration based on the physical quantity measured by an analytical such as ultraviolet (UV) or high-performance liquid chromatography (HPLC), and γ is the activity coefficient which is a measure of the nonideality of the saturated solution. A *γ-value* greater than 1 indicates a positive deviation from ideal behavior. This means that the cohesive solute-solute and solvent-solvent interactions are greater than the solute-solvent adhesive forces. When $\gamma = 1$, the system is considered to be ideal. A γ-value less than 1 represents a negative deviation from ideal behavior. Adhesive interactions between the solute-solvent are greater than the cohesive solute-solute and solvent-solvent interaction forces. For nonideal solutions, γ approaches 1 as the solute concentration approaches 0 as discussed in Section 6.9.3 Henry's Law.

As mentioned, $a_{nonelectrolyte}$ is the effective concentration of the drug or species. It accounts for intermolecular interactions that cause deviations from ideal behavior. Activity is a thermodynamic quantity that relates the chemical potential of the drug in a "real" solution. As a thermodynamic quantity, activity relates to the *effective behavior* of the drug or substance which may be different than expected based solely on its physical concentration.The actual or physical concentration is based on how the system was prepared. That is, what quantities of drug or excipients were weighed or measured. As mentioned earlier, the actual concentration can also be based on the physical quantity measured by UV or HPLC or other appropriate analytical instrument.

Example 6.2 Importance of Understanding the Difference Between Activity or Effective Concentration and Actual Concentration: Why Can Syrup USP (US Pharmacopeia) Be Stored at Room Temperature and Not Have Microbiological Growth?

Syrup USP (Syrup) is composed of only two ingredients, water (46% v/v) and sugar (85% w/v sucrose). Both of these ingredients are excellent substrates and nutrients for microorganisms to grow. Yet, unlike many products that contain water and sugar that are required to be refrigerated to prevent microbial growth, Syrup can safely be stored at room temperature. Why?

Water activity versus water content or actual water concentration is one reason for Syrup's microbiological preservative capability. High osmotic content is the second reason. For purposes of this example, the remainder of the discussion will focus on the important difference between water activity or effective concentration and water content or the actual amount of water in the Syrup. Water content is the "actual concentration" of water in the Syrup. Syrup was prepared by physically measuring the amount of water and adding it to the sucrose. Just as sucrose was physically weighed. The actual amount of water can also be determined by instrumental methods that can be used to specifically quantitate the quantity of water in the Syrup.

In the Syrup, the sucrose and water are attracted to each other and bind to each other. The extent of the water that is bound to the sucrose is no longer *free*. Microbes cannot use this *bound water*. The water must be free and available for them to use.

This is where activity, specifically water activity, comes into play. Water activity or the effective concentration of water is defined as the ratio of water vapor pressure in a substance or drug product to the vapor pressure of pure water at the same temperature under equilibrium conditions. When the water activity ratio is multiplied by 100%, it is called *percent water activity*. In this case, water activity or effective percent water measures the amount of water that is free to support microbe growth. There is a large body of research that has investigated the *critical water activity level* or *critical percent water activity level* that supports microbiological growth. Generally, a water activity above 0.90–0.95 is required for most bacteria to grow. In terms of percent, this means that for most bacteria to grow in an excipient or formulation or in this case Syrup, 90–95% of the water content must be free and not bound to support bacterial growth. Some pathogenic bacteria can grow at 86% water activity. A few yeasts and molds can grow at of 60% or higher water activity.

Syrup, USP water activity level is about 85–87%. This suggests that sucrose and water interact to bind some 13–15% of the water that was added to the formulation. That is, 13–15% of the water content is not available for microbe growth. The 85–87% water activity is too low to support the growth of most bacteria which require at least 90–95% free water.

In 2018, a new USP Water Activity Chapter <922> became available. The reader is encouraged to read this chapter to gain insight into how the USP views the importance of the measurement and use of water activity in the pharmaceutical industry.

6.10.1 Reference and Standard States

The component activity, *a*, is the solubility departure from the ideal solubility. As mentioned earlier, it is the effective concentration or observed concentration that is a deviation from the ideal solubility. To understand the extent of the deviation from the ideal, a *reference state* is established that defines the conditions for which the component exhibits ideal behavior.

The reference state for a solution is the condition where the activity equals its concentration (Eq. 6.54). The concentration can be expressed in mole fraction (*X*), molar (M), or molar (m) concentration scales. *A solution reference state is also defined as the condition where the activity coefficient equals 1* (Eq. 6.55).

<u>Reference State Conditions:</u>

$$\text{Solution Activity}, a = \text{Solution Concentration } (X, \text{M}, \text{m}) \tag{6.54}$$

$$\text{Activity Coefficient}, \gamma = 1 \text{ where Solution Activity/Solution Concentration} = 1 \tag{6.55}$$

Reference States:

1) Solvent – The reference state for a solvent using mole fraction scale is the pure solvent. In infinitely dilute solutions, the solvent behaves ideally and $a = X = 1$.
2) Miscible liquid solute – The liquid solute reference state is the pure liquid solute. Concentration is expressed in mole fraction.
3) Partially soluble liquid or solid solute – The reference state is an infinitely dilute solution where Eqs. (6.54) and (6.55) apply.

<u>Standard State Conditions:</u>

A *standard state* needs to be established for *activity*, like standard measures set for length (m), mass (kg), and time (s). The activity standard state for a solution component is set to unity, 1. (Eq. 6.56)

$$a_i = 1 \tag{6.56}$$

where a_i is the activity of the *i*th component. Using Eq. (6.52), a solution component standard state also occurs when the chemical potential of the component equals its standard chemical potential. Eq. (6.52a).

$$\mu_i = \mu_i^\circ + RT \ \ln\ (a_i = 1) \tag{6.52a}$$

In this case, using the logarithm operating Eq. (1.45), the second term on the right-hand side of Eq. (6.52a) becomes:

$$RT \ln\ (1) = 0 \tag{6.57}$$

And Eq. (6.52a) is:

$$\mu_i = \mu_i^\circ \tag{6.58}$$

Standard States:

1) Solvent – The solvent standard state is set as the *pure liquid at 1 atm where* $a = 1$.
2) Miscible liquid solute – The liquid solute standard state is *the pure liquid at 1 atm and* $a = 1$

3) Partially Soluble Liquid or Solid Solute – The definition of standard state of a partially soluble liquid or solid solute involves the definition of a hypothetical solution using X, M, or m concentration scales. The partially soluble liquid or solid solute standard state *is a hypothetical solution of unit concentration at infinitely dilute concentration that acts as an ideal solution.*

6.10.2 Activity Conventions (Adapted from Connors and Mecozzi)

1) A gas' fugacity is equal to its pressure or partial pressure at low to moderate pressures.
2) A solvent's activity equals its mole fraction. A pure liquid has an activity of 1.
3) A pure solid's activity equals 1.
4) A liquid or solid solute in an infinitely dilute solution ($\sim< 0.02\ X$ and $\sim<10^{-4}$ M) is equal to its concentration.

6.10.3 Activity Coefficients

The activity coefficient, γ, value can be less than, equal to, or greater than 1, depending on the relative strengths of the cohesion forces of the solute-solute and solvent-solvent interactions and the adhesion forces between the solute-solvent.

For liquid solutes, *Raoult's law* describes the partial pressure of a liquid solute over its solution. The relationship of the liquid solute's partial pressure to vapor pressure of the pure liquid and its composition is given by Eq. (6.59).

$$p_i = X_i \cdot P_{i,pure\ liquid} \tag{6.59}$$

where p_i the partial pressure of component i over its solution, X_i is the mole fraction of constituent i, and $P_{i,pure\ liquid}$ is the vapor pressure over pure liquid i. An ideal solution follows this linear relationship over the full composition range. The total pressure over an ideal solution of binary nonelectrolyte liquid constituents is:

$$P_{total} = X_1 P_{1,pure\ liquid} + X_2 P_{2,pure\ liquid} \tag{6.60}$$

which is the linear combination of the partial pressure of liquid 1 and the partial pressure of liquid 2. Figure 6.4 shows an ideal solution of two liquids that obeys Raoult's law. The partial pressures for both liquids are linear according to Eq. (6.59). The total pressure is given by Eq. (6.60). Benzene-toluene, hexane-heptane, and methanol-ethanol form virtually ideal solutions.

Example 6.3 Illustration of Nonideal Behavior of an Acetone-Ether Solution Exhibiting a Positive Deviation from Raoult's Law (Adapted from Daniels and Alberty, 1975)

Figure 6.5 is a plot of the nonideal partial pressures over an ether-acetone solution. The effective partial pressure curves are shown by the solid lines, and the dashed lines show the ideal partial pressures calculated from Raoult's law. The effective partial pressures deviate in a *positive direction* above the ideal Raoult's line. This is referred to as a *positive deviation* from Raoult's law and ideal behavior.

To reflect this nonideal behavior, an activity coefficient is added to Eq. (6.59). This is the same way that the nonideal solubility nature of a crystalline solute is handled as shown by Eq. (6.53), $a_{solubiity} = X_{ideal\ solubility} \cdot \gamma$. For nonideal liquids, Eq. (6.59) becomes:

$$p_i = \left(X_i \cdot P_{i,pure\ liquid}\right) \gamma \tag{6.61}$$

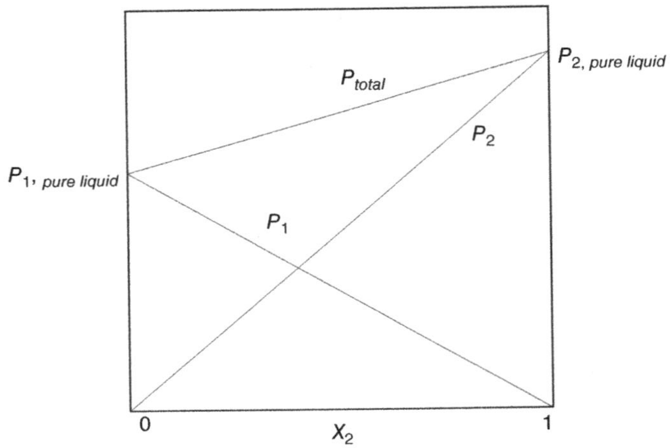

Figure 6.4 Ideal solution behavior of two liquids that obey Raoult's law. P_1 and P_2 are the partial pressures of liquid 1 and liquid 2. P_{total} is the total pressure over the binary solution. *Source:* Adapted from Connors and Mecozzi, 2010/John Wiley & Sons.

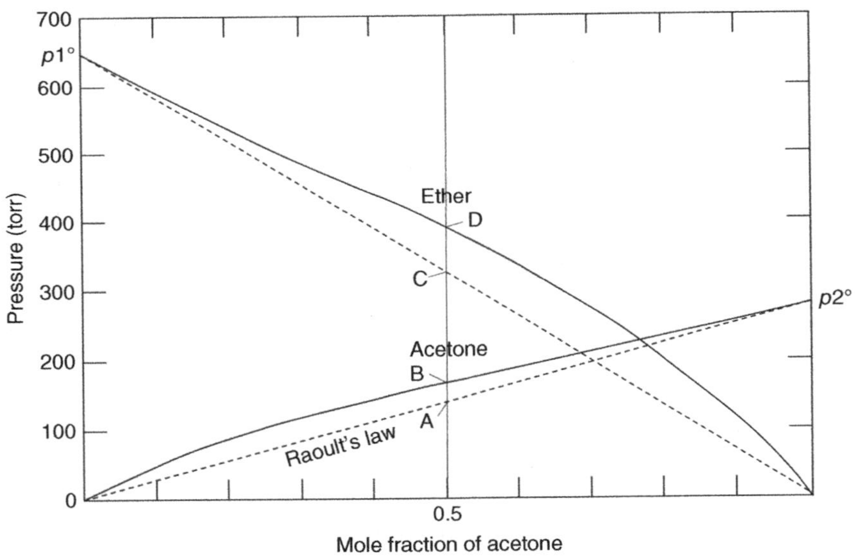

Figure 6.5 Nonideal partial pressures of acetone-ether solutions at 30 °C. *Source:* Adapted from Daniels and Alberty (1975).

In this case, p_i is the effective partial pressure, $(X_i \cdot P_{i,pure\ liquid})$ is the Raoult's law partial pressure, and γ is the activity coefficient. We can determine the magnitude of the nonideality of the acetone-ether solutions by calculating the activity coefficients of acetone and ether from Figure 6.5.

One way to calculate the acetone activity coefficient using Figure 6.5 is discussed next. If acetone and ether formed an ideal solution, the partial pressure of acetone at 0.5 mol fraction acetone would be at point A, where $X_{acetone} \cdot P_{pure\ acetone}$ is 142 Torr and represents Raoult's law partial pressure. The observed or effective acetone partial pressure, $p_{acetone}$, is at point B, 168 Torr. Next, we solve Eq. (6.61) for the acetone activity coefficient, $\gamma_{acetone}$ (see Eq. 6.62). Substituting the appropriate values for $p_{acetone}$ and $X_{acetone} \cdot P_{pure\ acetone}$ into Eq. (6.62) and solving for $\gamma_{acetone}$ gives:

$$\gamma_{acetone} = p_{acetone}/\left(X_{acetone} \cdot P_{pure\ acetone}\right) = 168\ \cancel{Torr}/142\ \cancel{Torr} = 1.18 \tag{6.62}$$

where the term $X_{acetone} \cdot P_{pure\ acetone} = 142$ Torr.

Similarly, the Raoult's law partial pressure for ether at 0.5 mol fraction ether at point C is 323 Torr. The effective ether partial pressure, p_{ether}, is D, 391 Torr. Solving for $\gamma_{acetone}$ gives:

$$\gamma_{ether} = p_{ether}/\left(X_{ether} \cdot P_{pure\ ether}\right) = 391\ \cancel{Torr}/323\ \cancel{Torr} = \underline{1.21} \tag{6.63}$$

Both activity coefficients for acetone and ether are greater than 1, meaning the acetone-ether solution acts in a nonideal manner. In addition, the activity coefficients of 1.18 for acetone and 1.21 ether indicate that the acetone-ether adhesion forces are weaker that the acetone-acetone and ether-ether cohesion forces. In simple terms, this means that acetone and ether "do not like each other and tend to push the other out of solution," creating a positive deviation from Raoult's law. Another way to think about it is that acetone or ether will find it easier to escape than in the pure state. This will increase the vapor pressure and result in positive deviation. Other examples of binary liquid solution systems that exhibit positive deviations are ethanol-cyclohexane, carbon tetrachloride-benzene, carbon tetrachloride-methanol, and carbon disulfide-acetone.

Example 6.4 Illustration of Nonideal Behavior of an Acetone-Chloroform Solution Showing a Negative Deviation from Raoult's Law (Adapted from Socratic Organic Chemistry)

An acetone-chloroform solution exhibits a negative deviation from Raoult's law (Figure 6.6).

The negative deviation from Raoult's law is evident from Figure 6.6. Equation (6.61) is used to reflect the nonideal deviation from Raoult's law which includes an activity coefficient term.

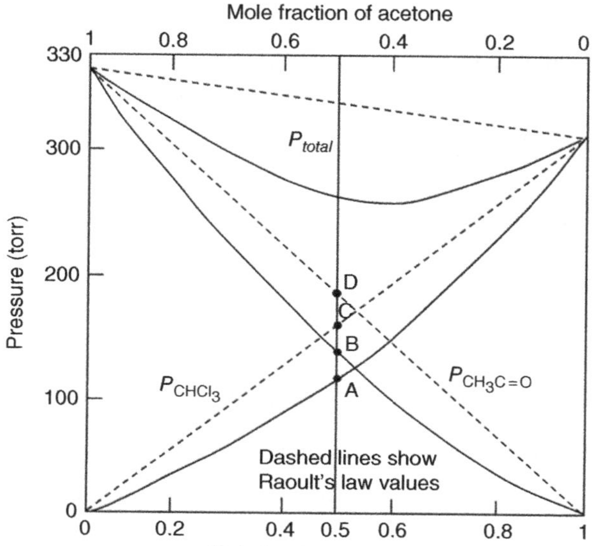

Figure 6.6 Nonideal partial pressures of chloroform-acetone solutions. *Source:* Adapted from Lowry (2023).

$$p_i = \left(X_i \cdot P_{i,pure\ liquid}\right) \gamma \tag{6.61}$$

An ideal chloroform solution at 0.5 mol fraction would have a partial pressure at point C where $X_{chlorform} \cdot P_{chloroform,\ pure\ liquid}$. The partial pressure at point C is 168 Torr. The effective partial chloroform pressure, $p_{chloroform}$, is at point A, 115 Torr. Eq. (6.61) can be solved for $\gamma_{chloroform}$.

$$\gamma_{chloroform} = p_{chloroform} / \left(X_{chlorform} \cdot P_{pure\ chloroform,}\right) = 115\ \text{\sout{Torr}}/168\ \text{\sout{Torr}} = \underline{0.685} \tag{6.64}$$

The activity acetone activity coefficient can be calculated in the same fashion.

For acetone, the Raoult's law pressure, $X_i \cdot P_{i,pure\ acetone}$, is at point D, and the effective partial pressure in at point B. Using Figure 6.6 the ideal acetone partial pressure is 197 Torr. Acetone's effective pressure is 142 Torr. The acetone activity coefficient is calculated from Eq. (6.61).

$$\gamma_{acetone} = p_{acetone} / \left(X_{acetone} \cdot P_{pure\ acetpme,}\right) = 142\ \text{\sout{Torr}}/197\ \text{\sout{Torr}} = \underline{0.721} \tag{6.65}$$

The activity coefficients for acetone and ether are less than 1, meaning the chloroform-acetone solution acts in a nonideal manner. Also, the chloroform activity coefficient of 0.685 and the acetone activity coefficient of 0.721 indicate that the chloroform-acetone adhesion forces are stronger than the chloroform-chloroform and acetone-acetone cohesion forces. These stronger chloroform-acetone adhesion forces decrease the amount of either compound from escaping to the vapor phase. In this case, the cohesion forces are due to hydrogen bonding. It is known that chloroform forms a hydrogen bond with acetone by sharing its lone hydrogen atom with acetone's ketone group. Other examples of binary solution systems that demonstrate negative deviations from Raoult's law are chloroform-acetic acid, chloroform-benzene, chloroform-diethyl ether, and acetic acid-methanol.

Example 6.5 Illustration of Nonideal Behavior of a Crystalline Solute Naphthalene in Chlorobenzene and Water Solutions

Given:

Naphthalene (N) ideal solubility is 0.265 mol fraction
Naphthalene's physically measured solubility in chlorobenzene (C) is 0.256 mol fraction.
Naphthalene's physically measure solubility in water (W) is 3.9E-6 mol fraction.

Calculate the activity coefficient for N-C and N-W solubility.
Activity Calculation:
At equilibrium, Eq. (6.51) describes the solubility process.

$$\text{Undissolved solute, } u \leftrightarrows \text{Dissolved solute, } d \tag{6.51}$$

and

$$\mu_u = \mu_d \tag{6.51a}$$

The activity or effective concentration of naphthalene solution is can be written as:

$$a_{naphthalene,effective\ concentration} = \gamma \cdot \left(\text{physical measured concentration}_{naphthalene}\right) \tag{6.66}$$

By convention, the activity of a solid and the solute in its saturated solution are equal to one. Substituting one into Eq. (6.66) results in:

$$1 = \gamma \cdot \left(\text{physical measured concentration}_{naphthalene}\right) \tag{6.67}$$

Now, solve Eq. (6.67) for γ which gives Eq. (6.68).

$$\gamma = 1/(\text{physical measured concentration}_{naphthalene}) \tag{6.68}$$

Equation (6.68) indicates that the solid solubility activity coefficient is inversely proportional to the *physical measured concentration*$_{naphthalene}$.

For the N-C solution, substituting for the (physical *measured concentration*$_{naphthalene}$) in Eq. (6.68) produces $\gamma_{N\text{-}C} = 1/0.256 = 3.9$. The $\gamma_{N\text{-}C}$ is reasonably close to one, suggesting that naphthalene solubility in chlorobenzene is nearly ideal. The comparison of naphthalene's 0.256 mol fraction solubility in chlorobenzene to naphthalene's ideal solubility of 0.265 mol fraction supports the idea that the N-C system exhibits near ideal solubility behavior.

The $\gamma_{N\text{-}W}$ can be calculated in a similar manner where $\gamma_{N\text{-}W} = 1/3.9\text{E-}6 = 2.6\text{E}5$. The $\gamma_{N\text{-}W}$ is much greater than 0. This means that a significant positive deviation from ideal behavior exists for the naphthalene-water system. The large $\gamma_{N\text{-}W}$ strongly suggests that the cohesive naphthalene-naphthalene and water-water interactions leave little possibility for naphthalene-water interactions leading to extremely poor solubility of the naphthalene in water.

6.11 Electrolyte Solutions

Electrolytes are compounds that form ions in solution. *Electrolytes* are classified as *strong electrolytes* and *weak electrolytes*. Strong electrolytes undergo near complete ionization in water. Weak electrolytes, on the other hand, only partially ionize in water.

Strong electrolytes are subdivided into strong acids, strong bases, and salts. The most common strong acid used in the product design is HCl. It is often used as a 0.1 M solution as dissolution medium to simulate HCl in the stomach. Hydrochloric acid is also used to adjust the pH of liquid formulations. Other strong acids are H_2SO_4 and HNO_3. A strong acid is often used to form a salt with a drug that is a weak base. Strong bases like NaOH, KOH, and LiOH are often used to form salts with weak acid drugs. Sodium hydroxide is also used to adjust the pH of liquid products. NaCl is the most common salt used in product design and manufacturing. It is commonly used to adjust tonicity of nasal, ophthalmic, and parenteral products. Potassium chloride, magnesium phosphate, and calcium phosphate are used primarily as mineral replacements.

Weak electrolytes are subdivided into weak acids and weak bases. Most drugs are weak acids or weak bases. About 75% of drugs are weak bases and 20% weak acids. The remainder 5% are nonelectrolytes. Weak acids such as acetic acid (CH_3COOH) and phosphoric acid (H_3PO_4) and their salts are often also used to create buffer systems. The most common weak bases are those containing a "nitrogen" atom. Tromethamine and histidine are common weak bases used to form basic buffer systems. Basic salts such as aluminum hydroxide ($Al[OH]_3$) and magnesium hydroxide ($Mg[OH]_2$) are over-the-counter antacids.

Salt crystal lattices consist of the individual salt ions. During solution formation with water, the salt crystal lattices dissolve in the water. The individual ions maintain their change in water as shown in Eq. (6.69).

$$H_2O_{(l)} + Na^+\,Cl^-_{(\text{crystal})} \rightarrow Na^+_{solution} + Cl^-_{solution} \tag{6.69}$$

Water and sodium chloride crystals participate in solution formation. Note that Eq. (6.69) represents a nonsaturated sodium chloride solution. Thus, the arrow points toward products Na^+, Cl^-, and H_2O.

Strong electrolytes like hydrochloric acid or sodium hydroxide dissociate in water to form hydronium ion or hydroxide ion. For the strong electrolyte, HCl, in relatively dilution solution, less than 0.1 M, the HCl is completely ionized.

$$H_2O_{(l)} + HCl_{(g)} \rightarrow H_3O^+ + Cl^- \tag{6.70}$$

For a weak electrolyte like acetic acid in an unsaturated solution, the weak electrolyte forms a hydronium ion and its conjugate anion that is equilibrium with acetic acid as shown in Eq. (6.71).

$$H_2O_{(l)} + CH_3COOH_{(l)} \rightleftharpoons H_3O^+ + CH_3COO^- \tag{6.71}$$

The equilibrium arrows indicate that acetic acid does not completely ionize in water.

6.12 Solubility of Slightly Soluble Strong Electrolyte Salts

Slightly soluble, strong electrolyte salts undergo near-complete ionization in water in a similar fashion as shown for sodium chloride in Eq. (6.69). The general equilibrium solubility equation is Eq. (6.72)

$$(\text{Cation ion}^+ \cdot \text{Anion}^-)_{solid} + H_2O \rightleftharpoons \text{Cation ion}^+{}_{(aqueous)} + \text{Anion}^-{}_{(aqueous)} \tag{6.72}$$

Silver chloride is a well-known example of a strong electrolyte of a slightly soluble salt. The equilibrium solubility equation is:

$$Ag^+ Cl^-{}_{(solid)} + H_2O \rightleftharpoons Ag^+{}_{(aqueous)} + Cl^-{}_{(aqueous)} \tag{6.73}$$

The silver chloride equilibrium solubility can be written as:

$$K = \left[Ag^+{}_{(aqueous)}\right]\left[Cl^-{}_{(aqueous)}\right] / \left[Ag^+ Cl^-{}_{(solid)}\right], \text{that is products/reactants} \tag{6.74}$$

where $K\,[H_2O] = K$, and $[Ag^+{}_{(aqueous)}]$, $[Cl^-{}_{(aqueous)}]$, and $[Ag^+Cl^-{}_{(solid)}]$ are the molar concentrations of the respective species when the activity coefficient is one.

The $Ag^+Cl^-{}_{(solid)}$ concentration is considered constant, giving a new equilibrium constant:

$$K\left[Ag^+ Cl^-{}_{(solid)}\right] = K_{solubility\ product} = \left[Ag^+{}_{(aqueous)}\right]\left[Cl^-{}_{(aqueous)}\right] \tag{6.75}$$

This equation holds when the ionic species are in low concentrations; the solution is deemed an ideal solution. As an ideal solution, the ionic species do not interact or form ion pairs or form other species due to hydrolysis in solution. The solubility product, $K_{solubility\ product}$ or more commonly reported as K_{sp}, *is the molar concentration of ionic constituents of the salt, each raised to the power of its equilibrium stoichiometric coefficient*.

For a generic salt, M_mX_n, the solubility product is:

$$K_{sp} = [M^+]^m\,[X^-]^n \tag{6.76}$$

where M is the cation and X is the anion, and m and n are the stoichiometric coefficients of the ions.

For example, the K_{sp} for calcium phosphate, $Ca_3(PO_4)_2$, which completely ionizes in water is:

$$Ca_3(PO_4)_{2\,(solid)} \rightleftharpoons 3Ca^{2+}{}_{aq} + 2PO_4{}^{3-}{}_{aq}$$
$$K_{sp} = \left[Ca^{2+}\right]^3\left[PO_4{}^{3-}\right]^2 \tag{6.77}$$

From Eq. (6.51a), the chemical potential of the undissolved solid, μ_u, and dissolved solid, μ_d, is equal.

$$\mu_u = \mu_d \tag{6.51a}$$

The development of Eqs. (6.78)–(6.85) follows the work of Connors and Mecozzi (2010). Equation (6.51a) can be extended to Eq. (6.72) for strong electrolytes giving Eq. (6.78):

$$\mu_{(cation\ ion \cdot anion)\ solid} = \mu_{cation\ ion\ (aqueous)} + \mu_{anion\ (aqueous)} \tag{6.78}$$

which can be written in terms of activities and standard chemical potentials:

$$\mu^{\circ}{}_{solid\ \pm} + RT\ \ln\ a_{solid\ \pm} = \mu^{\circ}{}_{cation} + RT\ \ln\ a_{cation} + \mu^{\circ}{}_{anion} + RT\ \ln\ a_{anion} \tag{6.79}$$

where μ° is the standard chemical potential of the identified species, $a_{solid\ \pm}$ is the activity of the strong electrolyte cation-anion solid, and a is the activity of the identified cation or anionic species. As noted in Section 6.10.2, the activity of a solid is 1. The second term on the left-hand side of Eq. (6.79) goes to zero since the ln 1 is 0. Then, the common terms can be combined to give:

$$\Delta\mu^{\circ} = \mu^{\circ}{}_{cation} + \mu^{\circ}{}_{anion} - \mu^{\circ}{}_{solid\ \pm} = -RT\ \ln\ a_{cation} \cdot a_{anion} \tag{6.80}$$

where $\Delta\mu^{\circ}$ is the standard chemical potential change for the standard chemical potentials of the cation ion and anion minus the standard chemical potential of solid electrolyte, R is the gas constant, T is the temperature in kelvin, and $a_{cation} \cdot a_{anion}$ is the product of the cation and anion activities. Note that that $-RT$ (ln a_{cation} + ln a_{anion}) terms becomes $-RT$ ln $a_{cation} \cdot a_{anion}$, using the logarithmic rule 1.46. The chemical potential is the molar Gibbs free energy as shown by Eq. (4.53).

$$\mu = G/n \tag{4.53}$$

A system at equilibrium is related to the Gibbs free energy by Eq. (5.3).

$$\Delta G^{\circ} = -\ (RT)\ \ln\ K \tag{5.3}$$

From Eqs. (6.79), (4.53), and (5.3), the solubility of a slightly soluble strong electrolyte is related to its solubility product by the change in its standard chemical potential, see Eq. (6.81). The

$$\Delta\mu^{\circ}{}_{strong\ electrolyte} = -RT\ \ln\ a_{cation} \cdot a_{anion} = -RT\ \ln\ K_{sp} \tag{6.81}$$

The $a_{cation} \cdot a_{anion}$ term is equivalent to the strong electrolyte's solubility product, $K_{sp.}$ The *mean ionic activity* is considered the *activity* of an electrolyte and is defined in Eq. (6.82).

$$a_{\pm} = (a_{+}{}^{m}\ a_{-}{}^{n})^{1/(m+n)} \tag{6.82}$$

$a_{\pm}$ is the *mean ionic activity* of the strong electrolyte salt, $a_{+}{}^{m}$ is the activity of the cation raised to the power of its stoichiometric number of ions (m), and $a_{-}{}^{n}$ is the anionic activity raised to the power of its stoichiometric quantity of ions (n).

The *mean ionic activity coefficient* for an electrolyte is:

$$\gamma_{\pm} = (\gamma_{+}{}^{m}\ \gamma_{-}{}^{n})^{1/(m+n)} \tag{6.83}$$

where $\gamma_{\pm}$ is the *mean ionic activity coefficient*, $\gamma_{+}{}^{m}$ is the activity coefficient for the cation raised to the stoichiometric value (m), and $\gamma_{-}{}^{n}$ is the activity coefficient of the of the anionic species raised to the stoichiometric number of anions (n). The mean activity coefficient is a function of the *ionic strength, I,* of the system. The mean activity coefficient is also a function of the valence charges on the ions and a factor, A, which is a function of temperature and the dielectric constant of the liquid. The logarithm of the mean activity coefficient is given in Eq. (6.84)

$$\text{Log}\,\gamma_{\pm} = -A z_{+} z_{-} I^{0.5} \tag{6.84}$$

where $\gamma_{\pm}$ is the mean ionic activity coefficient, A varies with the temperature and dielectric constant of the solvent, z_{+} z_{-} are the valence charges ignoring the algebraic signs on the charges, and I is the ionic strength of the solution.

The ionic strength of a solution is determined by Eq. (6.85).

$$I = 0.5 \sum M_i z_i^2 \tag{6.85}$$

The author directs the reader to Sinko (2006) and Connors and Mecozzi (2010) for further discussion and examples of mean ionic activity, mean ionic activity coefficients, and ionic strength.

Example 6.6 Silver Chloride's pK_{sp} is 9.80. What is AgCl's Molar Solubility? (Adapted from Chang, 1998)

Given:

AgCl pK_{sp} = 9.80 (note two significant figures in mantissa; see Section 1.8.4
pK_{sp} = −log K_{sp}
AgCl K_{sp} = $[Ag^+][Cl^-]$
Equilibrium concentration of Ag^+ = s (the molar solubility of the cation)
Equilibrium concentration of Cl^- = s (the molar solubility of the anion)

$$Ag^+Cl^-_{(solid)} + H_2O \rightleftharpoons Ag^+_{(aqueous)} + Cl^-_{(aqueous)} \tag{6.73}$$

$$K_{sp} = [Ag^+][Cl^-] = s \cdot s = s^2$$

First, determine the AgCl K_{sp} from the given pK_{sp} = 9.80.

$$-\log K_{sp} = 9.80$$

$$\underline{K_{sp}} = \text{antilog} - 9.80 = \underline{1.6\text{E} - 10\ \text{M}^2}\ (2\ \text{significant figures}) \tag{6.86}$$

Determine solubility of AgCl.

$$K_{sp} = s^2 = 1.6\text{E-}10\ \text{M}^2 \tag{6.87}$$

$$s = \left(1.6\text{E-}10\ \text{M}^2\right)^{0.5} = \underline{1.3\text{E-}5\ \text{M}} \tag{6.88}$$

$$\underline{\text{AgCl's molar solubility is 1.3E-5 M}} \tag{6.89}$$

Example 6.7 Common Ion Effect: Determine the Molar Solubility of Silver Chloride in 6.5E-3 M Silver Nitrate. (Adapted from Chang, 1998)

Given:

- Silver chloride solubility is 1.3E-5 M. Since silver chloride is a strong electrolyte, it completely dissociates in water to give 1.3E-5 M Ag^+ and 1.3E-5 M Cl^-
- $K_{sp} = [Ag^+][Cl^-] = 1.6\text{E-}10\ \text{M}^2$
- Silver nitrate molar solution is 6.5E-3 giving 6.5E-3 M Ag^+ and 6.5E-3 M NO_3^-

- Note: silver nitrate is a strong electrolyte that completely dissociates in water

$$\text{Initially } Ag^+Cl^-_{(solid)} + H_2O \rightleftharpoons Ag^+_{(aqueous)} + Cl^-_{(aqueous)} \tag{6.73}$$

- The addition of silver nitrate adds Ag^+ and NO_3^- ions to the product side of Eq. (6.73). The addition of Ag^+ causes stress on the product side. Applying Le Châtelier's principle, the stress is counteracted by the equilibrium moving toward the reactant side of Eq. (6.73), creating more solid Ag^+Cl^-. This should result in decrease in the silver chloride solubility.

Determine the molar solubility of silver chloride when a common silver ion is provided by the silver nitrate solution.

Let "s" equal the molar solubility of $Ag^+Cl^-_{(solid)}$ which equals the molar solubility of both silver ion and chloride ion.

The silver chloride solubility product is given earlier.

$$K_{sp} = [Ag^+][Cl^-] = 1.6\text{E-}10\ M^2$$

Accounting for the additional silver ion resulting from the addition of silver nitrate, the solubility product equation is given by Eq. (6.90).

$$1.6\text{E-}10\ M^2 = \left(*6.5\text{E-}3\ M\ Ag^+ + {}^{\#}s\ M\right)(s\ M) \tag{6.90}$$

where $*$ is the molar concentration of Ag^+ provided by the silver nitrate and $^{\#}s$ is the Ag^+ molar concentration from the silver chloride solubility.

Since the AgCl solubility is ~10^{-5} M (given earlier), the AgCl solubility is much smaller than the concentration of silver ion provided by silver nitrate. Therefore, the s in the first term on the right-hand side of Eq. (6.90) can be ignored giving Eq. (6.91).

$$1.6\text{E-}10\ M^2 = (6.5\text{E-}3\ M\ Ag^+)(s\ M) \tag{6.91}$$

Solving for "s" gives:

s = 2.5E-8 M which is the solubility of AgCl and the molar concentrations of silver and chloride ions.

The addition of 6.5E-3 M silver nitrate common ion decreased the AgCl solubility by 520-fold! (1.3E-5 M/ 2.5E-8 M = 520). This is consistent with Le Châtelier's principle.

6.13 Author's Anecdote – Common Ion Effect and the Ability of Thermodynamically Unstable Systems to be Manufactured and Used for Years Without Failure, Until One Day, Catastrophe

There never seems to be an easy drug product design project, even when it is supposed to be. This anecdote could be subtitled "A Slam Dunk Product Design Project Turns Ugly." The company I was working for licensed a drug for cystic fibrosis (CF) from a local internationally renowned university research group. The group had been developing amiloride-HCl, an approved diuretic, for CF for approximately three years. The researchers had developed a formulation which had been used in several small Phase I and Phase II clinical trials over that time.

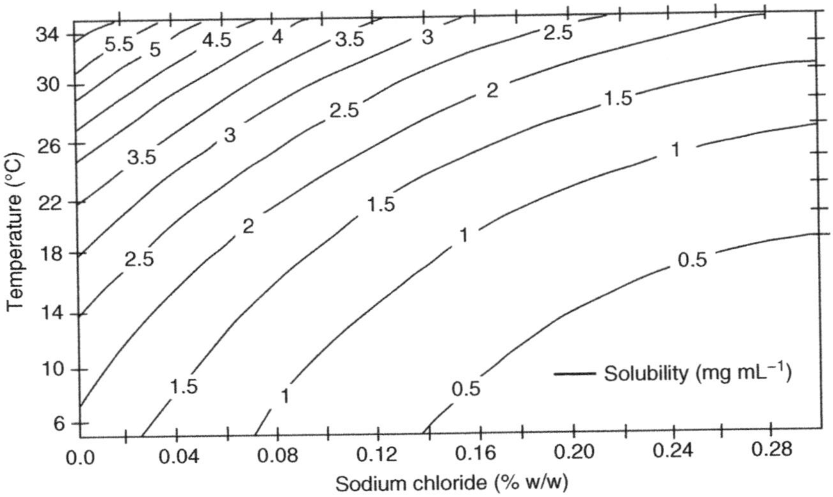

Figure 6.7 Contour plot of amiloride-HCl solubility as a function of NaCl concentration.

CF is a rare hereditary disease that is usually fatal. CF causes lung mucus to become thick. The thick mucus plugs vital lung structures and impedes normal respiration.

Our job was to replicate the established licensed formulation and make clinical supplies for Phase 2 and Phase 3 studies. The formulation was very simple. It was a sterile solution of amiloride-HCl and normal saline, 0.9% w/v NaCl. We contracted with a sterile manufacturing site that produced individual sterile blow-fill-seal plastic ampules of the formulation. The ampule solution would be used in a nebulizer. The 2.5 mg mL^{-1} drug product was manufactured without any issues. Once manufactured, the sterile ampules were placed on stability.

The early stability time points were uneventful. Then, the stability samples pulled at the 6-month time point had massive precipitation in all the ampules. We hypothesized that the issue was decreased solubility due to a common chloride ion effect. Of course, this hypothesis met serious push back from the inventors who had worked with the solution formulation for three plus years and had not seen any issues, including massive precipitation. None the less, we pursued investigating our hypothesis.

Figure 6.7 is a contour plot that shows the results of amiloride-HCl solubility study as a function of temperature and sodium chloride concentration.

Only at 34 °C and sodium chloride concentrations below ~0.22% w/w is the solubility of amiloride-HCL 2.5 mg mL^{-1} or higher. This is about 75% less NaCl than in the formulation.

These data initiated immediate discussion about using other tonicity adjusting agents or not adjust for tonicity at all. These suggestions raised grave concern about patient safety. It is well known that CF patients have what is called "reactive lungs." Any type of foreign material can cause the lungs to undergo bronchospasms. In addition, there was information available that CF patient lungs could be reactive to inhalation of water or nonisotonic inhalants. Sodium chloride was the only tonicity-adjusting agent reported for use in CF patients. The clinical community had to compromise. They wanted the highest possible dose so the patient inhalation times would not be excessively long, and they wanted to have largest amount of sodium chloride for tonicity without causing the drug to precipitate.

Using the above data, a dose of 1 mg mL^{-1} amiloride-HCl and 0.12% w/w NaCl was agreed to. This formulation was manufactured and placed on stability. The 1-, 3-, and 6-month points confirmed a stable product. Then unexpectedly at one year, several samples had one or two very fine crystals! What was this? Were we not conservative enough in picking the dose and the NaCl concentration for 15–30°C?

Our analytical department came to the rescue. They were able to isolate a few crystals and performed microscopy X-ray photoelectron spectroscopy (XPS) on single crystal samples. XPS can identify elements such as C, N, O, and S on the surfaces of materials. To everyone's surprise, the crystals were identified as (amiloride)$_2$ SO_4. It turns out that di-amiloride-sulfate is extremely insoluble. To put its solubility in perspective, it is ~2.8 times less soluble than barium sulfate and ~ 3.5 times less soluble than silver chloride. It also turned out that sulfate is ubiquitous even in the purest of water systems at single-digit ppm levels. At these very low levels of sulfate ion, the solubility of di-amiloride-sulfate was exceeded. Sadly, this seemingly simple project that held promise to treat the CF was discontinued with this new finding.

6.14 Solubility of Weak Acids and Bases

Before proceeding with this section, it may be helpful to review Section 5.9.1 Ionization Equilibrium of Weak Acids and Bases. Highly polar and ionized compounds have higher aqueous solubility than nonpolar and unionized species. In the case of weak acids and bases, the solution pH and the compound's pK_a(s) determine the percentage drug ionization. The drug's total solubility is reflected by the solubility of the unionized form and the drug percent ionized. The generalized Henderson-Hasselbach equation is given by Eq. (5.69).

$$pH = pK_a + \log\,[\text{conjugate base}]/[\text{conjugate acid}] \tag{5.69}$$

The total pH-solubility weak acid and base profile can be calculated by knowing the compound's pK_a, solubility of the unionized form, and the solubility of the ionized form(s). Since ~75% of drugs are weak bases, the examples in this chapter are for weak bases. Examples for weak acids predominate most textbooks and can be found in Sinko (2006) and Connors and Mecozzi (2010).

The solubility of a weak base and the ionization of its conjugate acid are shown in Scheme. (6.92).

$$\begin{array}{c} B_{solid} \\ \uparrow\downarrow \\ BH^+{}_{(aq)} + H_2O_{(l)} \rightleftharpoons B_{(aq)} + H_3O^+{}_{(aq)} \end{array} \tag{6.92}$$

Two equilibriums are operational in Eq. (6.92). The two equilibriums include the *intrinsic free base solubility* (solubility of the unionized free base) and the K_a of base conjugate acid, BH^+.

6.14.1 Solubility of a Weak Base as Function of the Drug's pKa, the System pH, and Solubility of the Unionized Base

Equation (5.68) explains the equilibrium process shown in (6.92).

$$K_a = \left[B_{(aq)}\right]\left[H_3O^+{}_{(aq)}\right]/\left[BH^+{}_{(aq)}\right] \tag{5.68}$$

where $B_{(aq)}$ and $BH^+{}_{(aq)}$ are the *unionized* and *ionized* species of the drug. The unionized *free base* form, $B_{(aq)}$ solubility is typically referred to as the *intrinsic solubility* and given the symbol S_o or $S_{unionized}$. The total solubility concentration from Eq. (6.92), S_{total} of the weak base, is equal to the solubility of the base plus the concentration of $BH^+{}_{(aq)}$ in solution. The concentration of $BH^+{}_{(aq)}$ is pH-dependent.

$$S_{total} = S_{unionized} + \left[BH^+{}_{(aq)}\right] \quad (6.93)$$

Using Eq. (5.68) and solving for $[BH^+{}_{(aq)}]$ gives:

$$\left[BH^+{}_{(aq)}\right] = \left[B_{(aq)}\right]\left[H_3O^+{}_{(aq)}\right]/K_a \quad (6.94)$$

The right-hand side of Eq. (6.94) can be substituted into Eq. (6.93)

$$S_{total} = S_{unionized} + \left[B_{(aq)}\right] H_3O^+{}_{(aq)}]/K_a = S_{unionized} + [S_{unionized}]\left[H_3O^+{}_{(aq)}\right]/K_a \quad (6.95)$$

where $[B_{(aq)}] = S_{unionized}$.

Combining common terms in Eq. (6.95) leads to Eq. (6.96) for the total solubility of a weak base:

$$S_{total} = S_{unionized}\left[1 + \left[H_3O^+{}_{(aq)}\right]/K_a\right] \quad (6.96)$$

Figure 6.8 shows the total solubility-pH profile using Eq. (6.96) when the pK_a of the weak base is 3, 5, or 7 and the intrinsic solubility (unionized solubility) is 1 $\mu g\ mL^{-1}$. The *y*-axis is total solubility, and the *x*-axis is pH. The weak base total solubility-pH profile can also be expressed in terms of pH and pK_a by using the definitions of pH and pK_a. This is performed later.

$$pH = -\log\left[H_3O^+{}_{(aq)}\right], \text{ and} - pH = \log\left[H_3O^+{}_{(aq)}\right] \quad (6.97)$$

Taking the antilog of both sides of the furthest right Eq. (6.97):

$$10^{-pH} = \left[H_3O^+{}_{(aq)}\right] \quad (6.98)$$

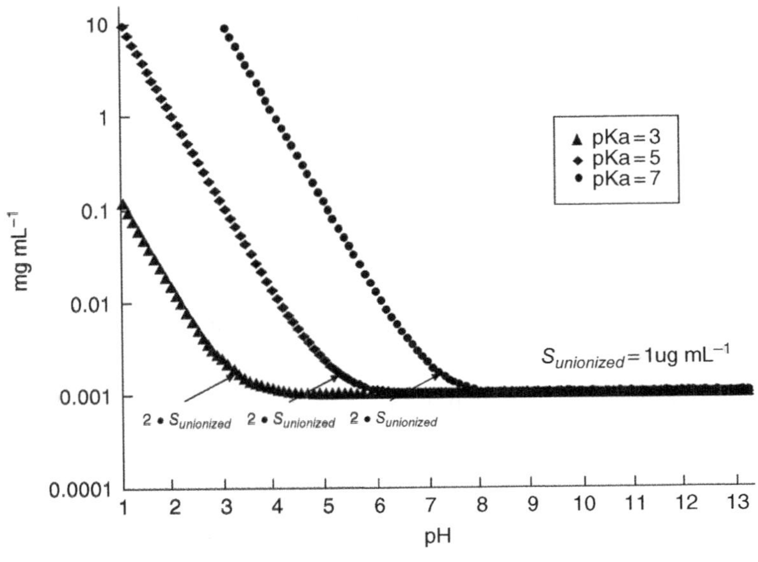

Figure 6.8 pH-solubility profile of a weak base having an intrinsic solubility of 0.001 mg mL^{-1} and hypothetical pK_a's of 3, 5, or 7 from Eq. (6.102). *Source:* Adapted from personal communication with Winnike (2014).

Using the definition of pK_a,

$$pK_a = -\log K_a, \text{and} - pK_a = \log K_a \tag{6.99}$$

Taking the antilog of both sides of the furthest right Eq. (6.99) gives:

$$10^{-pKa} = K_a \tag{6.100}$$

Substituting the respective terms into Eq. (6.96), $S_{total} = S_{unionized}\,[1 + [H_3O^+{}_{(aq)}]/K_a]$ gives:

$$S_{total} = S_{unionized}\left[1 + \left[10^{-pH}/10^{-pKa}\right]\right] \tag{6.101}$$

Using exponential operating rule 1.35 on Eq. (6.101) gives:

$$S_{total} = S_{unionized}\left[1 + \left[10^{(pKa-pH)}\right]\right] \tag{6.102}$$

This form of the total solubility makes calculations very simple if one has the drug's pK_a and the pH of interest.

6.14.2 When pH Equals pKa: The Total Weak Base Solubility is $2 \cdot S_{unionized}$

From Eq. (5.69), when the pK_a equals the pH, the ratio of the conjugate base and conjugate acid is one.

$$pH = pK_a + \log [\text{conjugate base}]/[\text{conjugate acid}] \tag{5.69}$$

$$pH - pK_a = 0 = \log [\text{conjugate base}]/[\text{conjugate acid}] \tag{6.103}$$

The antilog of 0 is 1. Therefore,

$$1 = [\text{conjugate base}]/[\text{conjugate acid}] \tag{6.104}$$

which means the concentrations of the conjugate base and acid are equal when the pH equals the pK_a of the base.

When the pK_a equals the pH, the total solubility of the weak base is two times the solubility of the unionized drug, $2 \cdot S_{unionized}$. This is can be determined using Eq. (6.102).

$$S_{total} = S_{unionized}\left[1 + \left[10^{(pKa-pH)}\right]\right] \tag{6.102}$$

When the pK_a equals the pH, 10^0 and gives Eq. (6.105).

$$S_{total} = S_{unionized} + [S_{unionized}] \cdot 1 = \underline{2 \cdot S_{unionized}} \tag{6.105}$$

showing the total weak base solubility at the point pH = pK_a is $2 \cdot S_{unionized}$.

6.14.3 A Weak Base is Nearly 100% Ionized at 2 pH Units Below its pK_a

From Figure 6.8, note that at 2 pH units below the weak base pK_a, the total drug solubility is 0.1 mg mL^{-1} for pK_a's of 3, 5, and 7. Since the drug intrinsic solubility (S_0 or $S_{unionized}$) is the same for each pK_a example, and the total solubility is the same at 2 pH units below each respective pK_a, this means that $S_{total} - S_{unionized}$. is the same. Equation (5.69) can also be used to show the ratio of the ionized to unionized species.

$$\text{pH} = \text{pK}_a + \log\ [\text{conjugate base, B}]/[\text{conjugate acid, BH}^+] \quad (5.69)$$

Rearrangement of Eq. (5.69) is:

$$\text{pH} - \text{pK}_a = \log\ [\text{B}]/[\text{BH}^+] \quad (6.106)$$

Taking the antilog of both sides gives:

$$10^{(\text{pH} - \text{pKa})} = [\text{B}]/[\text{BH}^+] \quad (6.107)$$

For a weak base with pK_a of 7, 2 pH units below the pK_a is 5. Substituting these values into Eq. (6.107) is:

$$10^{-2} = [\text{B}]/[\text{BH}^+];\ \text{and}\ [\text{BH}^+]/[\text{B}] = 10^2 \quad (6.108)$$

which shows that the ratio of the ionized form to unionized form is 100:1.

6.14.4 A Weak Base is Nearly 100% Unionized at 2 pH Units Above its pK_a

Employing Eq. (6.107), the ratio of unionized to ionized specie can be calculated in a similar way as mentioned earlier. In this example, the weak base has a $\text{pK}_a = 7$ and the solution pH is 9.

$$10^{(9-7)} = [\text{B}]/[\text{BH}^+];\ \text{and}\ 10^2 = [\text{B}]/[\text{BH}^+] \quad (6.109)$$

The ratio of the unionized drug to ionized drug is 100:1 or 99% unionized.

6.14.5 Similar Equations can be Developed for Weak Acids

The total solubility equations for a weak acid are provided.

$$S_{total} = S_{unionized}\left[1 + \text{K}_a/\left[\text{H}_3\text{O}^+{}_{(aq)}\right]\right] \quad (6.110)$$

where S_{total} is the total solubility of the weak acid, $S_{unionized}$ is the solubility of the free acid (unionized acid), K_a is the equilibrium constant of the free acid, and $[\text{H}_3\text{O}^+{}_{(aq)}]$ is the hydronium ion concentration. The total weak acid solubility, Eq. (6.111), can also be expressed in terms of pH and pK_a.

$$S_{total} = S_{unionized}\left[1 + 10^{(\text{pH} - \text{pKa})}\right] \quad (6.111)$$

6.14.6 Solubility-pH Profile for a Weak Acid and a Weak Base Drug

The total pH-solubility for a weak acid and a weak base drug is shown in Figure 6.9a,b.

Figure 6.9a,b illustrates the high predictability of the Henderson-Hasselbach equation for a weak monoacid or monobase having a single pKa. The confirmation of the theoretical equations from observed data required meticulous solubility data collection. Figure 6.9a, the limiting solubility of the weak acid is its unionized form, often referred to as the intrinsic solubility of the free acid (S_o or $S_{unionized}$). In this example, the pKa is 4.76. At two pH units below its pKa, the weak acid is nearly 100% (actually 99%) in its unionized free acid form. The free base form, unionized form, is the limiting solubility of the base. In Figure 9b, the base is almost completely ionized (99%) at two pH units below its pKa.

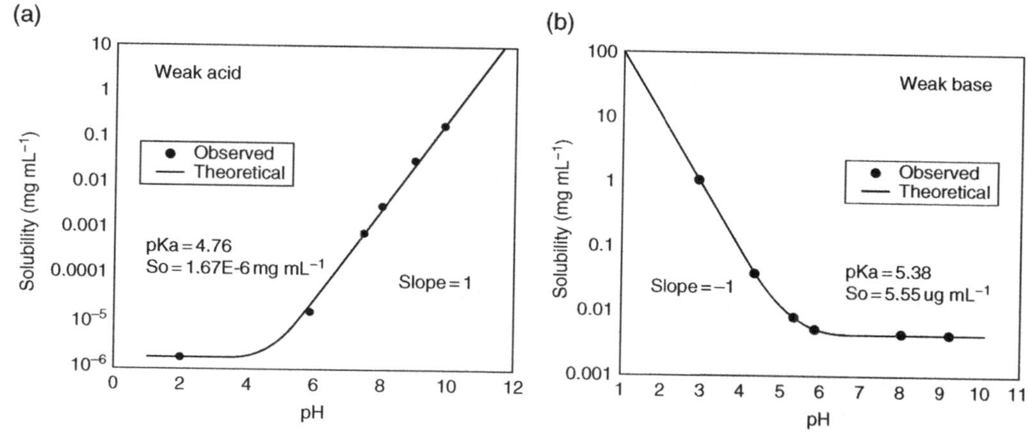

Figure 6.9a (the left figure) Comparison of observed solubility measurements and the predicted theoretical total solubility-pH profile for a weak acid using the Henderson-Hasselbach equation. **Figure 6.9b** (the right figure) Comparison of observed solubility measurements and the predicted theoretical total solubility-pH profile for a weak base using the Henderson-Hasselbach equation. *Source:* Adapted from personal communication with Winnike (2014).

6.14.7 Slope of the Ionized Portion of the Solubility Profile is Negative-One for a Weak Base and Positive-One for a Weak Acid

The weak base solubility profile is presented by Eq. (6.101)

$$S_{total} = S_{unionized}\left[1 + \left[10^{-\text{pH}}/10^{-\text{pKa}}\right]\right. \tag{6.101}$$

which can be rearranged.

$$S_{total} - S_{unionized} = S_{unionized}\left[10^{-\text{pH}}/10^{-\text{pKa}}\right] \tag{6.112}$$

The logarithm can be taken on both sides of Eq. (6.112). The logarithm operator, Eq. (1.46), was employed on the right-hand side of the equation to bring about:

$$\log\left(S_{total} - S_{unionized}\right) = \log S_{unionized} + \log\left[10^{-\text{pH}}/10^{-\text{pKa}}\right] \tag{6.113}$$

Using the logarithm operator, Eq. (1.47) for the second term on the right-hand side of Eq. (6.113) gives:

$$\log\left(S_{total} - S_{unionized}\right) = \log S_{unionized} + \log 10^{-\text{pH}} - \log 10^{-\text{pKa}} \tag{6.114}$$

Employing logarithm operator Eq. (1.48) to the last two terms on the right-hand side of the Eq. (6.114) is:

$$\log\left(S_{total} - S_{unionized}\right) = \log S_{unionized} - \text{pH} \log 10 - (-\text{pK}_a) \log 10 \tag{6.115}$$

and $\log 10 = 1$

$$\log\left(S_{total} - S_{unionized}\right) = \log\left(S_{unionized}\right) + \text{pK}_a - \text{pH} \tag{6.116}$$

The first two terms of Eq. (6.116) are collected and become:

$$\log\left(S_{total} - S_{unionized}\right) = \underline{-\text{pH}} + \log\left(S_{unionized}/\text{K}_a\right) \tag{6.117}$$

The $(S_{total} - S_{unionized})$ term represents the solubility of the ionized weak base line. As shown in Eq. (6.117) and Figure 6.9b, the slope of the weak base ionized species solubility line is negative 1.

A similar derivation may be done for a weak acid. In this case, the slope of the weak acid total solubility line is plus 1, Eq. (6.118):

$$\log(S_{total} - S_{unionized}) = +\text{pH} + \log(S_{unionized} \cdot \text{K}_a) \tag{6.118}$$

The slope of Figure 6.9a is consistent with Eq. (6.118).

6.14.8 The Solubility Product of a Weak Acid or Weak Base Counterion Limits the Total pH-Solubility of These Ionized Weak Electrolytes

It could be falsely concluded from Figures 6.8 and 6.9 that the solubility of the ionized forms of a weak acid or base is infinite. In fact, the solubility of the ionized forms is limited by the solubility product of the salt of the ionized form. Considering Figure 6.9b, if HCl was used to adjust the pH of the base below its pKa, the limiting solubility of this system would the ion product of the hydrochloride salt of the base.

As a reminder, when the free base is in solution equilibrium, the Ka describes the equilibrium processes for this system. These equilibria are given by Eq. (6.92)

$$\begin{array}{c} \text{B}_{solid} \\ \uparrow\downarrow \\ \text{BH}^+{}_{(aq)} + \text{H}_2\text{O}_{(l)} \rightleftharpoons \text{B}_{(aq)} + \text{H}_3\text{O}^+{}_{(aq)} \end{array} \tag{6.92}$$

The total solubility for this system is:

$$S_{total} = S_{unionized}\left[1 + \left[\text{H}_3\text{O}^+{}_{(aq)}\right]/\text{K}_a\right] \tag{6.96}$$

When the excess solid is the salt of the base and the available counterion, the solubility equilibrium is described by Eq. (6.119) and the solubility product, K_{sp}. The solubility product of slightly soluble electrolytes salts has already been discussed in Section 6.12.

$$(\text{BH}^+\text{Cl}^-)_{solid} \rightleftharpoons \text{BH}^+{}_{aq} + \text{Cl}^-{}_{aq} \tag{6.119}$$

A second portion of the total solubility profile accounts for this case when the salt of the free base is in excess is Eq. (6.120):

$$\text{Ksp} = \left[\text{BH}^+{}_{aq}\right]\left[\text{Cl}^-\right] \tag{6.120}$$

6.14.9 Weak Acid or Base Salt Total Solubility as a Function of pH, pKa, and K_{sp}

The total solubility of a weak acid or base salt is a function of its pH, pKa, and K_{sp}. The equation that describes the relationship of pH, pKa, and K_{sp} with the total solubility of a HCl salt of a weak base is shown in the derivation.

Given:

$$\text{Ksp} = \left[\text{BH}^+{}_{aq}\right]\left[\text{Cl}^-\right] \tag{6.120}$$

$$\text{K}_a = [\text{B}_{unionized}][\text{H}_3\text{O}^+]/[\text{BH}^+] \tag{5.68}$$

$$[B_{unionized}] = K_a\,[BH^+]/][H_3O^+]$$

$$S_{total} = \ [B_{unionized}] + [BH^+] \tag{6.93}$$

Substitution for $[B_{unionized}]$ into Eq. (6.93) gives:

$$S_{total} = [BH^+] + [B_{unionized}] + K_a[BH^+]/][H_3O^+] \tag{6.121}$$

and combining common terms gives:

$$S_{total} = [BH^+](1 + Ka/]/[H_3O^+]) \tag{6.122}$$

Solving for Eq. (6.122) for $[BH^+]$ is:

$$[BH^+] = S_{total}/(1 + Ka/]/[H_3O^+]) \tag{6.123}$$

Solving Eq. (6.120), $K_{sp} = [BH^+{}_{aq}]\ [Cl^-]$ for $[BH^+{}_{aq}]$ is:

$$[BH^+{}_{aq}] = Ksp/[Cl^-] \tag{6.124}$$

Substituting $[BH^+{}_{aq}]$ from Eq. (6.124) into Eq. (6.123) and solving for K_{sp} results in Eq. (6.125).

$$Ksp = [S_{total}/(1 + Ka/]/[\,H_3O^+])][Cl^-] \tag{6.125}$$

If $[Cl^-]$ only comes from dissolved B·HCl, then $[Cl^-] = S_{total}$. Substituting S_{total} for $[Cl^-]$ into Eq. (6.125) with S_{total} becomes:

$$Ksp = (S_{total})^2/(1 + Ka/]/[\,H_3O^+]) \tag{6.126}$$

and

$$(S_{total})^2 = Ksp\,(1 + Ka/]/[\,H_3O^+]) \tag{6.127}$$

Solving for the total solubility for Eq. (6.127) gives:

$$S_{total} = [Ksp\,(1 + Ka/]/[\,H_3O^+])]^{0.5} \tag{6.128}$$

and

$$S_{total} = \left[Ksp\left(1 + 10^{(pH-pKa)}\right)\right]^{0.5} \tag{6.129}$$

Equation (6.129) shows how the total solubility, that is the saturated solubility, of a salt of a weak base is a function of the base's pKa, pH, and the salt's K_{sp}. A similar equation can be derived for the salt of a weak acid.

6.14.10 Consolidated Solubility Profile for a Weak Base

The consolidated solubility profile combines Eqs. (6.111) and (6.120). An example of a consolidated solubility profile for an HCl salt of a weak basic drug is shown in Figure 6.10. This profile includes the weak base salt solubility portion reflecting the salt's K_{sp}. The portion is shown as the plateau region at pH values less than ~5.90. In this region, the excess solid is the drug salt which drives the

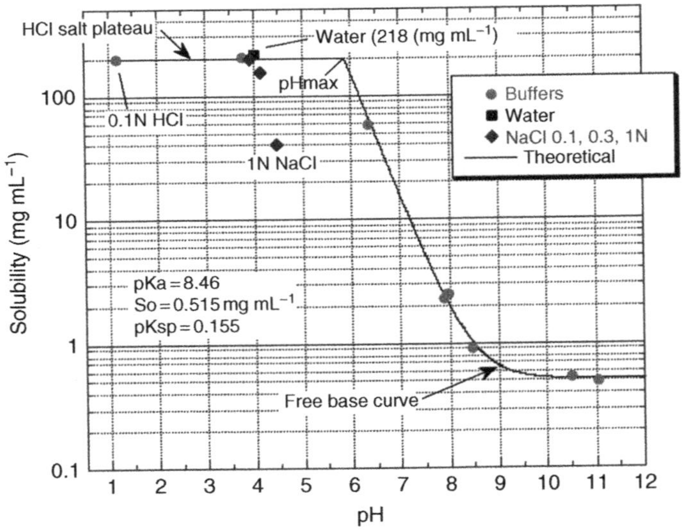

Figure 6.10 Complete solubility profile showing the excess salt-controlled solubility region (~pH 1 to ~5.9) and the unionized free-base-controlled region (pH ~5.9 to 11). *Source:* With permission from Winnike (2023).

solubility. The unionized free-base-driven solubility portion occurs between pH ~5.90 and 11.00. The excess solid in this region is the unionized free base. The experimental pHmax is graphically show to be ~5.90. The pHmax is where the two portions of the solubility profile intersect. At the pHmax, the excess solid is a mixture of the HCl salt and the unionized free base. The solubility at this pH is the sum of the solubilities of the unionized free base and the HCL salt. From Figure 6.10, the unionized free base's solubility is 0.515 mg mL^{-1} and the salt's solubility is ~200. mg mL^{-1} (determined from the graph). So, the total solubility at pHmax is the sum of these two values, 200.515 or 200.5 rounding to a one decimal point. In this case, the unionized free base contributes only ~0.3% of the total drug solubility at the pHmax. When the base contributes about 5 to 10% and higher to the total pHmax solubility, a peak can be observed at the pHmax.

The pHmax of a salt of a weak base can be calculated from Eq. (6.130).

$$\text{pHmax} = \text{pKa} + \log\left(\left[\text{B}_{\textit{unionized solubility}}\right]/\left[\text{BH}^{+}{}_{\textit{solubilty}}\right]\right) \tag{6.130}$$

Using information from the consolidated solubility profile of a weak-base HCl salt shown in Figure 6.10 and Eq. (6.130), the pHmax can be calculated.

Given:

Unionized free base (S_o) solubility = 0.515 mg mL^{-1}
HCl salt solubility = 200. mg mL^{-1} (read from graph)
Base pKa = 8.46

Substituting these values into Eq. (6.130) gives:

$$\text{pHmax} = 8.46 + \log\,(0.515\ \text{mg mL}^{-1}/200\ \text{mg mL}^{-1})$$
$$\text{pHmax} = 8.46 + \log 2.28\text{E-}3$$
$$\underline{\text{pHmax} = 8.46 \pm (-2.59) = 5.87}$$

From the graph, the intersection of the salt plateau solubility line and the unionized free solubility curved line is ~5.90 pH. This is in good agreement with the calculated pHmax value of 5.87.

The pHmax equation for a salt of a weak acid is:

$$\text{pHmax} = \text{pKa} + \log\left(\left[\text{BH}^{+}{}_{solubilty}\right] / \left[\text{B}_{unionized\ solubility}\right]\right). \quad (6.130a)$$

6.15 Weak Acid and Base Salt Disproportionation

Disproportionation is a reaction in which two entities such as free radicals, ions, atoms, and molecules interact with each other to give two or more different entities or species.

$$\text{A} + \text{B} \rightarrow \text{C} + \text{D} \quad (6.131)$$

It is well documented that salts of weak acid and base drugs undergo disproportionation in pharmaceutical liquid and solid dosage forms. On storage, disproportionation leads to the formation of the free acid and base forms which can have detrimental effects on drug solubility, dissolution, and bioavailability. Considering weak base salts, disproportionation entails the transfer of the drug's hydrogen ion to a hydrogen ion acceptor in the formulation matrix.

John et al. (2013) showed that solid matrix disproportionation is mediated by adsorbed water at the drug salt-excipient interface. Magnesium stearate was shown to be a good hydrogen ion acceptor that promoted disproportionation of a drug-HCl salt. The proposed disproportionation reaction scheme is shown:

$$2\,(\text{Drug} - \text{H}^{+}\text{Cl}^{-})\ [\text{drug} - \text{HCL salt}] + \text{R} - \text{COO}^{-}\text{Mg}^{+2-}\text{OOC} - \text{R}\ [\text{magnesium stearate}] \rightarrow$$
$$2\,(\text{R} - \text{COOH})\ [\text{stearic acid}] + \text{MgCl}_2 + 2\,(\text{Drug})\ [\text{unionized free drug base}]$$

Magnesium chloride deliquesces at 30% relative humidity which provides local water to mediate the reaction. The authors concluded that magnesium stearate is a "high-risk" excipient for weak base-HCl drugs and is capable of promoting salt disproportionation.

6.16 Application: Effect of Colloidal Sized Particles on Solubility

Colloidal particles have a size range from 1 nm to less than 1000 nm. The relationship between colloidal particle size and solubility was developed by Thomson (Lord Kelvin) (1871), Ostwald (1900) and Freundlich (1909). The *Kelvin and Ostwald-Freundlich* Eq. is given:

$$\text{Ln}\, S/S_{\infty} = 2\gamma M / r\rho RT \quad (6.132)$$

where S is the solubility of colloidal particles having a radius r and S_{∞} is the solubility of large, flat particles having an infinite radius, γ is the interfacial tension between the solid and solvent interface, M is the molecular weight of the solute, ρ is the solute density, the ratio M/ρ is the solute's molar volume, R is the gas constant, and T is in kelvin.

Solubility enhancement occurs because exceedingly small particles have greater specific surface area. This greater specific surface area results in greater surface free energy that leads to greater molecular escaping tendency at the surface of the particle.

The solubility enhancement, S/S_{∞}, of 10-nm, 100-nm, and 1000-nm particles can be calculated if the parameters in Eq. (6.132) are estimated (Bowman et al., 2006). Assuming $M = 500$ g mol^{-1}, $\rho = 1$ g mL^{-1}, $\gamma = 30$ erg cm^{-2}, $R = 8.314\text{E}7$ erg K^{-1} · mol, $T = 298$ K: the 10-nm, 100-nm, and 1000-nm particles were calculated to have 3.36- 1.13-, and 1.012-fold solubility increase.

References

Anonymous. (2022). *LibreTexts Chemistry.* Chapter 13.1: Factors affecting solution formation. https://chem.libretexts.org/Bookshelves/General_Chemistry/Book%3A_General_Chemistry%3A_Principles_Patterns_and_Applications_(Averill)/13%3A_Solutions/13.01%3A_Factors_Affecting_Solution_Formation (accessed April 2023).

Bowman, B.J., Ofner, C.M.I.I.I., and Shott, H. (2006). Colloidal dispersion. In: *Remington: The Science and Practice of Pharmacy*, 21e (ed. D.B. Troy), 293–318. Philadelphia: Lippincott Williams & Wilkins.

Chang, R. (1998). *Chemistry*, 6e. Boston: McGraw Hill.

Chen, L.-K., Cadwalladeer, D.E., and Jun, H.W. (1979). Nitrofurantoin solubility in aqueous urea and creatinine solution. *J. Pharm. Sci.* 89: 1389.

Connors, K.A. and Mecozzi, S. (2010). *Thermodynamics of Pharmaceutical Systems: An Introduction to Theory and Applications*, 2e. Hoboken, NJ: John Wiley & Sons.

Daniels, F. and Alberty, R.A. (1975). *Physical Chemistry*, 4e. New York: John Wiley & Sons.

Freundlich, H. (1909). *Kapillarchemie: Eine Darstellung der Chemie der Kolloide und verwandter Gebiete [Capillary Chemistry: A Presentation of Colloid Chemistry and Related Fields].* Leipzig, Germany: Akademische Verlagsgesellschaft.

Hancock, B.C. and Parks, M. (2000). What is the true solubility advantage for amorphous pharmaceuticals? *Pharm. Res.* 17 (4): 397–404.

John, C.T., Xu, W., Lupton, L.K., and Harmon, P.A. (2013). Formulating weakly basic HCl salts: Relative ability of common excipients to induce disproportionation and the unique deleterious effects of magnesium stearate. *Pharm. Res.*.

Liu, X.Y. (2000). Heterogeneous nucleation or homogenous nucleation? *J. Chem. Phys.* 112 (22): 9949–9955. https://doi.org/10.1063/1.481644.

Lowry, S. (2023). *LibreTexts Chemistry.* Chapter 8.8: Ideal vs. Real solutions. https://chem.libretexts.org/Bookshelves/General_Chemistry/Chem1_(Lower)/08%3A_Solutions/8.08%3A_Ideal_vs._Real_Solutions (accessed April 2023).

Martin, A., Swarbrick, J., and Cammarata, A. (1983). *Physical Pharmacy: Physical Chemical Principles in the Pharmaceutical Sciences*, 3e. Philadelphia: Lea & Febiger.

Ostwald, W. (1900). Ostwald's equation relating temperature, solubility, surface tension, and the radius of curvature of a phase boundary appears. *Z. Phys. Chem.* 34: 495–503.

Rathod, V., Stagner, W.C., Gajera, B., and Haware, R.V. (2020). Hybridized nanoamorphous micellar dispersion using a QbD-DM^3 Linked rational product design strategy for ritonavir: a BCS IV drug. *Int. J. Pharm.* 588: 119727. https://doi.org/10.1016/j.ijpharm.2020.119727.

Sinko, P. (2006). *Martin's Physical Pharmacy and Pharmaceutical Sciences*, 5e. Philadelphia: Lippincott Williams & Wilkins.

Thomson William (Lord Kelvin) (1871). On the equilibrium of vapour at a curved surface of liquid. *Philos. Mag.*, series. 1871(4) 42 (282): 448–452.

Winnike, R. (2023). Personal Communication. Winnike Consulting. Winnike.consultancy@gmail.com.

Chapter 6 Problem Set

6.1 Give two different but equivalent definitions of *solubility.*

6.2 Define unsaturated and supersaturated solutions.

6.3 What is the mole fraction, percent w/w, molarity, molality and percent w/v for lidocaine when 2.000 g lidocaine is dissolved in 100.0 g ethanol?
Givens:

Lidocaine (L) MW 234.3 g mol^{-1}
Ethanol (E) MW 46.07 g mol^{-1}; density 0.7892 g mL^{-1}

Using dimensional analysis, see Section 1.2 Physical Quantity Dimensions and Dimensional Analysis.

6.4 Name the six main intermolecular interaction forces and briefly describe each one and their relative strength.

6.5a $\Delta H_{solution} = \Delta H_1 + \Delta H_2 + \Delta H_3$ is an example of what law? Define that law.

6.5b For the enthalpy of solution model, what are $\Delta H_1 + \Delta H_2 + \Delta H_3$?

6.6 What is the heat of sublimation of naphthalene if the enthalpy of fusion is 19.01 kJ mol^{-1} and the enthalpy of vaporization is 47.44 kJ mol^{-1}?

6.7 When does $\Delta H_{solution} = \Delta H_{fusion}$?

6.8 Does Figure 6.1a represent an endothermic or exothermic heat of solution process? Explain the steps.

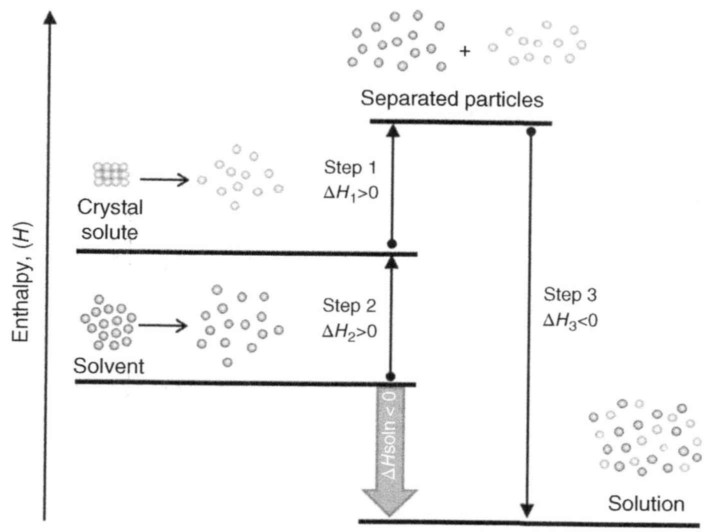

(a) Exothermic solution formation

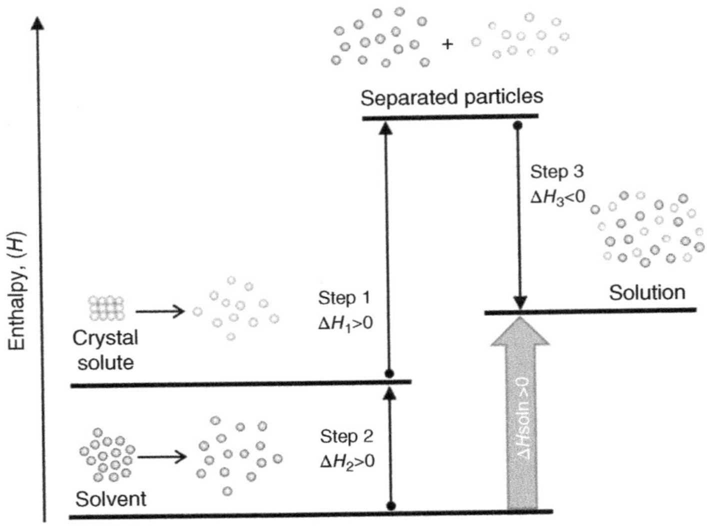

(b) Endothermic solution formation

6.9 Does Figure 6.1b represent an endothermic or exothermic heat of solution process? Explain the steps.

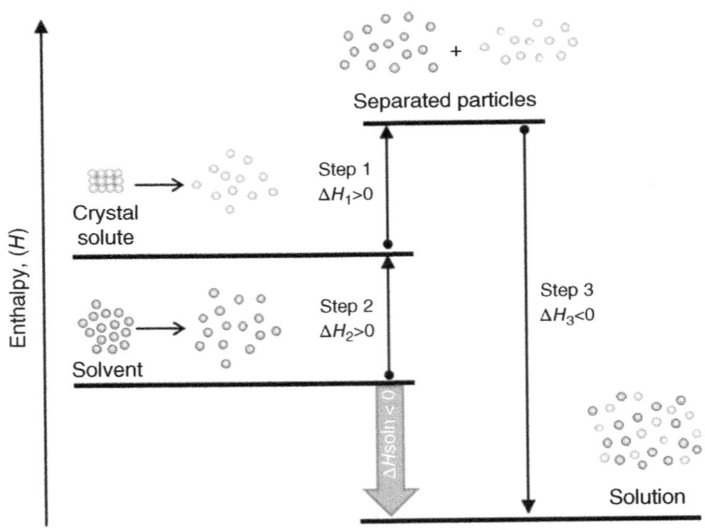

(a) Exothermic solution formation

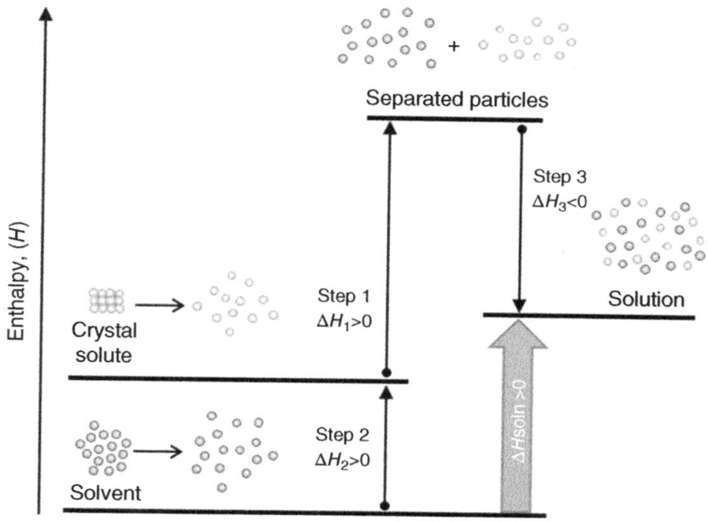

(b) Endothermic solution formation

6.10 Determination of $\Delta G^{\circ}_{solution}$ for $MgCl_2$ at 25 °C
Given:

$\Delta H^{\circ}_{solution} = -38.24$ kcal mol^{-1} (solvation has the largest impact on $\Delta H^{\circ}_{solution}$)
$\Delta S^{\circ}_{solution} = -0.0274$ kcal K^{-1} · mol (solvation and decreased order has the largest impact on $\Delta S^{\circ}_{solution}$)

6.11 Calculate naphthalene's solubility using the idea solubility equation (ΔH_{fusion} = 19.01 kJ mol^{-1})

6.12 Using the van't Hoff plot from the solubility data, calculate nitrofurantoin enthalpy of solution. Also calculate the entropy at 24 °C.
Given:

$$\text{Van't Hoff equation for solution: } \ln X_2 = -\Delta H^{\circ}_{solution}/RT + \Delta S^{\circ}_{solution}/R \quad (6.38)$$

Nitrofurantoin solubility data[a]

Temperature (°C)	Temperature (K)	Solubility, mole fraction (X_2)
24	297.15	6.01E-6
30	303.15	8.57E-6
37	310.15	1.32E-5
45	318.15	1.90E-5

Adapted from Chen et al. (1979) and Connors and Mecozzi (2010).

6.13 Provide the reference state and standard state conditions.

7

Surface Thermodynamics

7.1 Introduction to Interfacial Surface Region, Interfacial Tension, and Interfacial Energy

7.1.1 What Is a Surface or Interfacial Surface?

To talk about interfacial tension and interfacial free energy, we need to first discuss what is meant by a *surface* or *interfacial surface*. The reader is referred to Adamson and Gast (1997) for a thermodynamic development of the meaning of *interfacial surface region*. A "real" interfacial surface at equilibrium does not have a distinct boundary. The phases mix at the real interface and exhibit a gradual change in a species' density, concentration, and other extensive properties as illustrated in Figure 7.1. The interface occurs over a distance of 100 Angstroms or less. Following the development of Gibbs (1931), Adamson and Gast (1997) describe a convention that uses a convenient arbitrary surface or mathematical line to divide the interfacial surface between the two phases. The extensive property values for a "real" two phase system are determined by summing of the values associated with the bulk phases, α and β, and excess or deficiency values are assigned to the interfacial surface region.

According to this development, the following relationships hold (Adamson and Gast, 1997):

$$\text{Volume: } V^{\alpha} + V^{\beta} \tag{7.1}$$

$$\text{Internal energy: } U = U^{\alpha} + U^{\beta} + U^{\sigma} \tag{7.2}$$

$$\text{Entropy: } S = S^{\alpha} + S^{\beta} + S^{\sigma} \tag{7.3}$$

$$\text{Moles: } n_i = n_i{}^{\alpha} + n_i{}^{\beta} + n_i{}^{\sigma} \tag{7.4}$$

Here, α and β represent the two phases and σ indicates the excess or deficient quantity allocated to the interfacial surface region. It is interesting to note that the interfacial surface region is not assigned a volume implying the surface region is not three-dimensional. The interfacial surface region can be thought of as film-like phase region, as it is essentially two dimensional. This film-like region behaves differently than the bulk phases because the surface is at a higher energy state than either of the bulk phases.

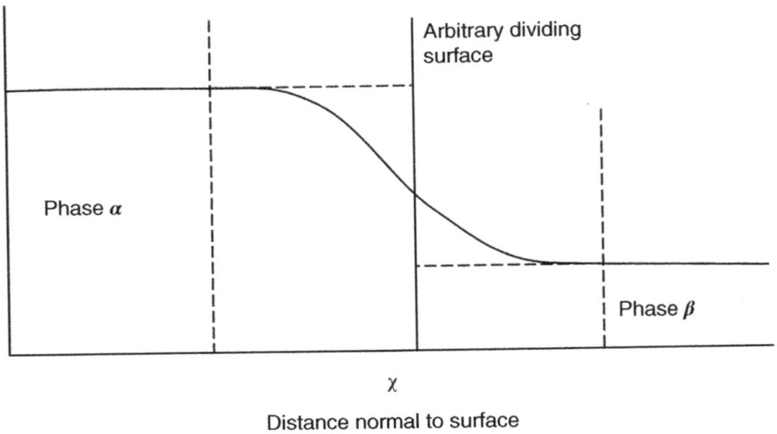

Figure 7.1 Schematic illustrating the surface or interfacial surface region. *Source:* Adapted from Adamson and Gast (1997).

7.1.2 Interfacial Tension

Interfacial tension results from the force imbalance which occurs at the interface. Interfacial tension is represented by the Greek letter gamma, γ or sigma, σ. In the case of a liquid-gas interface, the bulk liquid molecules experience no net forces because each molecule on average has the same number of neighboring molecules. At the interfacial surface region, the liquid molecules experience more intermolecular forces from the liquid than from the gas or vapor resulting in an inward and sideways pulling tension. The physics definition of *tension* is the pulling force transmitted axially away from the object. In this case, the bulk *cohesive* liquid intermolecular attractive "pulling" forces are greater than the *adhesive* intermolecular attractive "pulling" forces between the liquid and the gas, because the liquid is a condensed phase and the gas is a vapor phase. This means that the liquid-gas interfacial region has higher energy than the internal liquid, resulting in excess tension or potential energy. *Interfacial tension* is a measure of the *excess potential energy* at the liquid-gas interface and has units of N m^{-1}, or force per unit length. Interestingly, *interfacial tension* can also be defined in terms of either energy or work. In this case, interfacial tension is the energy or work required to increase the surface area of the interface. At constant temperature and pressure, this is also the definition of *interfacial Gibbs free energy*.

This raises some important questions. For example, how can the same physical phenomenon, interfacial tension, be defined in terms of force per unit length and energy per unit area? How can interfacial tension and interfacial Gibbs free energy have the same numerical value? The simple answer is interfacial tension and interfacial energy have the same fundamental units, kg s^{-2} (see Section 1.1.1). This will be shown in the next section (7.1.3 Interfacial Energy).

Surface tension is a subcategory of *interfacial tension*. When a gas or vapor is the interfacial phase in contact with a liquid or solid, the interfacial tension is designated as *surface tension*. By convention the liquid-vapor and solid-vapor interface is called a *surface*. This subcategorization is primarily based on the relative weak liquid-vapor and solid-vapor adhesive forces compared to relatively larger cohesive forces operating at the liquid-liquid, solid-liquid, and solid-solid interfaces. Table 7.1 summarizes the interfacial terminology.

Table 7.1 Example of Interfaces and Corresponding Interfacial Terminology.

Interfaces	Interfacial tension term	Example
Gas-gas	N/A	No interface exists
Liquid-gas (vapor)	Surface tension, γ_{lv}	Water-air interface
Solid-gas (vapor)	Surface tension, γ_{sv}	Hygroscopic drug, charcoal adsorption of an odor
Liquid-liquid	Interfacial tension, γ_{ll}	Two immiscible liquids or an oil-in-water emulsion
Solid-liquid	Interfacial tension, γ_{sl}	Suspension or a droplet of liquid on a solid surface
Solid-solid	Interfacial tension, γ_{ss}	Drug deposited on a lactose carrier for inhalation products

Please note, in this chapter, "interfacial tension" will be used occasionally to generically allude to both surface tension (γ_{lv} and γ_{sv}) and appropriate liquid-liquid, solid-liquid, and solid-solid interfacial tensions (γ_{ll}, γ_{sl}, and γ_{ss}), as interfacial tension is the more general term.

7.1.3 Interfacial Energy

From a thermodynamics perspective, it is easier to discuss interfacial tension in terms of interfacial energy. Interfacial energy is defined using the thermodynamic terms of energy and work. For a liquid, interfacial energy is the energy or work necessary to overcome the intermolecular attractive forces to increase the surface area of the liquid. Interfacial energy has units of J m^{-2}.

As mentioned earlier, it is worth noting that interfacial tension and interfacial energy have the same fundamental units and are in essence equivalent entities. This can be shown using dimensional algebra (see Section 1.2.5).

$$\text{Interfacial tension units: N m}^{-1} = (\cancel{\text{m}} \cdot \text{kg s}^{-2})/\cancel{\text{m}} = \underline{\text{kg s}^{-2}} \tag{7.5}$$

$$\text{Interfacial energy units: J m}^{-2} = \text{N} \cdot \text{m m}^{-2} = \{[(\cancel{\text{m}} \cdot \text{kg})\,\text{s}^{-2}] \cdot \cancel{\text{m}}\}\,\text{m}^{-2} = \underline{\text{kg s}^{-2}} \tag{7.6}$$

Comparison of Eqs. (7.5) and (7.6) show that interfacial tension and interfacial energy have the same fundamental units, kg s^{-2}. Therefore, the numerical values for interfacial tension and interfacial energy are equivalent. Water's surface tension is 71.8 mN m^{-1} (or 71.8 dynes cm^{-1}) in SI and cgs units, respectively. Its surface Gibbs free energy has the same numerical value, expressed as 71.8 mJ m^{-2} (or 71.8 ergs cm^{-2}). Since the interface in this case is water-air or water-vapor, the terms surface tension and surface free energy are employed.

The relationship between surface tension, γ, and *surface free energy*, $G^{surface\ free\ energy}$, is shown in Table 2.1 entitled "Relationship between Energy Form and Product of Intensive and Extensive Variables." Surface tension is an intensive variable like temperature and pressure and is independent of the size of the system. Surface free energy, on the other hand, is dependent on the magnitude of the system. Surface free energy is the product of surface tension and the change in surface area, $\gamma\ (\Delta A)$.

One might now ask when do I use interfacial tension and interfacial energy? Typically, surface tension or interfacial tension is used when the interfacial tension is measured as a force per unit length. Surface free energy and interfacial free energy, Gibbs free energy at constant temperature and pressure, are used to describe the energetics of the system. Here, interfacial energy is measured as the energy or work required to change a unit area of the interface.

7.2 Surface Tension of Liquid-Vapor Interfaces

As mentioned earlier, surface tension units are N m^{-1}, or force per unit length. The surface tensions of liquid-vapor systems can be measured by a number of techniques including capillary rise/fall, Du Noüy ring, Wilhelmy plate, and bubble pressure methods. The discussion of these methods follows.

7.2.1 Capillary Method

If a glass capillary is placed in a bulk liquid, the liquid in the tube may rise above or fall below the bulk liquid level. This spontaneous movement is known as *capillarity* and depends on the liquid's relative cohesion and adhesion forces. Capillarity is the result of the liquid's surface tension and the liquid-glass interfacial contact angle. We will discuss contact angle in more detail later in this chapter. Suffice it to say at this juncture, that a liquid's contact angle is related to γ_{lv}, γ_{sv}, and γ_{sl}. The capillary method is generally considered an equilibrium method (Henriksson and Eriksson, 2004; Tiberg et al. 2000). This relatively fast, simple, and inexpensive technique can be used for liquid-vapor interface systems. It can also be used for more complex systems like surfactant solutions. For example, Tilberg et al. showed that it can take 3–8 minutes for 0.1–1 mM aqueous surfactant solutions to reach capillary equilibrium. For systems such as these, an equilibrium in the position of the liquid is not reached until an equilibrium amount of surfactant has adsorbed to the glass. Thus, pretreating a hydrophilic capillary with the test surfactant solution significantly decreased the time required to reach an equilibrium surface tension value. A major limitation of this method is that it requires reasonably large amounts of test liquid.

Figure 7.2a–c shows different cases of capillarity depending of the liquid's surface tension and interaction with the glass surface. Figure 7.2a shows *capillary rise* for a liquid having a zero-contact angle with the glass surface. This is illustrated by the force arrows, *F*, drawn parallel to the capillary. Water rises in a capillary because it has strong adhesive interactions with a clean glass surface, and it readily spreads evenly over it. Stated another way, the surface free energy of the glass is much greater than the surface free energy of the liquid. This difference in surface free energies leads to complete spreading of the water on the glass surface, which is characterized by a zero-*contact angle.* We will also discuss spreading in more detail later in this chapter. The relationship that explains the upward force that leads to the column of water rising above the level of the bulk liquid is given by Eq. (7.7):

$$\text{Upward force} = (2\,\pi r)\cdot\gamma\cdot\cos\theta \tag{7.7}$$

where $2\pi r$ is the circumference of the capillary internal wall, r is the internal capillary radius, γ is the surface tension of the liquid, and θ is the water-glass contact angle. In Figure 7.2a, the contact angle, θ, is zero and cos θ is 1. In this case, the upward force is $(2\pi r)\,\gamma$.

The upward force must support the weight of the water column in the capillary. The weight of the water is equal to the water mass times the acceleration of gravity. The acceleration of gravity constant is 9.807 m s^{-2} at sea level gives the downward force caused by the column of water.

$$\text{Downward force} = \left(\pi r^2 h\right)\cdot g\cdot\rho \tag{7.8}$$

where $\pi r^2 h$ is the volume of the water column, r is the internal capillary radius, h is the height of the water column, g is the acceleration of gravity, and ρ is the liquid density. Please note that g is the

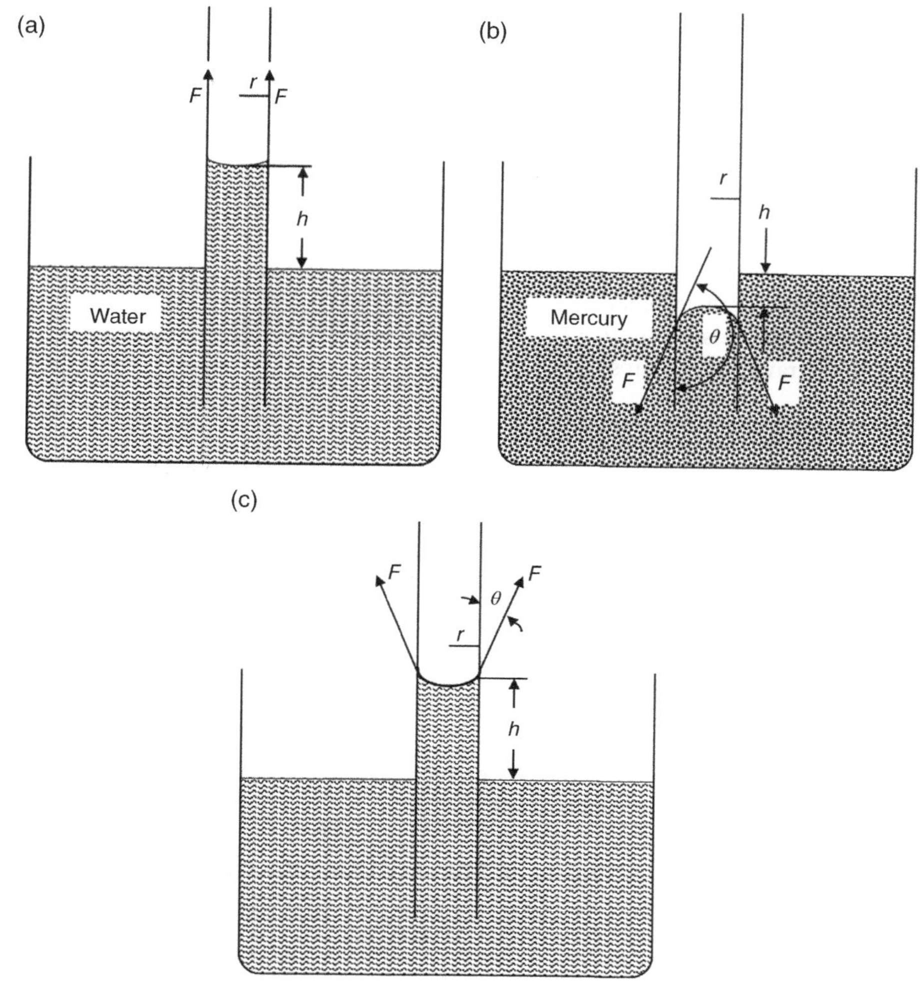

Figure 7.2 (a) Capillary surface tension method: capillary rise where liquid contact angle is zero. (b) Capillary surface tension method: capillary fall where the liquid contact angle is greater than 90°. (c) Capillary surface tension method: capillary rise where liquid contact angle is greater than 0° and less than 90°. *Source:* Adapted from Bummer (2006) who recognized the work of H. Semat (1957).

acceleration of gravity, not the gravitational constant, G, 6.674E-11 $\mathrm{N \cdot m \cdot kg^{-2}}$. Dimensional analysis shows that both Eqs. (7.7) and (7.8) are dimensionally consistent. Dimensional analysis for the upward force is:

$$(2\,\pi r) \cdot \gamma \cdot \cos\theta = \mathrm{m \cdot kg \cdot s^{-2}} = \mathrm{N\ (newtons)} \tag{7.9}$$

where γ units are kg $\mathrm{s^{-2}}$. Dimensional analysis for the downward force is:

$$(\pi r^2 h) \cdot g \cdot d = \cancel{\mathrm{m}}^{\cancel{3}} \cdot (\mathrm{m \cdot s^{-2}}) \cdot (\mathrm{kg} \cdot \cancel{\mathrm{m}}^{\cancel{-3}}) = \mathrm{m \cdot kg \cdot s^{-2}} = \mathrm{N} \tag{7.10}$$

Since Eqs. (7.7) and (7.8) describe two opposite and equal forces. We can use these equations to determine the surface tension of the liquid by setting them equal to each other and solving for γ.

$$(2\pi r)\cdot\gamma\cdot\cos\theta=(\pi r^2h)\cdot g\cdot\rho \tag{7.11}$$

$$\gamma=[(\pi r^2h)\cdot g\cdot\rho]/[(2\pi r)\cdot\cos\theta] \tag{7.12}$$

Example 7.1 Determine a Liquid's Surface Tension Using the Capillary Rise or Fall Method

Given:

Liquid density is 1.00 g cm^{-3}
Capillary radius is 0.0150 cm
Liquid rises 9.76 cm in the capillary
Liquid contact angle, θ, is 0°; $\cos\theta=1$
Acceleration of gravity is 981 cm s^{-2}

Calculate the surface tension of the liquid.
From Eq. (7.12)

$$\gamma=[(\pi r^2h)\cdot g\cdot\rho]/[(2\,\pi r)\cdot\cos\theta]$$
$$\gamma=[\cancel{3.14}\times(0.0150\,\cancel{cm})^2\cdot 9.76\,\cancel{cm})\cdot 981\,cm\,s^{-2}\cdot 1.00\,g\,\cancel{cm}^{-3}]/[2\cdot\cancel{3.14}\cdot 0.0150\,cm\cdot 1]$$

$\underline{\gamma=71.8\ (g\cdot cm\ s^{-2})\ cm^{-1}}$ or $\underline{71.8\text{ dynes } cm^{-1}}$. The numerical value which has three significant figures is consistent with the number of significant figures provided in the given section.

Note: (g·cm s^{-2}) = dyne. Therefore, the "derived units" (see Section 1.1.1) are consistent with the surface tension units given earlier as dynes cm^{-1}.

Figure 7.2b depicts the *capillary fall* for mercury which has a glass contact angle of 135°. Liquid contact angles greater than 90° exhibit capillary fall.

Example 7.2 Calculate the Surface Tension of Mercury Using the Capillary Rise or Fall Method

Given:

Mercury density is 13.6 g cm^{-3}
Capillary radius is 0.0150 cm
Mercury falls −3.29 cm in the capillary
Mercury contact angle, θ, is 135°; cos 135 = −0.707
Acceleration of Gravity is 981 cm s^{-2}

$$\gamma=[(\cancel{\pi} r^{\cancel{2}}h)\cdot g\cdot\rho]/[(2\cancel{\pi r})\cdot\cos\theta]$$
$$\gamma=\left(0.0150\,\cancel{cm}^2\cdot -3.29\,\cancel{cm}\cdot 981\,cm\,s^{-2}\cdot 13.6\,g\,cm^{-\cancel{3}}\right)/(2\cdot -0.707)$$
$$\underline{\gamma_{mercury}=466\,(g\cdot cm\ s^{-2})/cm=466\,\text{dyne}\,cm^{-1}}$$

Figure 7.2c illustrates a capillary rise for a liquid whose contact angle θ is greater than 0° and less than 90°. In this case, the upward force, indicated by the force arrows, is not parallel with the capillary wall as is the situation for complete wetting and a contact angle θ of zero. The cos θ in this range is positive which results in the rise of the liquid in the capillary.

7.2.2 Du Noüy Ring Method

The Du Noüy ring tensiometer method is used to measure a liquid's surface tension and the interfacial tension between two liquids. It is a dynamic technique and does not measure an equilibrium interfacial tension. It is a relatively simple and fast way to measure surface tension. To achieve more accurate surface tension values, it is recommended to use the equilibrium methods discussed in this section, as the necessary corrections for the Du Noüy ring method are significant. The main advantage of the Du Noüy ring method is its ability to measure liquid-liquid interfacial tensions. In that case, the material of the ring must be preferentially wetted by the lower liquid. Figure 7.3 depicts the experimental setup.

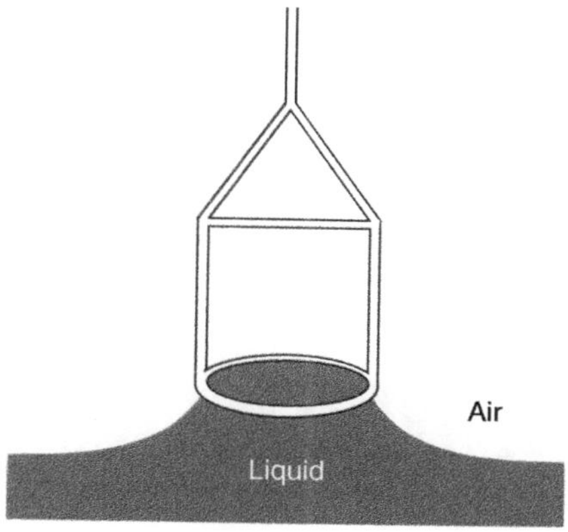

Figure 7.3 Du Noüy ring surface and interfacial tension method.

Typically, an ultraclean platinum ring is submerged below the interface. After immersion, the ring is pulled up through the interface. The maximum upward force is given by Eq. (7.13), and the interfacial tension can by calculated by Eq. (7.14).

$$F = w_{ring} + (4\pi R)\gamma \tag{7.13}$$

$$\gamma = \left(F - w_{ring}\right)/(4\pi R) \tag{7.14}$$

where γ is the interfacial tension, F is the maximum force required to pull the ring through the interface, w_{ring} is the weight of the ring minus the liquid buoyant force of the lower phase, R is the average of the inner and outer radius of the ring, and $4\pi R$ is the average perimeter of the ring. The constant "4" comes about because one is actually pulling two films. One film is being pulled by the inner surface of the ring ($2\pi(R-r)$), and one film is being pulled by the outer surface of the ring ($2\pi(R+r)$); r is the radius of the wire. The buoyant force is a function of the liquid density that the ring is immersed in, the volume displaced by the ring, and acceleration of gravity.

For more accurate measurements, correction factors need to be applied to account for the finite diameter of the ring. Eq. (7.14) becomes (7.15) when a correction factor is applied.

$$\gamma = \left(F - w_{ring}\right)/(4\pi R)f \tag{7.15}$$

where f is the correction factor. The reader is directed to discussions by Huh and Mason (1975, 1977) and Harkins and Huber (1930a, 1930b) regarding additional correction factors.

7.2.3 Wilhelmy Plate Method

The Wilhelmy plate method is a flexible, easy, accurate, and precise method for undertaking surface and interfacial tension measurements. In addition, the method requires no corrections. The experimental setup is shown in Figure 7.4.

The plate method can be performed in a static equilibrium or dynamic mode. The major disadvantage of this method is the relatively large quantity of liquid required. The plate is typically roughened platinum. Platinum has high surface free energy and generally leads to complete wetting by the liquid and a contact angle, θ, of 0°. In the static mode, the plate is lowered to "bridge" the two phases (liquid-vapor or liquid-liquid phases) and is not moved during testing. The bottom of

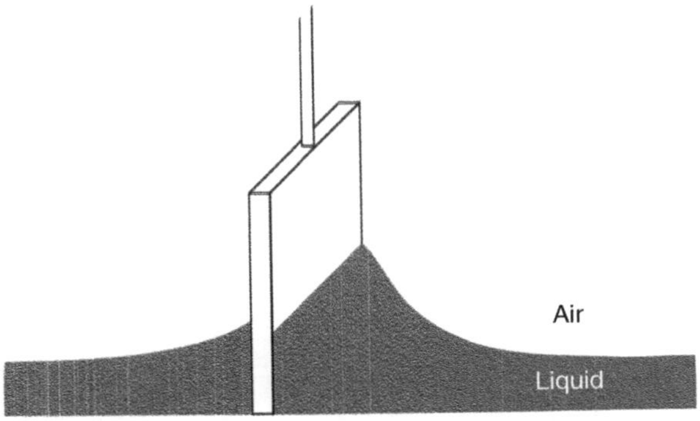

Figure 7.4 Wilhelmy plate surface and interfacial tension method.

the plate needs to be "caught" in the interface. A measured external force is applied to the plate to counterbalance the surface tension.

This static method also allows direct surface kinetic measurements over a wide range of timescales. Surface kinetic measurements are very useful for studying surfactant solutions and undertaking Langmuir trough studies where surface coverage, solute diffusion, and solute adsorption can be observed as a function of time.

In the dynamic mode, the plate is slowly lowered into the interface and then slowly withdrawn. This allows determination of the advancing contact angle during wetting and the receding contact angle during dewetting.

The measured force is a function of the system's interfacial tension and is given in Eq. (7.16).

$$F = (L \cos \theta)\, \gamma \tag{7.16}$$

where F is the tension or force acting on the plate, L is the *wetted length* (2·length + 2·width) of the three-phase perimeter boundary circumscribing the plate-liquid-air phases, θ is the contact angle between the plate and the liquid, and γ is the surface/interfacial tension. Equation (7.16) can be solved for the interfacial tension giving Eq. (7.17).

$$\gamma = F/(L \cos \theta) \tag{7.17}$$

7.2.4 Maximum Bubble Pressure Method

The maximum bubble pressure method is a dynamic method used for surface tension measurements. This method can only be used for determining a liquid-vapor surface tension, without using specialized equipment designed for liquid-liquid interfacial methods. Modern instruments can report quasistatic surface tensions or report dynamic surface tensions over time. The method it is simple, fast, accurate (± 0.25 nN m^{-1}), and precise (± 0.1 mN m^{-1}). It requires a minimum sample volume of 1 mL. One major advantage is the calculations do not depend on knowing the system's contact angle. However, it is required to know if the bubble forms from the inside diameter of the tube (the liquid wets the tube) or the outside diameter of the tube (the liquid does not wet the tube). The surface tension measurement is based on the Young-Laplace pressure equation, Eq. (7.18). This equation shows that the pressure inside a spherical bubble is a function of the bubble's surface tension and its curvature radius.

$$P_{bubble\ pressure} = (2\gamma)/r \tag{7.18}$$

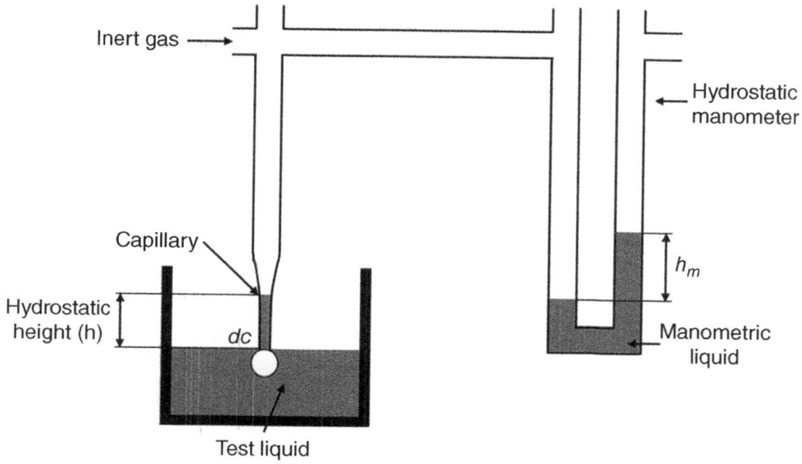

Figure 7.5 Schematic of a bubble pressure tensiometer. *dc* is the internal diameter of the capillary and h_m is manometric liquid height. *Source:* Adapted from Gasior and Debski (2019).

$P_{bubble\ pressure}$ is the pressure inside a spherical bubble, γ is the bubble film surface tension, and r is the radius of the bubble. Figure 7.5 shows a schematic of a bubble pressure tensiometer.

Figure 7.5 shows that the gas pressure required to form a bubble has two components. First, sufficient gas pressure is required to overcome the hydrostatic pressure resulting from the capillary rise of the test liquid. Second, additional gas pressure is required to form a hemispherical bubble shape. This is known as the *maximum pressure*. In this case, Eq. (7.18) becomes:

$$P_{bubble\ pressure} = \left(P_{max} - P_{hydrostatic\ pressure}\right) = (2\,\gamma)/r_{capillary} \tag{7.19}$$

where $P_{bubble\ pressure}$ is the pressure inside the bubble, P_{max} is the maximum pressure applied by the device to form a hemispherical bubble whose diameter is identical to the internal capillary diameter, $P_{hydrostatic\ pressure}$ is the upward force per unit area resulting from the capillary rise of the test liquid, γ is the liquid surface tension, and $r_{capillary}$ is the internal capillary radius.

From Figure 7.5, P_{max} and $P_{hydrostatic\ pressure}$ can be written in terms for measurable quantities.

$$P_{max} = g\left(\rho_{manometer\ liquid} \cdot h_{manometer\ liquid}\right) \tag{7.20}$$

where g is the acceleration of gravity, $\rho_{manometer\ liquid}$ is the density of the manometer liquid, and $h_{manometer\ liquid}$ is the manometer liquid height. The manometer liquid height is taken when the bubble emerging from the capillary reaches a hemispherical shape. When this occurs, the diameter of the hemisphere is equal to the internal or external diameter) of the capillary.

$$P_{hydrostatic\ pressure} = g\left(\rho_{test\ liquid} \cdot h_{hydrostatic\ height}\right) \tag{7.21}$$

where $\rho_{test\ liquid}$ and $h_{hydrostatic\ height}$ are test liquid density and the test liquid capillary rise height, respectively.

Solving Eq. (7.19) for γ results in Eq. (7.23).

$$\gamma = \left[\left(P_{max} - P_{hydrostatic\ pressure}\right) \cdot r_{capillary}\right]/2 \tag{7.23}$$

Table 7.2 Surface Tension of Pharmaceutical Liquids.

Liquid	Surface tension (mN m^{-1})
Water (25 °C)	71.8
PVP 1–5% w/v	65[a]
Glycerin (25 °C)	64.8
2% w/v A15LV, A4C, A15C, and A4M methylcelluloses, Dow Chemical (25 °C)	52[b]
PEG 200, 300, 400 (20 °C)	44.6
Dimethyl sulfoxide	42.9
Propylene carbonate	40.9
Castor oil	39.0
Propylene glycol (25 °C)	38.6
Triacetin (25 °C)	36
Oleic acid (20 °C)	32.8
Tricaprylin (20 °C)	30.0
Vegetable oil	30
Mineral oil	28
Ethanol	22
1-propanol	20

[a] Surface tension was independent concentration. This may be because a 1% solution saturated the aqueous surface.
[b] Surface tension was independent of viscosity (Nassatto et al., 2014).

The calculated values for P_{max} and $P_{hydrostatic\ pressure}$ and the known $r_{capillary}$ value are substituted into Eq. (7.23) to determine the test liquid's surface tension. Commercial bubble point tensiometers automatically report the surface tension.

Additional corrections for gravity flattening of the bubble and viscosity effects can improve the accuracy of the reported surface tension. Further discussion of these corrections is available through SINTERFACE Technologies (2003).

7.2.5 Surface Tension of Selected Pharmaceutical Liquids

The surface tension of several pharmaceutical liquids is listed in Table 7.2. The values are provided for general comparison purposes because the surface tensions were determined at different temperatures. The effect of temperature on surface tension will be discussed in Section 7.3: Effect of Temperature on Surface Tension.

7.3 Effect of Temperature on Surface Tension

Surface tension decreases with an increase in temperature because the increase in thermal energy increases molecular motion and decreases cohesion. There are a number of empirical equations that describe this phenomenon (Eötvös, 1886; Guggenheim, 1945). The surface tension of a liquid

Table 7.3 Surface Tension of Selected Liquids, N m^{-1} as a Function of Temperature[a].

Temperature (°C)	H_2O	CCl_4	C_6H_6	$C_6H_5NO_2$	C_2H_5OH	CH_3COOH
0	0.07564	0.0290	0.0316	0.0464	0.0240	0.0295
25	0.07197	0.0261	0.0282	0.0432	0.0218	0.0271
50	0.06791	0.0231	0.0250	0.0402	0.0198	0.0246
75	0.0635	0.0202	0.0219	0.0373	—	0.0220

[a] Adapted from Daniels and Alberty (1975).

goes to zero at its critical temperature. The critical temperature for a liquid is the point for which no amount of pressure is capable of condensing the vapor to a liquid. Water's critical temperature is about 374 °C. Table 7.3 illustrates that the surface tension decreases with the increase in temperature.

Water's surface tension decreased about 19% over the temperature range from 0 °C to 75 °C. Acetic acid's surface tension decreased approximately 34%, while chloroform and benzene showed a 44% decrease over the same temperature range. This limited set of data suggests that the compounds demonstrating hydrogen bonding are affected less by increased temperature than those compounds that do not hydrogen bond.

7.4 Surface Tension as Force Per Unit Length and as Energy per Unit Area

Figure 7.6a–c shows a three-sided wire frame. This frame is enclosed by a theoretical friction-free movable bar. The sides of the wire frame have hooks at the bottom to prevent the movable bar from sliding off the frame. This apparatus can be dipped into a soap solution to form a soap film on the frame.

From Section 7.1.2: Interfacial Tension, the liquid-vapor interfacial region has higher free energy than the internal liquid, resulting in excess potential energy. This potential energy is the surface free energy of the liquid-vapor interfacial region. Since surface free energy is the product of surface tension and the change in surface area (see Section 7.1.3: Interfacial Energy), the soap film system seeks to minimize its potential energy by minimizing its surface area.

The apparatus is hung allowing gravitational force, $F_{gravitational\ force}$ to be applied perpendicular to the soap film. When $F_{gravitational\ force}$ is smaller than the soap-film force, $F_{soap\ film}$, the soap film retracts and the bar is drawn upward by $F_{soap\ film}$ to minimize the film's surface area. Figure 7.6a,b shows the film at its equilibrium position. Here, the gravitation force, $F_{gravitational\ force}$, equals $F_{soap\ film}$. In Figure 7.6c, the bar is weighted sufficiently so that $F_{gravitational\ force}$ exceeds $F_{soap\ film}$ and the film breaks.

Equation (7.24) describes the equilibrium film position shown in Figure 7.6a,b.

$$F_{gravitational\ \mathrm{force}} = F_{soap\ film} \tag{7.24}$$

Equation (7.25) can now be used to calculate the surface tension of the soap film, where $F_{soap\ film}$ is directly proportional to the soap-film surface tension.

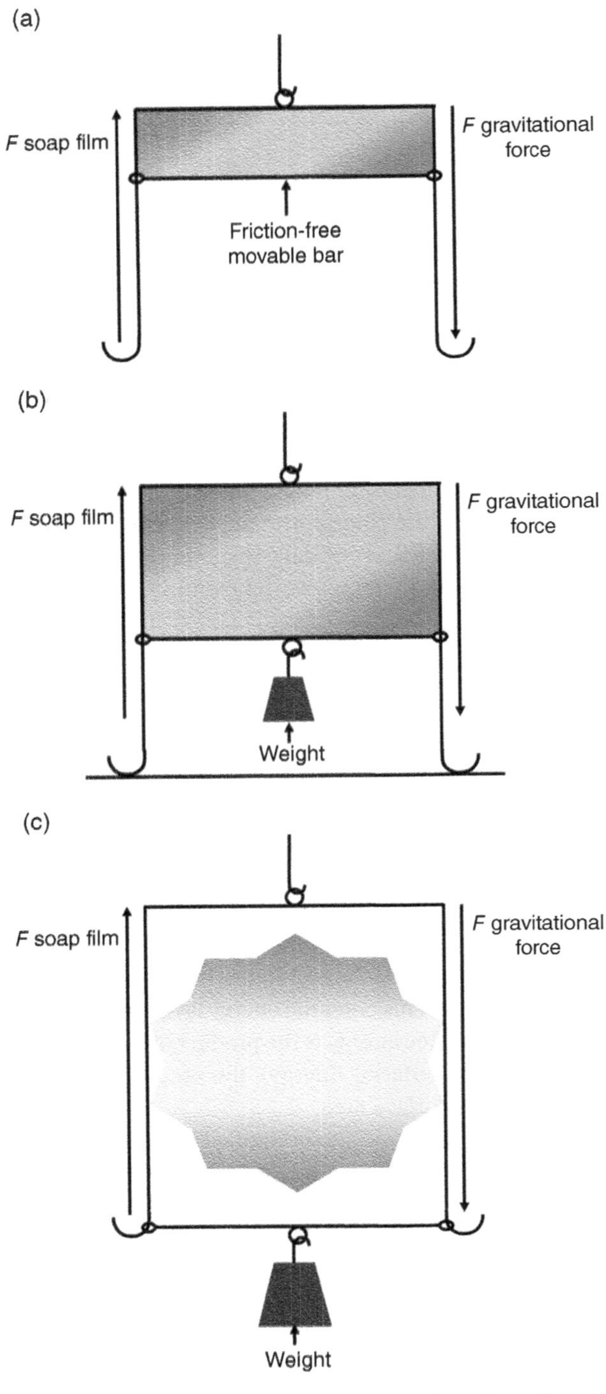

Figure 7.6 (a) Soap-film equilibrium position where the soap-film force equals the gravitational force. The soap film minimizes its surface area to minimize its surface free energy. (b) Soap-film equilibrium position where the soap-film force equals the gravitational force. The soap-film surface area is increased by the added weight. (c) The larger weight and gravitational force exceed the soap-film force and the film breaks.

$$F_{soap\ film} = 2 \cdot \gamma \cdot l \tag{7.25}$$

$F_{soap\ film}$ is the surface-film force, γ is the soap-film surface tension, and l is the movable bar length at equilibrium. The number 2 appears in the equation because the film has two liquid-vapor surfaces, one on the "back" and "front" of the soap film.

Example 7.3 Determination of a Soap's Surface Tension and Surface Energy Using a Wire Frame Apparatus with a Movable Bar.

Given:

Bar length = 5.0 cm
Weight of bar = 0.080 g
Added weight = 0.40 g

1) First calculate $F_{gravitational\ force}$.

$$F_{gravitational\ force} = (\text{bar weight} + \text{added weight}) \cdot \text{acceleration of gravity} \tag{7.26}$$

$$\underline{F_{gravitational\ force}} = (0.080\text{ g} + 0.40\text{ g}) \cdot 981\text{ cm sec}^{-2}) = \underline{470.88\text{ g} \cdot \text{cm} \cdot \text{sec}^{-2}} \tag{7.27}$$

2) From Eq. (7.24),

$$F_{gravitational\ force} = F_{soap\ film,\ so}$$

$$F_{soap\ film} = 470.88\text{ g} \cdot \text{cm} \cdot \text{sec}^{-2} = 470.88\text{ dynes} \tag{7.28}$$

3) From Eq. (7.25),

$$F_{soap\ film} = 2 \cdot \gamma \cdot l\text{, where}$$

$$\gamma = (470.88\text{ dynes})/(2 \cdot 5.0\text{ cm}) = \underline{47.088\text{ dynes cm}^{-1}}(\text{cgs units}) \tag{7.29}$$

4) Using dimensional algebra (see Section 1.2.5) and the conversion ratio (1E-5 N)/(1 dyne) = 1, the soap surface tension in SI units is:

$$47.088\ \cancel{\text{dynes}} \cdot \cancel{\text{cm}^{-1}}\left(1\text{E-5}\ \cancel{\text{N}} \cdot 1\ \cancel{\text{dyne}^{-1}}\right) \cdot \left(1\text{E2}\ \cancel{\text{cm}} \cdot \text{m}^{-1}\right) \cdot \left(10^3\text{ mN} \cdot \cancel{\text{N}^{-1}}\right) = \underline{47.088\text{ mN m}^{-1}} \tag{7.30}$$

5) Rounding to the correct significant figures, the soap's surface tension is:

$$\underline{47\text{ dynes cm}^{-1}}\text{ and }\underline{47\text{ mN m}^{-1}}\text{; force per unit length} \tag{7.31}$$

The soap film's surface free energy units is $\underline{47\text{ ergs m}^{-2}}$ or $\underline{47\text{ mJ m}^{-2}}$ (7.32). This is the surface free energy required to extend the soap film.

7.5 Surface Tension as a Measure of Intermolecular Cohesive Force and Work of Cohesion

Surface tension is the measure of the net effect between a liquid's intermolecular cohesive forces and liquid-gas adhesive forces. In general, the liquid-vapor adhesion is weak compared to the liquid-liquid cohesive forces, because gas-phase molecules are much farther apart than molecules in a condensed phase. The intermolecular cohesive forces depend on the liquid's chemical structure. Mercury, a metallic liquid, has a high surface tension of 466 mN m^{-1} or 466 mJ m^{-2} and quite strong cohesive forces as a result of mercury's metallic bonds. Water has a high surface tension of 71.8 mN m^{-1}, or 71.8 mJ m^{-2}, and a relatively high cohesiveness due its ability to form four hydrogen bonds. As a comparison, ethanol's surface tension is 22.3 mN m^{-1} or 22.3 mJ m^{-2}. Ethanol is capable of forming only three hydrogen bonds.

The *work of cohesion* is the measure of energy or work required to create a new surface by overcoming the liquid's intermolecular cohesive forces. *Cohesion* is defined as the intermolecular attractive forces between the *same* molecules.

Example 7.4 Determine Water's Work of Cohesion

Let us imagine a cylinder of water having a cross-sectional area of 1 cm^2. Now picture the water cylinder being pulled apart to form two new cylinders, creating a new "bottom" for the top cylinder and a new "top" for the bottom cylinder (see Figure 7.7).

This splitting of the cylinder creates two new surfaces each having a surface tension of γ. Energy or work was required to create these two new surfaces. This work is called the *work of cohesion,* w_c. The work of cohesion general equation is provided by Eq. (7.33).

$$w_c = 2\gamma \tag{7.33}$$

Since we are interested in the work of cohesion, we will use water's surface tension in units of mJ m^{-2}. From Eq. (7.33), water's work of cohesion can be calculated.

$$w_c = 2\ (71.8\ \text{mJ m}^{-2}) = \underline{144\ \text{mJ m}^{-2}}\ \text{(with correct significant figures)} \tag{7.34}$$

Table 2.1 shows that the Gibbs surface free energy, $\Delta G^{surface\ free\ energy}$ is the product of, γ, and the change in surface area, $\Delta A^{surface}$.

$$\Delta G^{surface\ free\ energy} = \gamma \cdot \Delta A^{surface} \tag{7.35}$$

If $\Delta A^{surface}$ is positive, $\Delta G^{surface\ free\ energy}$ is positive and work must be done to expand the surface area. The process is nonspontaneous. If $\Delta A^{surface}$ is negative, $\Delta G^{surface\ free\ energy}$ is negative, and the surface area decreases spontaneously.

The reader is directed to Sinko (2023) and Connors and Mecozzi (2010) for further discussions regarding Sections 7.5, 7.6, 7.7, and 7.8.

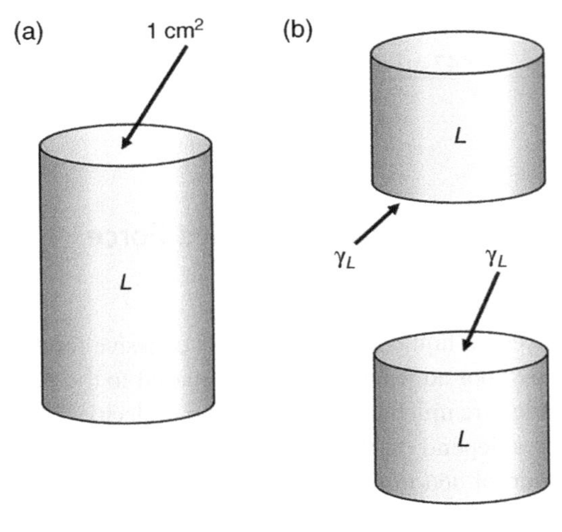

Figure 7.7 Creating a new surface from a cylindrical column of water.

7.6 Interfacial Tension and Work of Adhesion

When two mutually saturated immiscible liquids form an interface, the system possesses an interfacial tension, γ_{12}, where 1 and 2 represent the respective liquids. The magnitude of the interfacial tension depends on the strength of the intermolecular cohesion forces between like molecules and the intermolecular adhesion between unlike molecules. *Interfacial tension* results from the force per unit length existing at the interface between two immiscible phases. *Interfacial tension* can also be defined in energy or work units and is the energy or work required to increase the interfacial surface area. The work required to increase such an interface is called the *work of adhesion,* w_a. Because w_a involves work or energy, the process is also referred to as *thermodynamic adhesion*.

Let us imagine two cylinders of mutually saturated immiscible liquid 1 and liquid 2 which are in intimate contact over a well-defined area. This interface is formed by intermolecular *adhesive* attractive forces between the two liquids. *Work of adhesion* reflects the energy necessary to separate a *unit area* of interface. When the interface is pulled apart, two new surfaces are generated containing pure liquid 1 and pure liquid 2 which then form interfaces with the vapor phase. These new surfaces have surface tensions of γ_1 and γ_2. In this final state, the liquid-liquid interface no longer exists and this needs to be accounted for in the work of adhesion equation. Therefore, the overall energy gain or energy loss is given by the Dupré (1869),

$$w_a = (\gamma_1 + \gamma_2) - \gamma_{12} \tag{7.36}$$

where w_a, is the work of adhesion or energy per unit area (ergs cm^{-2} or mJ m^{-2}), γ_1 and γ_2 are the surface tensions of pure liquids 1 and 2 reported as energy per unit area, and γ_{12} is the interfacial tension between liquids 1 and 2 as energy per unit area.

7.6.1 Interfacial Tension of Pharmaceutical Liquids

Table 7.4 lists the interfacial tensions between water and some selected liquids.

7.7 Liquid-Liquid Cohesive and Adhesive Forces: Spreading of Two Immiscible Liquids

Figure 7.8a and b illustrates the spreading of one liquid on another.

Figure 7.8a shows a small drop of the *"test liquid"* that has been placed on a *"substrate liquid"* which has been deformed into a lens by gravity. In this instance, the test liquid has not spread across

Table 7.4 Interfacial Tension of Selected Liquids and Water

Liquid-liquid interface	Interfacial tension (mN m^{-1})
Water-vegetable oils	23–26
Water-medium chain triglycerides	17
Water-olive oil	17
Water-castor oil	17
Water-oleic acid	12
Water-mineral oil	4

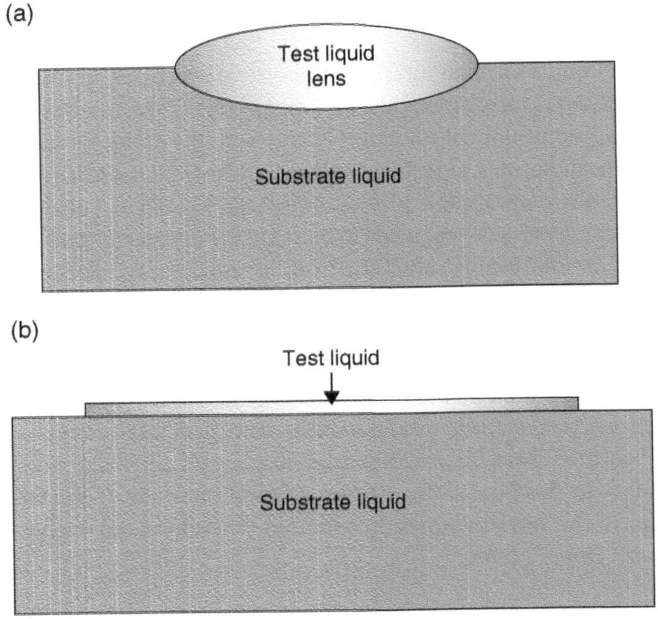

Figure 7.8 (a) The test liquid does not spread across the substrate liquid. The test liquid forms a "lens" on the surface of the substrate. (b) The test liquid spreads across the substrate liquid forming a thin film on the substrate.

that substrate liquid. In Figure 7.8b, a small drop of test liquid has been placed on a substrate liquid and has spread across the substrate to form a thin film. The spreading of a liquid upon another liquid depends on the work of adhesion between the two liquids and the work of cohesion of the test liquid. The equation that describes spreading of the test liquid on a substrate liquid is:

$$S_{test\ liquid/substrate\ liquid} = w_a - w_{c,test\ liquid} \tag{7.37}$$

where $S_{test\ liquid/substrate\ liquid}$ is the *initial spreading coefficient* for the spreading of the test liquid on the substrate, w_a is the work of adhesion between the test liquid and the substrate liquid, and $w_{c,test\ liquid}$ is the work of cohesion of the test liquid. Equation (7.37) indicates that if the spreading coefficient is positive, spreading of the test liquid is spontaneous. This makes sense since the adhesion force between the two liquids is stronger than the cohesion force of the test liquid.

Equations (7.33) and (7.36) can be substituted into Eq. (7.37) to give the spreading coefficient in terms of surface and interfacial tension instead of work of adhesion and work of cohesion. This is useful since surface and interfacial tensions can be measured experimentally. Equation (7.38) shows the substituted relationships:

$$S_{test\ liquid/substrate\ liquid} = \left(\gamma_{test\ liquid} + \gamma_{substrate\ liquid} - \gamma_{test\ liquid/substrate\ liquid}\right) - 2\gamma_{test\ liquid} \tag{7.38}$$

where $S_{test\ liquid/substrate\ liquid}$ is the initial spreading coefficient, $\gamma_{test\ liquid}$ is the surface tension of the test liquid, $\gamma_{substrate\ liquid}$ is the surface tension of the substrate liquid, and $\gamma_{test\ liquid/substrate\ liquid}$ is the interfacial tension between the test and substrate liquids. Terms in Eq. (7.38) can be consolidated to give Eq. (7.39).

$$S_{test\ liquid/substrate\ liquid} = \gamma_{substrate\ liquid} - \gamma_{test\ liquid} - \gamma_{test\ liquid/substrate\ liquid} \tag{7.39}$$

The subtracted terms can be collected resulting in Eq. (7.40).

$$S_{test\ liquid/substrate\ liquid} = \gamma_{substrate\ liquid} - \left(\gamma_{test\ liquid} + \gamma_{test\ liquid/substrate\ liquid}\right) \tag{7.40}$$

The test liquid will spontaneously spread over the substrate liquid when the initial spreading coefficient is positive.

Example 7.5a Calculate the Spreading Coefficient When a Drop of Oleic Acid (test liquid) Is Placed on Water (substrate liquid)

Given:

Surface tension of oleic acid is 32.8 mN m^{-1}
Surface tension of water is 71.8 mN m^{-1}
Interfacial surface tension of water-oleic acid is 12 mN m^{-1}
$S_{test\ liquid/substrate\ liquid} = \gamma_{substrate\ liquid} - (\gamma_{test\ liquid} + \gamma_{test\ liquid/substrate\ liquid})$

Substitute for the appropriate variables into the spreading coefficient equation.

$$S_{test\ liquid/substrate\ liquid} = \gamma_{substrate\ liquid} - \left(\gamma_{test\ liquid} + \gamma_{test\ liquid/substrate\ liquid}\right)$$

$$S_{test\ liquid/substrate\ liquid} = 71.8\ \text{mN m}^{-1} - \left(32.8\ \text{mN m}^{-1} + 12\ \text{mN m}^{-1}\right) = \underline{27\ \text{mN m}^{-1}} \tag{7.41}$$

The spreading coefficient is 27 mN m^{-1} and is strongly positive. Therefore, oleic acid will spread across the water.

Example 7.5b Calculate the Spreading Coefficient When a Drop of Water (Test Liquid) Is Placed on Oleic Acid (Substrate Liquid).

$$S_{test\ liquid/substrate\ liquid} = \gamma_{substrate\ liquid} - \left(\gamma_{test\ liquid} + \gamma_{test\ liquid/substrate\ liquid}\right)$$

$$\underline{S_{test\ liquid/substrate\ liquid}} = 32.8\ \text{mN m}^{-1} - \left(71.8\ \text{mN m}^{-1} + 12\ \text{mN m}^{-1}\right) = \underline{-51\ \text{mN m}^{-1}} \tag{7.42}$$

The spreading coefficient is – 51 mN m^{-1}, indicating that water will not spread on oleic acid.

Interestingly, oleic acid will spread on water, but water will not spread on oleic acid. This means that the selection of the liquid substrate is a key determinant to spreading. Equations (7.37) and (7.40) are importantly designated as initial spreading coefficients. This highlights the fact that the spreading coefficient is not an equilibrium value. As the test liquid spreads over the substrate liquid, the two liquids eventually become saturated with each other. This creates new surface tensions for each liquid which may impact the calculation of the spreading coefficient and even result in a change from a positive spreading coefficient to a negative spreading coefficient. If the primary interest is in spreading at equilibrium, it is suggested that the surface tensions of mutually saturated test liquid and substrate liquid be used in Eq. (7.40). The interfacial tension value is by definition reported for mutually saturated liquids.

7.8 Wetting and Spreading of Liquids on Solids, Work of Solid-Liquid Adhesion, and Solid-Liquid Interfacial Tension

The phenomenon of liquids *wetting* or *spreading* on solid surfaces has been observed for centuries. The phenomenon of liquids *beading* on solid surfaces is also well known. Figure 7.9 depicts an equilibrium system where a liquid drop is in contact with a flat solid surface.

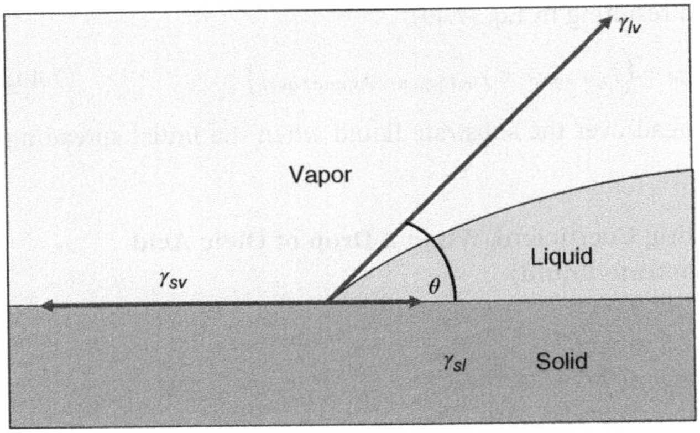

Figure 7.9 Equilibrium contact angle θ of liquid drop on a solid.

In 1805, Thomas Young developed this vector diagram to explain wetting, spreading, and beading. This figure shows the relationship between the solid surface tension, γ_{sv}, liquid surface tension, γ_{lv}, and solid-liquid interfacial tension, γ_{sl}. The vector arrows are proportional to their respective surface and interfacial tensions. At equilibrium, the γ_{sv} vector is counterbalanced by the sum of the γ_{sl} and γ_{lv} cos θ vectors. These relationships are described by Young's equation:

$$\gamma_{sv} = \gamma_{sl} + \gamma_{lv} \cos\theta \tag{7.43}$$

where θ is the *contact angle* formed when a liquid drop contacts a solid surface, γ_{sv} is the surface tension of the solid-vapor interface, γ_{sl} is the interfacial tension between the solid-liquid, and γ_{lv} is the surface tension of the liquid-vapor. Angle θ defines the shape of the drop. The contact angle, θ, is measured at the point where the liquid, solid, and vapor phases meet. For illustration, Table 7.5 gives cos θ for several contact angles.

When a liquid forms a thin film which completely spreads across the solid, the contact angle is 0°. This solid is said to be completely wetted by the liquid. When θ approaches 180°, the surface is not wetted and a spherical drop forms at a single point of contact on the solid. Intermediate contact angles lead to varying degrees of wetting and spreading. As a point of correctness, a contact angle of 180° has never been experimentally observed. This is because the London dispersion force operates between all materials and across phase boundaries. A contact angle of 180° implies that there is no interaction between the liquid and the solid. As was just explained, that is not the case for any real system.

Table 7.5 Cos θ Values for Selected Contact Angles.

Contact angle (θ°)	cos θ
0	1
45	0.7071
90	0
135	−0.7071
180	−1

7.8.1 Spreading of a Liquid on a Solid

The liquid-liquid spreading equation (Eq. 7.39) can also be used to describe the spreading of a liquid on a solid. The liquid-solid spreading equation becomes:

$$S_{test\ liquid/substrate\ solid} = \gamma_{substrate\ solid} - \gamma_{test\ liquid} - \gamma_{test\ liquid/substrate\ solid} \quad (7.44)$$

Since the surface tension of the solid and the interfacial tension between the solid and the liquid are difficult to determine experimentally, Eq. (7.44) has practical limitations. Young's equation in combination with Eq. (7.44) can be used to derive a solid-liquid *spreading coefficient* in terms of the liquid's surface tension and the solid-liquid contact angle. Both these terms can be readily determined experimentally.

Young's equation, Eq. (7.43), is:

$$\gamma_{sv} = \gamma_{sl} + \gamma_{lv}\cos\theta \quad (7.43)$$

Equation (7.44) can be rewritten using subscripts consistent with those given in Young's equation and shown in Eq. (7.44a).

$$S_{test\ liquid/substrate\ solid} = \gamma_{sv} - \gamma_{lv} - \gamma_{sl} \quad (7.44a)$$

The right-hand side of Eq. (7.43) is substituted for solid-vapor surface tension, γ_{sv}, in Eq. (7.44a) to give:

$$S_{test\ liquid/substrate\ solid} = S = \gamma_{sl} + \gamma_{lv}\cos\theta - \gamma_{lv} - \gamma_{sl} \quad (7.45)$$

Cancelling common terms and rearranging for common terms gives Eqs. (7.46) and (7.47).

$$S_{test\ liquid/substrate\ solid} = \cancel{\gamma_{sl}} + \gamma_{lv}\cos\theta - \gamma_{lv} - \cancel{\gamma_{sl}} \quad (7.46)$$

$$S_{test\ liquid/substrate\ solid} = S = \gamma_{lv}(\cos\theta - 1) \quad (7.47)$$

The spreading coefficient, S, of a liquid on a solid can now be calculated, using Eq. (7.47), by determining the experimental values for the surface tension of the test liquid and the contact angle between the liquid and solid. The latter can be determined using a contact angle goniometer. Note that in the case of a liquid spreading over a solid, S equals 0 or is less than 0. Zero indicates complete spreading and negative S values indicate varying degrees of partial wetting. The closer the S value is to zero the better the spreading.

Example 7.6 Wetting and Spreading of Water Saturated with Tablet Excipients on Tablets Prepared with Different Binders

The speed of tablet disintegration and dissolution often depends on the wetting and spreading of the disintegration and dissolution fluid on the tablet surface. Esezobo et al. (1989) studied the influence of contact angle on disintegration time of three tablet formulations containing 5% of three different binders. Water saturated with all the tablet excipients except the binders was used as the disintegration medium. Table 7.6 reports the surface tension of the water saturated with all the tablet excipients except the binders, contact angle, spreading coefficient, and disintegration time.

Spreading coefficient, S, sample calculations for povidone using Eq. (7.47) are provided.

The reported γ_{lv} and θ values are substituted into $S = \gamma_{lv}(\cos\theta - 1)$ to give:

$$S = 71.23\ \text{mN m}^{-1}(\cos 41.8 - 1) = 71.23\,(0.7455 - 1) = 71.23\,(-0.2545) = \underline{-18.1\ \text{mN m}^{-1}} \quad (7.48)$$

All the spreading coefficients in Table 7.6 are negative. According to the authors, better spreading occurred with lower negative S values. The rank order of binder formulations having the best

Table 7.6 Effect of Wetting and Spreading on Tablet Disintegration Time[a].

Binder	Surface tension ($mN\ m^{-1}$)[a]	Contact angle (°)	Spreading coefficient ($mN\ m^{-1}$)	Disintegration time (min)
Povidone	71.23	41.8	−18.1	17.0
Gelatin	71.23	43.7	−19.7	23.5
Tapioca	71.23	40.8	−17.3	2.0

[a] Water saturated with all the tablet excipients except the binders.

wetting and spreading was tapioca > povidone > gelatin, and shorter disintegration times follow this same rank order: tapioca > povidone > gelatin. Although the disintegration time of tapioca starch exhibits consistent rank order with the spreading coefficient values, the significantly shorter disintegration time of 2.0 minutes suggests that other factors besides spreading coefficient may be at play. For example, tapioca starch is a known disintegrant while povidone and gelatin are not considered disintegrants.

Example 7.7 Spreading and Adhesion Application: Wet Granulation

Wet granulation uses the principle of spreading and adhesion of a granulating liquid onto a powder bed with the aim of enlarging the overall particle size distribution of the final dried powder granulation. Rumpf (1962) described the mechanisms of interparticle interaction and bonding that occur during wet granulation. There are four mobile granulation film states that relate to the amount and the type of liquid distribution: *pendular, funicular, capillary, and droplet.* Pendular, funicular, and capillary stages are important in the formation of acceptable granules. The droplet granulation state is overwetted, and an unusable wet doughy liquid mass is formed.

Equations (7.49) and (7.50) describe the parameters critical for successful wet granulation. The granulation states are related to how much granulation liquid is in the granule pores. The relative granule pore liquid saturation, S, is given by:

$$S = \frac{M_l}{M_p}\left(\frac{1-\varepsilon}{\varepsilon}\right)\rho \tag{7.49}$$

where S is the relative liquid saturation of the granule pores, M_l the granulating solution mass, M_p the powder blend mass, ε the granule porosity or voids, and ρ the powder blend true density.

During the pendular granulation stage, liquid bridges cause adhesion through granulation liquid surface tension and capillary force. The powder starts to increase in particle size as powder particles start to adhere to one another.

The funicular state is an intermediate state between the pendular and capillary stages. As more liquid is added, the liquid bridges coalesce. This forms a continuous liquid phase that connects multiple particles, but the air pockets remain. Mechanical kneading and consolidation forces increase the density of the granules and force air out of the void spaces. Decreasing the void spaces decreases the overall material porosity which leads to the $(1-\varepsilon)/\varepsilon$ term increasing. This kneading process decreases the powder porosity and decreases the distance between two powder blend particles.

When the granule voids are filled with granulation liquid, the capillary stage has been reached. As shown in Eq. (7.50), the larger the value of S, the relative saturation of granule pores, the greater the wet granule tensile strength.

The strength of a moist idealized spherical monodispersed granulation is given by

$$\sigma = S\frac{(1-\varepsilon)}{\varepsilon}\frac{\gamma}{d}\cos\theta \tag{7.50}$$

where σ is the moist granule tensile strength, S is the relative liquid saturation level, γ is the surface tension of the granulating liquid, d is the diameter of the capillary wall, or in this case the distance between two particles, and θ is the contact angle of the granulating liquid and powder blend. The capillary suction pressure is given by the $\gamma \cos \theta$ term, which is the same term given in Eq. (7.7) for determining the capillary upward force.

To increase granule strength, the formulator can increase the relative saturation of granule pores by adding more granulation liquid, decreasing the granule porosity and distance between particles by kneading time and energy, increasing the granulating liquid's surface tension, and achieving a contact angle between the powder blend and the granulating liquid of 0.

There may be a trade-off between increasing the liquid's surface tension and decreasing the contact angle. Surfactants can be added to the granulating liquid to decrease the contact angle to improve the wetting of the powder. Adding surfactants will also decrease the surface tension of the granulating liquid, which has a negative effect on granule strength. One way to decrease θ without adding surfactants to the granulating liquid is to add more hydrophilic excipients, such as mannitol, pregelatinized starch, and microcrystalline cellulose, to the powder blend (see Example 7.6). Binders also affect the surface tension of the binder liquid. See Table 7.2 Surface Tension of Pharmaceutical Liquids.

7.8.2 Work of Solid-Liquid Adhesion

Young's equation, Eq. 7.43, $\gamma_{sv} = \gamma_{sl} + \gamma_{lv} \cos \theta$, is useful to conceptually understand the wetting and spreading of a liquid on a solid resulting from solid-liquid adhesive forces, but the determination of γ_{sv} is difficult and time-consuming and γ_{sl} cannot be measured directly. Pockels (1914), addressed the limitations of the Young equation by unifying the Young and the Dupré equations. For a solid-liquid interface system, the general work of adhesion term w_a becomes w_{sl} to represent the adhesive work of a solid-liquid interface.

The Dupré work of adhesion equation ($w_a = (\gamma_1 + \gamma_2) - \gamma_{12}$, Eq. 7.36) is restated in Eq. (7.51) using the Young equation surface tension terms. Here, $w_a = w_{sl}, \gamma_1 = \gamma_{lv}, \gamma_2 = \gamma_{sv}$, and $\gamma_{12} = \gamma_{sl}$ for the solid-liquid system.

$$w_a = w_{sl} = \gamma_{lv} + \gamma_{sv} - \gamma_{sl} \tag{7.51}$$

Substituting for γ_{sv} from the Young equation into Eq. (7.51) results in Eq. (7.52).

$$w_{sl} = \gamma_{lv} + \gamma_{sl} + \gamma_{lv} \cos \theta - \gamma_{sl} \tag{7.52}$$

Cancelling common terms and rearranging for common terms gives Eq. (7.53). This contribution by Agnes Pockels, a *citizen scientist,* led to what is known as the Young-Dupré equation some 109 years after Young's equation and 45 years after the Dupré equation had been published.

$$w_{sl} = \gamma_{lv} + \cancel{\gamma_{sl}} + \gamma_{lv} \cos \theta - \cancel{\gamma_{sl}}$$

$$w_{sl} = \gamma_{lv} \ (\cos \theta + 1) \ \left(\text{Young-Dupré equation}\right) \tag{7.53}$$

where w_{sl} is the reversible work of adhesion of the solid-liquid interface. The liquid surface tension, γ_{lv}, and θ can be measured experimentally, and w_{sl}, the solid-liquid work of adhesion, can be calculated from Eq. (7.53). In Section 7.9, the Young-Dupré equation is an important component in determining the surface tension/energy of solids.

7.9 Solid Surface Tension/Energy

The surface tension, γ_{sv}, or surface free energy of solid drugs and excipients has broad implications for drug product design. Surface tension/energy can explain many physical and performance differences observed during drug product design and solid material evaluation. The surface tension/energy of a material may differ due to powder formation route, such as spray-drying, solvent crystallization, crystallization, or amorphization using supercritical fluid techniques and others; and crystal facet preference or dependence. Polymorphs, as well as R and L chiral material forms, have different surface tension/energies. Surface tension/energy can impact wettability, dissolution and bioavailability, cohesion, and adhesion of inhalation dry powder formulations, and subsequent drug delivery performance. Affected performance factors include powder flow, powder mixing uniformity, wet granulation effectiveness, tabletability, film-coating adhesion and quality, capsule filling, dry powder inhaler drug deposition, and batch-to-batch variability. External factors like humidity may also affect solid surface tension/energy.

As important and ubiquitous as surface tension/energy is to drug product development, it has only been since the 1960s that the theory and analytical methods have become available to more fully understand and study the influence of solid surface tension/energy on drug product design. Since the 1960s, solid surface tension/energy research and its application have increased exponentially. The development of current solid surface tension/energy theory relies on the fact that the solid-vapor surface energy is equivalent to the solid-vapor surface tension. The chronological development of solid surface tension/energy theory and applicable analytical methods is summarized in Table 7.7.

The underlying surface energy theory will now be presented followed by the development of the OWRK, Wu, and vOCG theories and equations. The development of this theory applies the Dupré work of adhesion (Eq. 7.36) and the Young-Dupré (Eq. 7.50) equations which have been discussed earlier.

7.9.1 Underlying Solid Surface Tension/Energy Theory: Combining Berthelot's Rule, Dupré Work of Cohesion, and Anges Pockles' Unifying Young-Dupré Equations

Berthelot (1889) suggested that the adhesive work between a solid and a liquid was equal to the geometric mean of the cohesive work of the solid and of the liquid (Eq. 7.54).

$$w_{sl} = (w_{ss}w_{ll})^{0.5} \tag{7.54}$$

where w_{sl} is the solid-liquid work of adhesion and w_{ss} and w_{ll} are the work of cohesion for the solid and for the liquid. Combining the work of cohesion from Eq. (7.33), $w_c = 2\gamma$, and Eq. (7.54) leads to Eq. (7.55):

$$w_{sl} = (2\gamma_{sv} \cdot 2\gamma_{lv})^{0.5} = 2(\gamma_{sv} \cdot \gamma_{lv})^{0.5} \tag{7.55}$$

Both Eq. (7.55) and the Young-Dupré equation (Eq. 7.53), $w_{sl} = \gamma_{lv}(\cos\theta + 1)$, express solid-liquid adhesive work and can be set equal to one another giving Eq. (7.56).

$$\gamma_{lv}(\cos\theta + 1) = 2(\gamma_{sv} \cdot \gamma_{lv})^{0.5} \tag{7.56}$$

Dividing both sides of Eq. (7.56) by 2 results in Eq. (7.57).

$$0.5\,\gamma_{lv}(\cos\theta + 1) = (\gamma_{sv}\gamma_{lv})^{0.5} \tag{7.57}$$

Table 7.7 Chronological Development of Solid Surface Tension/Energy Theory and Supporting Analytical Methods.

Year	Scientist(s)	Contribution	Equation(s)
1805	Thomas Young	Young's equation describing a vector diagram to explain wetting, spreading, and beading.	$\gamma_{sv} = \gamma_{sl} + \gamma_{lv} \cos \theta$[b]
1805	***Thomas Young***	***Contact Angle Measurement***[a]	
1869	Athanase Dupré	Described thermodynamic adhesion or work of adhesion equation for a two-component system.	$w_a = \gamma_1 + \gamma_2 - \gamma_{12}$[c]
1889	Marcellin Berthelot	Berthelot's rule states that adhesive work of a solid-liquid system equals the geometric mean of the work of cohesion of the solid and of the liquid. This equation is the foundation for most modern surface tension/energy theories.	$w_{sl} = (w_{ss}w_{ll})^{0.5}$[d]
1914	Anges Pockles	Unified the work of Young and Dupré by deriving what is known as the Young-Dupré equation. This equation finds important applications for most of the contemporary surface tension/energy theories.	$w_{sl} = \gamma_{lv} (\cos \theta + 1)$[e]
1962	**Böhme et al.**	***Reported the use of Centrifugal Particle Detachment Method to measure surface energy***[f]	
1964	Fowkes	Stated the solid surface tension/energy can be approximated by the sum of each type of interaction force (dispersive, polar, hydrogen, induction, acid-base, and others).	$\gamma_{sv}^{Tot} = \gamma_{sv}^{d} + \gamma_{sv}^{p} + \gamma_{sv}^{h} + \gamma_{sv}^{i} + \gamma_{sv}^{ab} + \gamma_{sv}^{o}$[g]
1967	***Kiselev***	***Reported the use of Inverse Gas Chromatography for surface energy measurements***[h]	
1969[i] 1970[j] 1971[k]	Owen-Wendt Kaelble Rabel (OWRK)	The OWRK surface tension/energy theory describes the total solid surface energy as a linear combination of dispersive and polar attractive forces (Eq. 1).	(Eq. 1) $\gamma_{sv}^{Tot} = \gamma_{sv}^{d} + \gamma_{sv}^{p}$
		The final form of OWRK assimilates the Young-Dupré equation, Fowkes' concept of first-order linear combinations for energy components, and Berthelot's rule concept of mean geometric mean. The model is given by Eq. (2).	(Eq. 2) $0.5 \cdot \gamma_{lv} (\cos \theta + 1) = (\gamma_{sv}^{d} \gamma_{lv}^{d})^{0.5} + (\gamma_{sv}^{p} \gamma_{lv}^{p})^{0.5}$

(*Continued*)

Table 7.7 (Continued)

Year	Scientist(s)	Contribution	Equation(s)
1973	Wu	Wu followed the ideas of OWRK but used the harmonic mean instead of the geometric mean of OWRK for work of adhesion.	$0.25(\cos\theta + 1) = [(\gamma_{sv}{}^{d}\,\gamma_{lv}{}^{d}/\gamma_{sv}{}^{d} + \gamma_{lv}{}^{d}) + (\gamma_{sv}{}^{P}\,\gamma_{lv}{}^{P}/\gamma_{sv}{}^{P} + \gamma_{lv}{}^{d})$[l]
1986	***Binnig et al.***	***Initial report of Atomic Force Microscopy.***[m]	
1988	Van Oss-Chaudhury-Good (vOCG)	In the vOCG theory, the Lifshitz-van der Waals interactions are considered the same as dispersive attractive forces. The model equation incorporates the Young-Dupré equation, Fowkes' concept of first order linear combinations for components, and Berthelot's rule concept of mean geometric mean.	$0.5(\cos\theta + 1) = (\gamma_{sv}{}^{d}\,\gamma_{lv}{}^{d})^{0.5} + (\gamma_{sv}{}^{acid}\gamma_{slv}{}^{base})^{0.5} + (\gamma_{sv}{}^{base}\,\gamma_{lv}{}^{acid})^{0.5}$[n]
1993	***Ohnesorge and Binnig***	***Reported the ability to achieve atomic resolution with Atomic Force Microscopy.***[o]	

[a] ***Bold and italics indicates the introduction of a new surface energy analytical method.***
[b] Young (1805).
[c] Dupré (1869).
[d] Berthelot (1898).
[e] Pockels (1914).
[f] Böhme (1962).
[g] Fowkes (1964).
[h] Kiselev (1967).
[i] Owens and Wendt (1969).
[j] Kaelble (1970).
[k] Rabel (1971).
[l] Wu (1973).
[m] Binnig et al. (1986).
[n] Van Oss et al. (1988).
[o] Ohnesorge and Binig (1993).

Equation (7.57) can now be used to calculate the surface tension/energy of a solid, γ_{sv}, if γ_{lv} and the contact angle θ are determined experimentally.

Also note: Eq. (7.57) is the general form of the OWRK equation $[0.5\,\gamma_{lv}(\cos\theta + 1) = (\gamma_{sv}{}^{d}\,\gamma_{lv}{}^{d})^{0.5} + (\gamma_{sv}{}^{P}\,\gamma_{lv}{}^{P})^{0.5})]$.

7.9.2 Underlying Solid Surface Tension/Energy Theory: Fowkes Concept

Fowkes (1964) posited that the solid surface tension/energy was the sum of independent interactive forces such as dispersion, polar, hydrogen, induced, acid-base, and other interactions. This is shown in Eq. (7.58).

$$\gamma_{sv}{}^{Tot} = \gamma_{sv}{}^{d} + \gamma_{sv}{}^{P} + \gamma_{sv}{}^{h} + \gamma_{sv}{}^{i} + \gamma_{sv}{}^{ab} + \gamma_{sv}{}^{o} \tag{7.58}$$

where $\gamma_{sv}{}^{Tot}$ is the total surface tension/energy, $\gamma_{sv}{}^{d}$ is the dispersive energy component, $\gamma_{sv}{}^{p}$ is the polar component, $\gamma_{sv}{}^{h}$ is the hydrogen-bonding component, $\gamma_{sv}{}^{i}$ is an inductive attractive effect, $\gamma_{sv}{}^{ab}$ is the acid-base electron donating and accepting mechanism, and $\gamma_{sv}{}^{o}$ represents other unaccounted interactive forces. For a review of interaction forces, see Section 6.3.2.

7.9.3 Owen, Wendt, Rabel, and Kaelble (OWRK) Surface Tension/Energy Equation

Owen and Wendt (1969), Rabel (1971), and Kaelble (1970) proposed that all the terms of the Fowkes concept except the dispersive component could be grouped under a polar term. The total surface tension/energy becomes:

$$\gamma_{sv}{}^{Tot} = \gamma_{sv}{}^{d} + \gamma_{sv}{}^{p} \tag{7.59}$$

Using Berthelot's rule and expanding the general form of the Eq. (7.57), $0.5\gamma_{lv}(\cos\theta + 1) = (\gamma_{sv}\gamma_{lv})^{0.5}$, to include dispersive attractive forces and polar attractive forces leads to the OWRK equation (see Eq. 7.60)

$$0.5\gamma_{lv}\ (\cos\theta + 1) = \left(\gamma_{sv}{}^{d}\ \gamma_{lv}{}^{d}\right)^{0.5} + \left(\gamma_{sv}{}^{p}\ \gamma_{lv}{}^{p}\right)^{0.5}\quad \text{(OWRK Equation)} \tag{7.60}$$

Equation (7.60) has two unknowns, $\gamma_{sv}{}^{d}$ and $\gamma_{sv}{}^{p}$. To solve this equation for the unknowns, two liquids that have known dispersive and polar components can be chosen as *test liquids*. Common polar test liquids are water, glycerol, and formamide. Diiodomethane and α-bromonaphthalene are often used as dispersive test liquids. A drop of the test liquid is placed onto a flat solid surface that can be cast, molded, or compressed. Once the contact angles for the two test liquids are measured, $\gamma_{sv}{}^{d}$ and $\gamma_{sv}{}^{p}$ can be calculated using the surface tension/energy components shown in Table 7.8.

Example 7.8 Calculate a Solid Polymer's Surface Tension/Energy Using the OWRK and Fowkes Equations (Adapted from Laureen, 2024)

Given:

Diiodomethane (Dim)

$\gamma_{Dim,lv}{}^{Tot} = 50.8\ \text{mN m}^{-1}$
$\gamma_{Dim,lv}{}^{d} = 50.8\ \text{mN m}^{-1}$
$\gamma_{Dim,lv}{}^{p} = 0\ \text{mN m}^{-1}$
$\theta = 39°$

Water (W)

$\gamma_{W,lv}{}^{\text{Tot}} = 72.8\ \text{mN m}^{-1}$
$\gamma_{W,lv}{}^{\text{d}} = 21.8\ \text{mN m}^{-1}$

Table 7.8 Surface Tension/Energy Components of Common Test Liquids[a].

Liquid	Total surface energy (mJ m^{-2})	Dispersive component (mJ m^{-2})	Polar component (mJ m^{-2})
Diiodomethane	50.8	50.8	0
Formamide	58.0	39.0	19.0
Glycerin	64.0	34.0	30.0
Water	72.8	21.8	51.0

[a] Adapted from https://www.accudynetest.com/surface_tension_table.html.

$\gamma_{W,lv}{}^{p} = 51.0\ \mathrm{mN\ m^{-1}}$
$\theta = 83°$

OWRK Equation

$$0.5\,\gamma_{lv}\,(\cos\theta + 1) = \left(\gamma_{sv}{}^{d}\,\gamma_{lv}{}^{d}\right)^{0.5} + \left(\gamma_{sv}{}^{p}\,\gamma_{lv}{}^{p}\right)^{05}$$

Fowkes Equation

$$\gamma_{sv}{}^{Tot} = \gamma_{sv}{}^{d} + \gamma_{sv}{}^{p}$$

Calculation for the *dispersive component* of the polymer's total surface tension/energy using the Dim liquid data.

1) Substitute the givens into the OWRK equation.

$$\left(\gamma_{sv}{}^{d}\mathrm{mN\,m^{-1}}\cdot 50.8\,\mathrm{mN\,m^{-1}}\right)^{0.5} + \left(\gamma_{sv}{}^{P}\mathrm{mN\,m^{-1}}\cdot 0\,\mathrm{mN\,m^{-1}}\right)^{0.5} = 0.5\cdot 50.8\,\mathrm{mN\,m^{-1}}\cdot(\cos 39° + 1) \tag{7.61}$$

Equation (7.61) has consistent units on each side of the equation.

2) Restating Eq. (7.61) after math operations.

$$\left(\gamma_{sv}{}^{d}\ \mathrm{mN\ m^{-1}}\cdot 50.8\ \mathrm{mN\ m^{-1}}\right)^{0.5} + 0 = 0.5\cdot 50.8\ \mathrm{mN\ m^{-1}}\ (0.7771 + 1) \tag{7.62}$$

$$\left(\gamma_{sv}{}^{d}\mathrm{mN\ m^{-1}}\cdot 50.8\ \mathrm{mN\ m^{-1}}\right)^{0.5} = \ = 45.138\ \mathrm{mN\ m^{-1}}$$

Square each side

$$\gamma_{sv}{}^{d}\,\mathrm{mN\ m^{-1}}\cdot 50.8\ \mathrm{mN\ m^{-1}} = 2037.44\ (\mathrm{mN\ m})^{-2}$$

$$\gamma_{sv}{}^{d}\,\mathrm{mN\ m^{-1}} = 2037.44\ (\mathrm{mN\ m})^{-2}/50.8\ \mathrm{mN\ m^{-1}}$$

$$\underline{\gamma_{sv}{}^{d}\,\mathrm{mN\ m^{-1}} = 40.1\ \mathrm{mN\ m^{-1}}\ \text{with correct significant figures and units}} \tag{7.63}$$

The dispersive surface tension/energy, $\gamma_{sv}{}^{d}$, for the solid polymer is <u>40.1 mN m^{-1}</u>.

3) A similar procedure and calculations can be performed to determine the polar surface tension/energy component of the polymer. These calculations result in a polar surface tension/energy component of <u>2.5 mN m^{-1}</u>.
4) The Fowkes equation, $\gamma_{sv}{}^{Tot} = \gamma_{sv}{}^{d} + \gamma_{sv}{}^{P}$, is used to calculate the total polymer surface tension/energy.

$$\gamma_{sv}{}^{Tot} = \gamma_{sv}{}^{d} + \gamma_{sv}{}^{P} = 40.1\ \mathrm{mN\ m^{-1}} + 2.5\ \mathrm{mN\ m^{-1}} = \underline{42.6\ \mathrm{mN\ m^{-1}}} \tag{7.64}$$

The total polymer surface tension/energy is <u>42.6 mN m^{-1}</u>.

7.9.4 Wu Surface Tension/Energy Equation Using Harmonic Mean for Adhesive Work

Wu (1973) accepted the Fowkes concept but chose to use the harmonic mean to describe the adhesive work where w_{sl} is given by Eq. (7.65)

$$w_{sl} = 4\left[\left(\gamma_{sv}{}^{d}\ \gamma_{lv}{}^{d}\right)/\left(\gamma_{sv}{}^{d} + \gamma_{lv}{}^{d}\right) + \left(\gamma_{sv}{}^{P}\ \gamma_{lv}{}^{P}\right)/\left(\gamma_{sv}{}^{P} + \gamma_{lv}{}^{P}\right)\right] \tag{7.65}$$

Setting the Young-Dupré equation (Eq. 7.53), $w_{sl} = \gamma_{lv}(\cos\theta + 1)$, equal to the Wu equation gives:

$$\gamma_{lv}\ (\cos\theta + 1) = 4\left[\left(\gamma_{sv}{}^{d}\ \gamma_{lv}{}^{d}\right)/\left(\gamma_{sv}{}^{d} + \gamma_{lv}{}^{d}\right) + \left(\gamma_{sv}{}^{p}\ \gamma_{lv}{}^{p}\right)/\left(\gamma_{sv}{}^{p} + \gamma_{lv}{}^{p}\right)\right] \quad (7.66)$$

Dividing both sides of Eq. (7.66) by 4 gives the simplified Wu surface energy equation which can be used to calculate the dispersive and polar adhesive interactions.

$$0.25\,\gamma_{lv}\ (\cos\theta + 1) = \left[\left(\gamma_{sv}{}^{d}\ \gamma_{lv}{}^{d}\right)/\left(\gamma_{sv}{}^{d} + \gamma_{lv}{}^{d}\right) + \left(\gamma_{sv}{}^{p}\ \gamma_{lv}{}^{p}\right)/\left(\gamma_{sv}{}^{p} + \gamma_{lv}{}^{p}\right)\right] \quad (7.67)$$

The Wu equation also requires at least two *test liquids* with known dispersive and polar components.

7.9.5 Van Oss-Chaudhury-Good (vOCG) Surface Tension/Energy Equation Using Acid-Base Adhesion Theory

Van Oss-Chaudhury-Good (vOCG) postulated that the important surface energy components were dispersion forces and acid-base interactions. vOCG adopted Berthelot's and Fowkes concepts and employed the geometric mean and linear combination of forces. The vOCG equation takes a similar form to the OWRK equation.

$$0.5\,\gamma_{lv}\ (\cos\theta + 1) = \left(\gamma_{sv}{}^{d}\gamma_{lv}{}^{d}\right)^{0.5} + \left(\gamma_{sv}{}^{acid}\gamma_{lv}{}^{acid}\right)^{0.5} + \left(\gamma_{sv}{}^{base}\gamma_{lv}{}^{base}\right)^{0.5} \quad (7.68)$$

Equation (7.68) has three unknowns, $\gamma_{sv}{}^{d}$, $\gamma_{sv}{}^{acid}$, and $\gamma_{sv}{}^{base}$, which require three *test liquids* with known dispersive, acid, and base surface tension/energy components to solve the equation.

7.9.6 Solid Surface Tension/Energy Analytical Methods

Table 7.7 lists the chronological introduction of centrifugal particle detachment (CPD), inverse gas chromatography (IGC), and atomic force microscopy (AFM) surface energy methods. Discussion of these methods is outside the scope of this textbook. However, several references are provided for further reading: CPD (Podczeck, 1998; Tran et al., 2021), IGC (Kiselev, 1967; Grimsey et al., 2002), and AFM (Lamprou and Smith, 2015).

7.9.7 Pharmaceutical Application: Prediction of Dry Powder Inhaler Formulation Performance Using Surface Tension/Energy Measurements

Most dry powder inhaler formulations incorporate a *drug carrier* that is used to improve powder flow for filling into the device, capsules, or aluminum pouches. The carrier, typically lactose, often has a particle size greater than 10 μm. Particles of that size are deposited in the oral pharyngeal cavity and do not reach the lungs. The carriers can also help improve the powder content uniformity of low-dose inhalation medications, decrease powder aggregation, and improve powder inhalation performance compared to drug alone. For acceptable performance, a sensitive balance between the drug and carrier cohesive and adhesive forces is required.

A delivered dose of 1–5 μm sized drug is often considered an acceptable delivered particle size range. The fraction of the delivered drug dose that is in the 1–5 μm range is referred to as the *fine particle fraction* (FPF). The larger the FPF, the larger the amount of drug that is expected to reach the lungs and provide a therapeutic effect. In this regard, FPF is considered a critical drug product performance attribute.

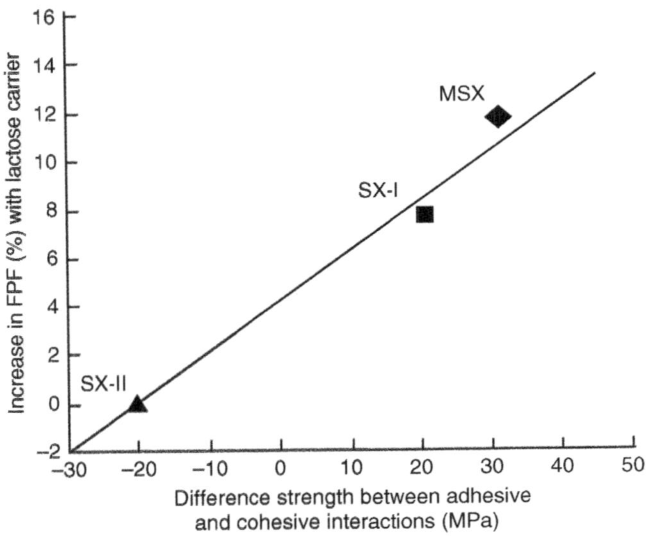

Figure 7.10 Effect of the difference between adhesive and cohesive attractive forces on percent fine powder fraction. MSX (micronized salmeterol xinafoate) is the stable polymorph and the form used in most commercial products. SX-I (supercritical fluid processed Form I salmeterol xinafoate polymorph) and SX-II (supercritical fluid processed Form I salmeterol xinafoate polymorph) are unstable polymorphs prepared by solution enhance dispersion with supercritical fluids (SEDS™). *Source:* Adapted from Grimsey et al. (2002).

In this application example, IGC was used to assess the cohesive and adhesive interactions between lactose and salmeterol xinafoate which was prepared by three different processes.

Figure 7.10 indicates that larger differences between the adhesive force between the drug and carrier and the drug and carrier cohesive forces result in an increased percentage of the drug fine particle fraction. It is generally accepted that an increase in fine particle fraction should lead to an increased amount of drug delivered to the lungs, which in turn is expected to improve the therapeutic effect.

This application also shows that IGC adhesion-cohesion data can be used to predict dry powder aerosol performance. IGC may also be used to monitor batch-to-batch drug substance, excipients, and drug formulation surface tensions/energies.

7.10 Relationship Between Surface Curvature, Surface Tension/Energy, and the Pressure Difference Across a Bubble's Curved Surface

A bubble is spherical because this shape minimizes its surface free energy. The surface free energy of a bubble is its surface area times the surface tension of the film, $G^{surface\ free\ energy} = (4\pi r^2)\,\gamma$ where $G^{surface\ free\ energy}$ is the Gibbs surface free energy, r is the sphere radius, and γ is the surface tension of the liquid. The following discussion will show that the pressure inside a bubble, P_{int}, is greater than the outside pressure, P_{ext}. The outside of the bubble is also referred to as its convex surface side. The concave bubble surface is viewed from the bubble's center. The bubble's size is constant when the external work, w_{ext}, performed on the bubble film and the internal work, w_{int}, are equal.

The external work is the sum of two work components: $P_{ext}V$ work and γA surface free energy work. The internal work arises from PV work, $P_{int}\,\Delta V$. For any incremental change in the bubble size, there will be incremental changes in external work (w_{ext}), internal work (w_{int}), and bubble radius (r). The relevant equations are provided.

Incremental external work resulting from a change in bubble size is given by Eq. (7.69).

$$dw_{ext}/dr = (P_{ext}\cdot dV)/dr + (\gamma\cdot dA)/dr = \left[P_{ext}\, d\left(4\pi r^3/3\right)\right]/dr \;+ \left[\left(\gamma\, d\left(4\pi r^2\right)\right]/dr \quad (7.69)$$

The derivative of Eq. (7.69) is (see differential Eq. (1.85)):

$$P_{ext}\left(4\pi r^2\right) + \gamma(8\pi r) \quad (7.69a)$$

Incremental internal work resulting from a change in bubble size is:

$$dw_{int}/dr = (P_{int}\cdot dV)\; dr = \left[P_{int}\, d\left(4\pi r^3/3\right)\right]\; dr \quad (7.70)$$

Taking the derivative of $[P_{int}\; d\; (4\pi r^3/3)]\; dr$ with respect to r gives:

$$P_{ext}\left(4\pi r^2\right) \quad (7.70a)$$

At equilibrium, Eqs. (7.70a) and (7.69a) are equal and:

$$P_{int}\left(4\pi r^2\right) = P_{ext}\left(4\pi r^2\right) + \gamma\;(8\pi r) \quad (7.71)$$

The pressure difference across the bubble film is:

$$P_{int} - P_{ext} = (2\gamma)/\; r = \Delta P \quad (7.72)$$

Equation (7.72) shows that the pressure inside the bubble, $P_{int,}$ is $(2\gamma)/r$ greater than the outside bubble pressure, P_{ext}. This is the pressure difference, ΔP, across the film. This equation also shows that the smaller the bubble, the greater the pressure difference. That is, the concave side of the bubble (looking from center of the bubble) has higher pressure than the convex side of the bubble (looking from the outside of the bubble).

Equation (7.72) is the reduced form of a more general equation, the Young-Laplace equation, that is used to determine the surface pressure change for both spherical and nonspherical surfaces and is shown in Eq. (7.73).

$$\Delta P = \gamma\;(1/R_1 + 1/R_2) = \gamma\, H \quad (7.73)$$

where ΔP is the pressure difference between the inside and outside a nonspherical surface such as a pendant drop, R_1 is the radius of a circle tangent to a point on the nonspherical surface, R_2 is the radius of a second circle that is tangent to the same point and is perpendicular to circle created by R_1, and H *is* the *mean surface curvature*. Adamson and Gast (1997) provide a more detailed description of these radii. Except for a sphere, there is a need to describe a curved surface using two radii of two orthogonal circles tangent to a point on the curved surface of interest. For spheres, Eq. (7.73) reduces to Eq. (7.72) because the two radii of the two orthogonal tangent circles are equal.

7.11 The Effect of Curvature and Surface Tension on Vapor Pressure

The effect of curvature and surface tension on vapor pressure and solubility is described by the *Kelvin* (1871, see Sir William Thomson) and *Ostwald* (1900)-*Freundlich* (1900) *equations*. The derivation of the Kelvin equation follows that of Wurster (2024), to provide insight into its formation and fundamental basis.

The Kelvin equation can be derived by envisioning two closed systems at equilibrium at the same constant volume and temperature. One closed system is a *water droplet* having a radius of less than one micron which is in equilibrium with its vapor. Since the water droplet is in equilibrium with its vapor, the chemical potential of the condensed water droplet surface equals the water vapor chemical potential. For convenience, the chemical potential of the condensed water droplet surface will be referred to as the *chemical potential of the curved surface*, μ_{curved}. A *flat-water surface phase* in equilibrium with its vapor comprises the second closed system. The chemical potential of the flat-water surface will be referred to as the *chemical potential of the flat surface*, μ_{flat}.

The question can now be asked: "how much work would be done if dn water molecules are transferred from the interior of the water having a flat surface to the interior of the water droplet when the transfer occurs at constant temperature (T), constant total volume (V), and constant surface area ($\mathcal{A}^{surface}$)"? The Helmholtz free energy equation can describe this work.

$$d(\text{Helmholtz free energy}_{curved}) = -SdT - PdV - \gamma d\mathcal{A}^{surface} + \mu_{curved}\, dn \quad (7.74)$$

$$d\left(\text{Helmholtz free energy}_{flat}\right) = -SdT - PdV - \gamma d\mathcal{A}^{surface} + \mu_{flat}\, dn \quad (7.75)$$

The work done in the molecular transfer process is equal to the difference in the chemical potential between the two systems times the number of molecules transferred (Eq. 7.76).

$$d(\text{Helmholtz free energy}_{curved}) - d\left(\text{Helmholtz free energy}_{flat}\right) = \left(\mu_{curved} - \mu_{flat}\right) dn \quad (7.76)$$

Since the system was initially at equilibrium and an infinitesimally small change has been made to the system, the work from Eq. (7.76) must be the same as if the dn molecules had been used to expand the surface. In this case,

$$\left(\mu_{curved} - \mu_{flat}\right) dn = \gamma d\mathcal{A}^{surface} \quad (7.77)$$

From right side of Eqs. (7.69) and (7.69a)

$$\gamma d\mathcal{A}^{surface} = \left[\gamma\, d\left(4\pi r^2\right)\right]/dr = \gamma(8\pi r)dr \quad (7.78)$$

Substituting the right side of Eq. (7.78) into Eq. (7.77) for $\gamma\ d\mathcal{A}^{surface}$ gives Eq. (7.79).

$$\left(\mu_{curved} - \mu_{flat}\right) dn = \gamma(8\pi r)\, dr \quad (7.79)$$

The next step in deriving the Kelvin equation requires dn to be defined in experimentally measurable terms. The change in drop volume, dV, when the water molecules are added is equal to molecular volume (υ) times dn (see Eq. 7.80)

$$dV = \upsilon\ dn \quad (7.80)$$

From the right side of Eqs. (7.69) and (7.69a)

$$dV = 4\pi r^2 dr \quad (7.81)$$

Substituting the right side of Eq. (7.81) into Eq. (7.80) for dV gives Eq. (7.82).

$$4\pi r^2 dr = \upsilon\ dn \quad (7.82)$$

Rearranging for *dr* gives:

$$dr = (v\ dn)/\left(4\pi r^2\right) \tag{7.83}$$

Now, we can substitute for *dr* in Eq. (7.79).

$$\left(\mu_{curved} - \mu_{flat}\right)\ \cancel{dn} = [\gamma(8\pi\cancel{r})]\ (v\ \cancel{dn})/\left(4\pi\cancel{r^2}\right) \tag{7.84}$$

and

$$\left(\mu_{curved} - \mu_{flat}\right) = 2\gamma v/r \text{ (the Kelvin equation)} \tag{7.85}$$

This is the general form of the Kelvin equation. It is valid for liquids and solids. Please note that there is a *positive difference* in chemical potentials between vapors over a *convex* curved surface and a planar flat surface where $\mu_{curved} > \mu_{flat}$. On the other hand, there is a *negative difference* in chemical potentials between vapors over a *concave* surface and a planar flat surface, and $\mu_{flat} > \mu_{curved}$.

For ideal vapors, the chemical potential is related to vapor pressure at standard temperature and pressure as indicated in Eqs. (7.86) and (7.87). For review, see the development of Eq. (4.74), showing the relationship between the chemical potential of an idea gas and its vapor pressure at standard temperature and pressure where P_1 is 1 atm Equations (7.86) and (7.87) are obtained by substituting *k* for *R* in Eq. (4.74).

$$\mu_{curved} = \mu^\circ + kT \ln\ (P_{curved}) \tag{7.86}$$

$$\mu_{flat} = \mu^\circ + kT \ln\ \left(P_{flat}\right) \tag{7.87}$$

where μ_{curved} is the chemical potential of the water vapor above the curved surface, μ_{flat} is the chemical potential of the water vapor over the flat surface, μ° is the chemical potential at standard temperature and pressure, *k* is the Boltzmann's constant, P_{curved} is the vapor pressure over the curved surface, and P_{flat} is the vapor pressure over the flat surface. The difference in the chemical potentials is:

$$\mu_{curved} - \mu_{flat} = kT\ \left[\ln\ (P_{curved}) - \ln\ \left(P_{flat}\right)\right] = kT \ln\ \left(P_{curved}/P_{flat}\right) \tag{7.88}$$

(see logarithm operation Eq. [1.47])

Substituting $2\gamma v/r$ from the Kelvin equation (Eq. 7.85) for $\mu_{curved} - \mu_{flat}$ in Eq. (7.88) gives:

$$2\gamma v/r = kT\ \ \ln\ \left(P_{curved}/P_{flat}\right) \tag{7.89}$$

Dividing both sides of Eq. (7.89) by kT *gives* Eq. (7.90).

$$\ln\ \left(P_{curved}/P_{flat}\right) = (2\gamma v/)/(rkT) \tag{7.90}$$

Molecular volume (v) data are less commonly available than molar volume, V_m, data. To convert molecular volume to molar volume, the numerator and denomination of the right side of Eq. (7.90) is multiplied by Avogadro's number (N_A) giving Eq. (7.91).

$$\ln\ \left(P_{curved}/P_{flat}\right) = (2\gamma\, V_m/)/(rRT) \tag{7.91}$$

where $v\, N_A = V_m$, and $k\, N_A = R$. Equation (7.91) becomes the Kelvin equation expressed in terms of vapor pressure as function of surface tension and the radius of the particle.

This form of the Kelvin-Ostwald-Freundlich equation is also key to understanding capillary condensation that can occur during vapor adsorption onto solids. Capillary condensation will be discussed in more detail in Chapter 8: Adsorption Phenomena.

7.12 The Effect of Curvature on Solubility

The Kelvin-Ostwald-Freundlich equation shows the effect of curvature and surface tension on solubility:

$$\ln S_r/S_\infty = (2\gamma V_m)/(rRT) = (2\gamma M)/(r\rho RT) \tag{7.91}$$

where S_∞ is the solubility of a particle with an infinite radius (a flat surface), S_r is the solubility of a particle of radius r, γ is the interfacial tension between the particle and the solution media, V_m is the molar volume, M is the molecular mass of the material of the particle, ρ is the density of the material of the particle, R is the gas constant, and T is the temperature in kelvin. Equation (7.91) was used to explain the theoretical solubility differences of 10 nm, 100 nm, and 1000 nm particles in Section 6.15.

This equation states that drug solubility will increase as the particle size decreases. This logarithmic relationship is observed for drug particles 1 μm and smaller and is illustrated in the application.

7.12.1 Pharmaceutical Application of the Kelvin-Ostwald-Freundlich Equation: Increasing Drug Solubility and Dissolution Rate by Decreasing Drug Particle Size to One Micron and Smaller

Equation (7.91) shows that as the radius of curvature decreases, drug solubility increases. As the interfacial tension between the drug particle and the dissolution media increases, the drug solubility also increases.

The relative solubility of a drug, S_r/S_∞, increases as the particle size decreases. This phenomenon is observed at nanometer sizes. For example, the relative drug solubilities of particles having a radius = 1000 nm, 100 nm, and 10 nm are 1.012, 1.13, and 3.35, respectively.

At nanometer size ranges, drug dissolution is increased by the drug's solubility increase and also an increase in specific surface area (SSA) for the same mass of particles. The Nernst (1904) and Brunner (1904) equation describes the effect of drug solubility and particle surface area on the drug's dissolution rate:

$$dC/dt = (DA/hV)C_s \tag{7.76}$$

where C is the concentration of drug in solution at time t, D is the drug-solute diffusion coefficient, A is the surface area of the drug particle or particles, h is the concentration gradient thickness of the stagnant diffusion layer, V is the volume of the dissolution media, and C_s is the drug saturation or equilibrium solubility.

Nanometer-sized drug particles are defined as particles with one or more dimensions of 100 nm or less (Nanotechnology Act, 2003). Nanometer-sized drug particles increase drug dissolution by two mechanisms. First, as noted earlier, nanodrug solubility, C_s, is increased. From Eq. (7.76), the dissolution rate is directly proportional to this solubility increase. The dissolution rate is also directly proportional to the powder's surface area. A drug's specific surface area, SSA ($m^2\,g^{-1}$ of drug) for particles of 2000 nm, 200 nm, and 20 nm diameter and of unit density is 3 $m^2\,g^{-1}$, 30 $m^2\,g^{-1}$, and 300 $m^2\,g^{-1}$, respectively. To put these SSAs in perspective, talc's SSA ranges between 2.41 $m^2\,g^{-1}$ and 2.42 $m^2\,g^{-1}$ and magnesium stearate SSA ranges from 1.6 to 14.8 $m^2\,g^{-1}$.

Therefore, the reduction in particle size to a nanosized range leads to increased drug solubility and greater surface area. These changes can lead to potentially significantly increased dissolution rates, which in turn may lead to much improved bioavailability.

7.12.2 Author's Anecdote – Albuterol Sulfate Pressurized Suspension Inhaler: Catastrophic Crystallization in Chlorofluorocarbon Propellant

Ostwald ripening is a dispersed-system phenomenon involving the redistribution of molecules from small dispersed particles to larger dispersed particles at the expense of the smaller particles. The Young-Laplace equation and the Kelvin-Ostwald-Freundlich equations support the idea that large particles are energetically more stable and thermodynamically favored. Drug particles in a dispersion are constantly dissolving and recrystallizing. The Kelvin (also Thomson, 1871)-Ostwald (1900)-Freundlich (1909) equation and the Nernst and Brunner equation tells us that the smaller particles, especially those in the 100 nm range and smaller, dissolve faster because of higher solubility and higher SSA. Since the larger particles are thermodynamically favored, the dissolved drug in solution preferentially deposits on the larger particles. Over time, the larger particles grow at the expense of the smaller particles.

Many pressurized metered-dose inhalers have been formulated as a drug dispersion in chlorofluorocarbon and hydrofluorocarbon propellants. Ostwald ripening of aerosol dispersions has been well documented and can lead to catastrophic product failure. Inhalation aerosols must deliver drug over a narrow particle size range of 0.5–10 μm and a mass median aerodynamic diameter (MMAD) of around 3–5 μm. Two issues can arise from Ostwald ripening. The MMAD may increase to the point that the drug cannot reach the critical areas in the lungs and the product is ineffective. Additionally, the particle size may increase to the point that the product no longer meets its particle size product specification.

In the 1980s, albuterol sulfate (AS) pressurized inhalers were failing a dose per actuation product specification. This was very unusual since the product had been on the market for several years. When the albuterol aluminum cans were opened and inspected, it was found that large AS crystals were adhering to the aluminum sides. Imperfections on the aluminum can serve as high energy sites for crystallization and crystal growth. As smaller drug particles dissolve, the drug tends to redeposit/recrystallize on the can sidewalls. This deposited drug is no longer available to be delivered to the lungs and drug dose delivery can fall below an acceptable level.

What seemed unusual was the fact that the aluminum can supplier had never changed. In addition, there was no record that the aluminum quality or can manufacturing had changed.

It was noted that the water content in the chlorofluorocarbon propellant dispersion was higher than normal, but well within the product specification. After an exhaustive investigation, it was finally determined that the elastomer ring sealing formulation had been changed slightly. This change led to the ability of atmospheric moisture to diffuse through the elastomer, dissolve in the CFC propellant, and then dissolve AS. It was speculated that *Ostwald ripening* was accelerated by the higher moisture content and more AS being dissolved to then recrystallize. This catastrophic failure was caused by a miniscule part per million water increase.

The problem was solved by using the "old" ring elastomer formulation. To minimize potential future crystal growth and drug loss on the side of the cans, the aluminum cans were also coated with a polymer that significantly lowered the AS aluminum surface adhesion energy.

References

Adamson, A.W. and Gast, A.P. (1997). *Physical Chemistry of Surfaces*, 6e, 58–60. New Delhi: Wiley India.

Berthelot, D. (1898). Sur le mélange des gaz. *Comptes rendus hebdomdaires des séances de l'Académie des Sciences* 126: 1703.

Binnig, G., Quate, C.F., and Gerber, C. (1986). Atomic force microscope. *PRL* 56: 930–936.

Böhme, G., Krupp, H., Rabenhorst, H., and Sandstede, G. (1962). Adhesion measurements involving small particles. *Trans. Inst. Chem. Eng.* 40: 252–259.

Bummer, PM. Interfacial phenomena. In: Troy BD, (ed.) (2006). *Remington: The Science and Practice of Pharmacy*, 21e. Philadelphia: Lippincott Williams & Wilkins: pp284–285.

Connors, K.A. and Mecozzi, S. (2010). *Thermodynamics of Pharmaceutical Systems: An Introduction to Theory and Applications*, 2e. Hoboken, NJ: John Wiley & Sons.

Daniels, F. and Alberty, R.A. (1975). *Physical Chemistry*, 4e, 254. New York: John Wiley & Sons.

Dupré, A. (1869). *Theorie Mechaniquie de la Chaleur*, 369ff. Paris: Gauthier-Villars.

Eötvös, R. (1886). *Weid Ann.* 27: 456.

Esezobo, S., Zubair, S., and Pilben, N.J. (1989). Effects of tapioca obtained from cassava (Manihot utilissima) on the disintegration and dissolution rates of paracetamol tablets. *Pharm. Pharmacol.* 41: 7–10. https://doi.org/10.1111/j.2042-7158.1989.tb06319.x.

Fowkes, F.M. (1964). Attractive forces at interfaces. *Ind. Eng. Chem.* 56: 40.

Freundlich, H. (1909). *Kapillarchemie: Eine Darstellung der Chemie der Kolloide und verwandter Gebiete [Capillary Chemistry: A presentation of colloid chemistry and related fields]*. Leipzig, Germany: Akademische Verlagsgesellschaft.

Gasior, W., Debski, A. Surdat 3: Data base of physicochemical properties of alloys. 2019. https://www.researchgate.net/publication/338008876_SURDAT_3_Database_of_physicochemical_properties_of_alloys.

Gibbs, J.W. (1931). *The Collected Works of J. W. Gibbs*, vol. 1, 219. New York: Longmans, Green.

Grimsey, I.M., Feeley, J.C., and York, P. (2002). Analysis of the surface energy of pharmaceutical powders by inverse gas chromatography. *J. Pharm. Sci.* 91: 571–583.

Guggenheim, E.A. (1945). *J. Chem. Phys.* 13: 253.

Harkins, W.D. and Huber, J.A. (1930a). Method for the determination of surface and interfacial tension from the maximum pull on a ring. *J. Am. Chem. Soc.* 52: 1751–1772.

Harkins, W.D. and Huber, J.A. (1930b). Surface tension by the ring method. *Science* 72: 73–775.

Henriksson, U. and Eriksson, J.C. (2004). Thermodynamics of capillary rise: Why is the meniscus curved. *J Chem. Ed.* 81: 150–154.

Huh, C. and Mason, S.G. (1975). A rigorous theory of ring tensiometry. *Colloid Polym. Sci.* 253: 566–580.

Huh, C. and Mason, S.G. (1977). A rigorous theory of ring tensiometry: addendum on the wall effect. *Colloid Polym. Sci.* 255: 460–467.

Kaelble, D.H. (1970). Dispersion-polar surface tension properties of organic solids. *J. Adhesion* 2: 66–81.

Kiselev, A.V. (1967). In advances. In: *Chromatography* (ed. J.C. Giddings and R.A. Keller). New York: Marcel Dekker Co.

Lamprou, D.A. and Smith, J. (2015). Chapter 21: Applications of AFM in pharmaceutical sciences. In: *Analytical techniques in pharmaceutical sciences* (ed. A. Müllertz, Y. Perrie, and T. Rades). New York: Springer.

Laureen, S. Surface free energy — what is it and how to measure it? White Paper, Biolin Scientific. info@biolinscientific.com. https://content.biolinscientific.com/wp-surface-free-energy (accessed April 2024).

Nanotechnology Act. S.189 – 21st century nanotechnology research and development act. 2003.

Nassatto, P.L., Pignon, F., Silveira, J.L.M. et al. (2014). Interfacial properties of methylcelluloses: the influence of molar mass. *Polymers* 6: 2961–2673.

Nernst, W. (1904). Theorie der Reaktionsgeschwindigkeit in heterogenen Systemen. *Z. Phys. Chem.* 1904 (47U): 52–55.

Ohnesorge, F. and Binig, G. (1993). True atomic resolution by atomic force microscopy through repulsive and attractive forces. *Science* 260 (5113): 1451–1456. https://doi.org/10.1126/science.260.5113.1451.

Ostwald, W. (1900). Über die vermeintliche Isomerie des roten undgelben Queck-silberoxyds und die Oberlfächenspannung festerKörper. *Z. Phys. Chem.* 34: 495–503.

Owens, D. and Wendt, R. (1969). Estimation of the surface free energy of polymers. *J. Appl. Polym. Sci.* 13: 1741–1747.

Pockels, A. (1914). "Über Randwinkel und Ausbreitung von Flüssigkeiten auf festen Körpern" (on contact angles and the flow of fluids on solid bodies). *Physikalische Zeitschrift* 15: 39.

Podczeck, F. (1998). *Particle-Particle Adhesion in Pharmaceutical Powder Handling*, p143. London: World Scientific Publishing Co.

Rabel, W. (1971). Einige Aspekte der Benetzungstheorie und ihre Anwendung auf die Untersuchung und Veränderung der Oberflächeneigenschaften von Polymeren. *Farbe und Lack* 77: 997–1005.

Rumpf, H. (1962). The strength of granules and agglomerates. In: *Agglomeration* (ed. W. Knepper), 379–414. New York: Interscience.

SINTERFACE *Technologies*, 2003. https://www.mpikg.mpg.de/4862784/BPA-Info.pdf (accessed June 2023).

Sinko, P.J. (2023). *Martin's Physical Pharmacy and Pharmaceutical Sciences*, 8e. Baltimore, MD: Lippincott Williams & Wilkins.

Thomson, W (Lord Kelvin) (1871). On the equilibrium of vapour at a curved surface of liquid. *Philosophical Magazine, Series* 42 (282): 448–452.

Tiberg, F., Zhmud, B., Hallstensson, K., and von Bahr, M. (2000). Capillary rise of surfactant solutions. *Phys. Chem. Chem. Phys.* 2: 5189–5196.

Tran, D.T., Bittner, R., and Zámostný. (2021). Adhesion force measurement by centrifuge technique as a tool for predicting interactive mixture stability. *Chem. Eng. Res. Design.* 165: 467–476. https://doi.org/10.1016/j.cherd.2020.10.012.

Van Oss, D.J., Chaudhury, M.K., and Good, R.J. (1988). Interfacial Lifshitz-van der Waals and polar interactions in macroscopic systems. *Chem. Rev.* 88: 927.

Wu, S. (1973). Polar and nonpolar interactions in adhesion. *J. Adhes.* 5: 5.

Wurster DE. *Personal Communication.* Apr 2024.

Young, T. (1805). An essay on the cohesion of fluids. *Philos. Trans.* 95: 65–87. https://doi.org/10.1098/rstl.1805.0005.

Chapter 7 Problem Set

7.1 Describe the interfacial surface region.

7.2 Define *interfacial tension.*

7.3 Define *interfacial energy.*

7.4 Show why interfacial tension and interfacial energy values can be interchanged.

7.5 What is the difference between surface tension and interfacial tension?

7.6 Name four techniques that can be used to measure the surface tension of liquid-vapor interfaces and briefly describe them.

7.7 How is surface tension effected by temperature?

7.8 Using surface tension and interfacial tension terms, what are the expressions for cohesive and adhesive work?

7.9 What is the equation that describes the spreading of two immiscible liquids one (test liquid) on the other (substrate liquid)?

7.10 What is the spreading coefficient when a drop of water is placed on mineral oil?
Given:

Surface tension of water is 71.8 mN m^{-1}
Surface tension of mineral oil is 28 mN m^{-1}
Interfacial tension of water-mineral oil is 4 mN m^{-1}
$S_{test\ liquid/substrate\ liquid} = \gamma_{substrate\ liquid} - (\gamma_{test\ liquid} + \gamma_{test\ liquid,\ substrate\ liquid})$

7.11 What is the spreading coefficient of a film-coating solution consisting of 9% hydroxypropyl methylcellulose (HPMC) and 1% polyethylene glycol (PEG) 400 on a compressed tablet consisting of 75.2% microcrystalline cellulose (MCC), 24.2% lactose, 0.4% magnesium stearate, and 0.2% silicon dioxide (all w/w %).
Given:

Film coating surface tension is 45.6 mN m^{-1}
Contact angle is 43°
$S_{test\ liquid/substrate\ solid} = S = \gamma_{lv}\ (\cos\theta - 1)$

7.12 What is the adhesive work equation?

7.13 Using the OWRK model equation, calculate a polymer's polar surface energy effects.
Given:

Water

$\gamma_{Wlv}{}^{Tot} = 72.8$ mN m^{-1}
$\gamma_{Wlv}{}^{d} = 21.8$ mN m^{-1}
$\gamma_{Wlv}{}^{P} = 51.0$ mN m^{-1}
$\theta = 83°$

Glycerin

$\gamma_{Wlv}{}^{Tot} = 64.0$ mN m^{-1}
$\gamma_{Wlv}{}^{d} = 34.0$ mN m^{-1}
$\gamma_{Wlv}{}^{P} = 30.0$ mN m^{-1}

OWRK Eq.

$$0.5\ \gamma_{lv}\ (\cos\theta + 1) = (\gamma_{sv}{}^{d}\ \gamma_{lv}{}^{d})^{0.5} + (\gamma_{sv}{}^{P}\ \gamma_{lv}{}^{P})^{0.5}$$

7.14 (a) Calculate the pressure difference across a 10.0-mm-diameter bubble. (b) Calculate the pressure inside the 10 mm bubble. The bubble film is comprised of soapy water, $\gamma = 25.0$ mN m^{-1}.

7.15 Calculate the relative solubility increase predicted for a 10-nm-radius particle (20 nm diameter).

Given:

$\gamma = 30.0$ mN m^{-1} = 0.0300 N m^{-1}
$M = 500$ g mol^{-1} = 0.500 kg mol^{-1}
$r = 10.0$ nm = 1.00E-8 m
$\rho = 1.00$ g cm^{-3} = 1000 kg m^{-3}
$R = 8.314$ N·m·mol^{-1}·K^{-1}
$\ln S/S_\infty = (2\gamma M)/(r\rho RT)$

7.16 Calculate the SSA (m^2 g^{-1}) of a 20.0-nm-diameter particle.

Given:

Material density = 1.00 g cm^{-3}
Specific surface area = 6/(density diameter)
1 nm = 1E-9 m
1 cm^3 = 1E-6 m^3

8

Adsorption Phenomena

8.1 Introduction

Adsorption phenomena play key roles in drug product design. Gas adsorption studies are frequently used to characterize a pharmaceutical material's surface area and porous nature. The surface area and porous properties impact compressibility, flowability, and dissolution, to name a few. Drug adsorption to container closure systems can affect delivered dose or stability. Adsorption of proteins to glass and plastic container walls is known to lead to drug dose loss and protein denaturation. Drug adsorption to plastic on metered dose inhalers needs to be accounted for. Water vapor adsorption to drugs and excipients can result in powder agglomeration, flow issues, softening of tablets, and drug stability issues.

Adsorption occurs onto solids as the result of their surface energy, which leads to the adhesion of atoms, molecules, and ions. In addition, the Lennard-Jones (1931) intermolecular interaction energy curve shows that as a neutral atom or molecule approaches the solid surface, the potential energy of the system decreases.

$$V = 4\varepsilon\left[(\sigma/r)^{12} - (\sigma/r)^{6}\right] \tag{8.1}$$

Where V is the intermolecular potential of the two approaching particles, ε is the depth of the potential energy well, σ is the distance between particles when $V = 0$, and r is the distance between the particles. The $(\sigma/r)^6$ term originates from the van der Waal's attraction theory. The $(\sigma/r)^{12}$ is associated with atom or molecule repulsion. The exponent, 12, is an empirical approximation of the rapid increase in repulsion. The power of 12 is convenient because it is the square of 6, which is the exponent of the attractive van der Waal's term. This equation assumes that the solid surface is homogeneous and the only interaction is via London dispersion force. As discussed later, most solid surfaces are heterogeneous, and the adsorption depends on the physicochemical nature of the adsorbate and adsorptive.

The adsorbed atoms, molecules, or ions form a *two-dimensional phase* or *film*. This two-dimensional film retains translational freedom and is referred to as a *mobile* film. When a film loses its mobility, it is called an *immobile film*. The film is comprised of an *adsorbate*, which is adhered to the solid surface *substrate*, also known as the *adsorbent*. Finally, *adsorptives* are the atoms, molecules, and ions that undergo the adsorption process to become adsorbates.

Thermodynamics for Drug Product Design: A Guide for Pharmaceutical Scientists, First Edition. William Craig Stagner.

8.1.1 Physisorption and Chemisorption

Atoms, molecules, or ions can adhere to solid surfaces by two general mechanisms: physical adsorption or *physisorption,* and chemical bonding or *chemisorption*. This chapter is focused on physisorption. Although the distinctions between the two processes can overlap, Table 8.1 highlights the common differences between the two adsorption mechanisms.

8.1.2 Adsorption and Absorption

Adsorption and *absorption* only differ by one letter, but they are very different processes. Adsorption involves the interaction of atoms, molecules, and ions to the surface of a solid substrate. Absorption is the mass transfer of atoms, molecules, and ions into the interior, as opposed to the surface, of a material. Table 8.2 provides a summary of the differences between physical adsorption and absorption. In the case where it is not clear-cut if the mechanism is adsorption

Table 8.1 Typical Characteristics of Physisorption and Chemisorption.

Properties	Physisorption	Chemisorption
Nature	Reversible	Nonreversible
Bonding interaction	van der Waals forces (London dispersion force, Debye force, and Keesom force), hydrogen bonding, and ion–ion interactions	Chemical interactions involving orbital overlap and charge transfer (that is, covalent bond formation)
Adsorption enthalpies	5–50 kJ mol^{-1}[a]	40–800 kJ mol^{-1}
Saturation uptake	Mono and multilayer	Monolayer
Surface specificity	Often, but not required in all cases	Required
Temperature range	Favored at low temperature. Near or below condensation of gas. Adsorption is exothermic and its extent decreases with increase in temperature	Increasing temperature increases the extent of adsorption, and effects may be complex.[b]
Adsorption kinetics	Fast	Variable. Activated process.

[a] Related to mass and polarity, similar to heats of liquefaction.
[b] Increase in temperature increases the kinetic energy that acts as the activation energy for the chemical reaction to occur.

Table 8.2 Differences between Adsorption and Absorption.

Properties	Adsorption (physical)	Absorption
Phenomenon	Surface	Bulk
Enthalpy	Exothermic	Endothermic
Temperature	Favored at low temperatures	Favored at higher temperatures
Reaction Rate	Fast	Diffusion controlled
Concentration	Surface concentration differs from bulk concentration	At equilibrium, uniform concentration in bulk

or absorption, or that both adsorption and absorption occur in the same material, the terms *sorbate, sorbent,* and *sorptive* are used.

8.1.3 Adsorption and Desorption Profiles

Adsorption designates the continual increase in adsorbate amount with an increase in adsorptive exposure. *Desorption* conveys the counter process where the adsorbate amount progressively decreases. An adsorption process or desorption process indicates the experimental direction from which the data were obtained. An *adsorption–desorption isothermic profile* is obtained experimentally when a pretreated degassed sample undergoes onward adsorptive interaction with a predesignated adsorptive exposure level. It should be noted that degassing includes all gases, including water vapor. The desorption process is then initiated and is stopped when the original low adsorptive exposure level is reached. A constant temperature plot of the adsorption and desorption data results in *adsorption–desorption profile.*

It is not uncommon for the adsorption and desorption isotherms to coincide with each other. *Adsorption–desorption hysteresis* (ADH) occurs when the adsorption and desorption curves are not superimposable. There are several possible reasons for this hysteresis that will be discussed later.

8.1.4 Porous Absorbents

Porosity is the ratio of the total pore volume to the volume of the material. Porous adsorbents are divided into *internal* and *external surface* materials. *Internal surfaces* contain *macropores, mesopores,* and *micropores* having pores and widths of greater than 50 nm, between 2 nm and 50 nm, and not exceeding 2 nm, respectively. External surfaces are considered to be nonmicroporous in nature. The above-mentioned definition of an internal surface is convenient but seldom completely accurate. Typically, even a very porous material has some external surface (Thommes et al. 2015).

8.2 Adsorption Data

8.2.1 Data Reporting

The most convenient method to report adsorption–desorption isotherms is graphically. Figure 8.1 shows the updated 2014 IUPAC Physisorption Isotherm Classification System, consisting of six vapor-solid isothermal adsorption–desorption profiles. The *x*-axis is recorded as relative pressure, P/P^o, where P is the equilibrium vapor pressure and P^o is the saturation pressure of the vapor at the experimental temperature. The dependent variable is the amount adsorbed in mg or the concentration of moles of adsorbate per gram of degassed adsorbent. Moisture adsorption–desorption profiles often report the independent variable as percent relative humidity (%RH) and the dependent variable as percent weight change. %RH is simply $(P/P^o) \cdot 100$.

8.2.2 Reversible Type I Isotherms

Type I isotherms are reversible, and the curvature is concave to the *x*-axis. The shape is the result of microporous solids that have a small portion of nonmicroporous external surface or solids, which only have external surfaces. The limited amount of adsorption is governed by accessible pore

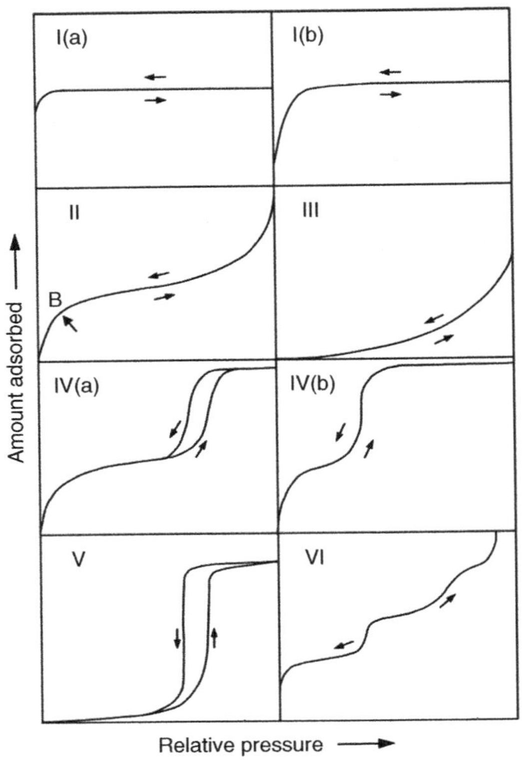

Figure 8.1 Physisorption isotherm classification system. *Source:* Thommes et al. (2015)/Walter de Gruyter GmbH.

volume and external surface. The rapid uptake at low relative pressures is due to strong adsorbent–adsorbate interactions. Slow uptake of adsorptive gas is demonstrated for adsorbents having a pore size that covers a broader range.

8.2.3 Reversible Type II Isotherm

Nonporous or microporous solids are responsible for reversible Type II isotherms. A "sharp" concave knee at point B, usually represents the completion of the monolayer coverage and the transition from monolayer adsorption to multilayer adsorbate formation that continues throughout the almost linear portion of the curve. A more gradual concave curvature suggests that monolayer and multilayer formation coincides. At P/P^o approaching or equal to 1, the adsorbed multilayer seems to increase without limit.

8.2.4 Reversible Type III Isotherm

Nonporous or microporous solids are also responsible for the reversible Type III isotherms. The initial portion of the curve is convex to the P/P^o axis and exhibits a gradual increase in amount adsorbed. It is posited that the slow increase of amount adsorbed at low P/P^o is due to weak adsorbate–adsorbent interactions. Past monolayer coverage, the rate of surface coverage increases with each increment of P/P^o because the adsorbate interacts with itself more strongly than it interacts with the solid surface. At the saturation pressure, $P/P^o = 1$, the amount adsorbed reaches a limit.

8.2.5 Reversible Type IV Isotherm

Reversible Type IV isotherms arise from 2 nm to 50 nm mesoporous substrate materials. The adsorption phase is similar in shape to the Type II isotherms, which suggests monolayer formation before multilayer formation occurs. The overall adsorption–desorption profile is a function of the adsorbate–adsorbent interaction energy and the appearance of condensed adsorbate in the mesopores at pressures less than the saturation pressure, P^o. The plateau region beyond the convex curvature seen from the *x*-axis is characteristic of the Type IV profile.

Type IVa isotherms show hysteresis due to capillary condensation in mesopores wider than 4 nm. In general, capillary condensation occurs due to increased attractive forces among multilayer adsorption gas molecules in a confined space. This condensation occurs below P^o and is explained by the Kelvin equation (see Section 7.11) The hysteresis occurs because the pore filling and pore emptying occur by two different processes. In general, for a closed pore, condensation is initiated by a nucleation step. Condensation is thought to take place from the closed end toward the opening. During pore desorption, the evaporation of a reversible liquid–vapor transition starts at the pore opening and continues to evaporate at the meniscus surface. Section 8.6 will discuss hysteresis in more detail.

Absorbents having smaller widths than 4 nm do not exhibit condensation or hysteresis (Type IVb isotherms). Note that isotherm IVb is still distinguished by an amount adsorbed plateau region of high relative pressure.

8.2.6 Reversible Type V Isotherm

Isotherm V's shape at low relative pressure is similar to isotherm III. As noted with Type III, the slow increase in amount adsorbed is due to weak adsorbent–adsorbate interactions and a lack of favorable surface interaction sites. At higher relative pressures, the slope of the adsorption curve increases sharply. The adsorption shape is most likely due to delayed condensation in the micropores and mesopores. The high P/P^o plateau is due to ultimate adsorption being limited. The sharp desorption is explained by the "narrow-neck ink bottle effect," where the neck diameter is uniform. Once evaporative conditions are met, the condensed phase undergoes rapid and uniform evaporation.

8.2.7 Reversible Type VI Isotherm

Type VI isotherms show superimposable adsorption–desorption as seen in Types I, II, III, and IVb. The height of each step function is related to the capacity of each adsorbed layer. At high P/P^o less than 1, the amount adsorbed does not reach a limit.

8.2.8 General Observations

A favorable adsorption isotherm has a concave curvature as seen from the *x*-axis. Favorable adsorption means that a relatively large amount of adsorption occurs at low relative pressures of less than 0.25 P/P^o or so. The concave curves are shown for Type I, II, IV, and VI isotherms.

Convex curvature to the *x*-axis represents unfavorable adsorption. The unfavorable adsorption is attributed to weak adsorbate–adsorbent interaction forces. Adsorption occurs at higher relative pressures of around 0.25 to 0.50 P/P^o.

Hysteresis shown in Type IVa and V is the result of the condensation of the vapor in the confined small spaces of mesopores and micropores.

8.3 Freundlich, Langmuir, and Brunauer, Emmett, and Teller Adsorption Isotherm Equations

A *gas/vapor adsorption study* involves exposing a degassed adsorbent to increasing relative gas/vapor pressures at constant temperature. A plot of the amount gas adsorbed versus adsorptive relative gas pressure gives the adsorption profile. A *solute adsorption study* involves adding a known amount of adsorbent to the adsorptive solution and measuring the equilibrium adsorptive concentration. In this case, a plot of the amount adsorptive adsorbed per amount adsorbent versus equilibrium adsorptive concentration gives the adsorption profile.

There is an important distinction to be made between vapor/solid adsorption systems and solution/solid adsorption systems. A vapor/solid system is, at minimum, a two-component system. The solid surface can have a certain fraction of the solid surface covered by the adsorptive gas/vapor and a certain fraction of the surface being bare solid or unbound surface.

Adsorption from solution involves, at minimum, a three-component system, consisting of solute, solvent, and solid surface. The surface of the solid is never bare; all of the solid surface must be covered by either solute, solvent, or a mixture of both.

The foregoing distinction is important in terms of the model equations to be used. This can best be illustrated with the Langmuir equation. The Langmuir (1916a, 1916b, 1918) equation was originally derived for vapor/solid adsorption, and the derivation involves a kinetic argument for a two-component system. The derivation depends upon a certain fraction of the solid surface being covered and a certain fraction of the surface being bare or unbound solid.

As stated earlier, adsorption from solution involves at a minimum, three components and a solid surface that is never bare. Thus, a thermodynamic argument employing an equilibrium constant must be employed. The reactant side of the equilibrium is adsorbed solvent and solute in solution. The product side is solvent in solution and adsorbed solute. The final equation looks like the Langmuir equation, and the variables have close to the same meaning. However, since the assumptions between the kinetic argument and the thermodynamic argument are different, there are slight differences in the meanings of the parameters, which will be discussed later. Thus, when the Langmuir equation is derived using a thermodynamic argument for the adsorption of an adsorptive solute from solution, it is referred to as the "Langmuir-like equation." These distinctions follow that of Wurster (2024)

The fundamental differences between the two adsorption systems have been ignored by many authors. Given the number of equations used to model adsorption data that will be presented, this textbook will not attempt to clarify the record for all of them. The selection of a truly applicable equation will be left to the reader. There are two, three, four, and five-parameter adsorption isotherm equations that have been used to model the adsorption process.

William Henry's linear isotherm adsorption model, which was developed in the early 1800s, had been used to describe a partitioning phenomenon. Henry's isotherm equation has served as a prototypical equation for those that followed (Eq. 8.2). This early model was limited to explaining low pressure or solute adsorption phenomena.

$$q_e = K_{HE}\ P/P^o = K_{HE}\, C_e \tag{8.2}$$

For solution adsorption studies, q_e is the extent or amount adsorbed as adsorptive mass/adsorbent mass. q_e increases with increasing adsorptive-solute solution concentration. K_{HE} is Henry's adsorption constant, and it gives a measure of the *adsorbent's adsorption capacity* C_e is the equilibrium concentration of the *adsorptive solute solution* after exposure to a known amount of adsorbent.

The specific units for solute adsorption study equations are dependent on the adsorption capacity of the system, which depends on the adsorbate–adsorbent interactions. For example, for unfavorable adsorption processes, it may take grams of adsorbent to adsorb only milligrams of adsorptive. On the other hand, highly favorable adsorption may result in grams of adsorptive adsorbing to a gram of adsorbent.

The *generic units* and *commonly used units* for Henry's equation and other adsorption isotherms are: q_e (adsorptive mass adsorbed/mass adsorbent and $mg_{adsorptive\ adsorbed}/g_{adsorbent}$), K_{HE} is Henry's constant (specific units should provide dimensional consistency), and C_e is the equilibrium concentration of the adsorptive-solute solution after exposure to the adsorbent. Generic C_e units are in mass adsorptive solute or moles adsorptive solute per volume of solution. The commonly used terms are $mg_{adsorptive\ solute}/L$ solution. K_{HE} units are commonly L/g adsorbents.

The three most commonly used adsorption isotherm equations are the Freundlich, Langmuir, and Brunauer, Emmett, and Teller (BET) models. These are discussed in the next sections.

8.3.1 Freundlich Adsorption Isotherm Equation

In 1907, Herbert Freundlich developed an empirical exponential equation that modeled experimental adsorption data. Later, this empirical model was derived from thermodynamic principles by Adamson and Gast (1997). The Freundlich equation was first developed for vapor adsorption onto solid heterogenous surfaces. It has, however, been successfully employed for adsorptive-solute solution solid surface adsorption processes. This section's discussion will be limited to solution/solid adsorption. Freundlich's exponential equation is provided in (Eq. 8.3)

$$q_e = K_F\, C_e^{1/n} \tag{8.3}$$

where q_e is the adsorption extent ($mg_{adsorptive\ adsorbed}/g_{adsorbent}$) and K_F is the Freundlich isotherm constant. K_F has units $L_{solution}/g_{adsorbent}$ and is related to adsorbent–adsorption capacity, C_e ($mg_{adsorptive\ (solute)}/L_{solution}$) is the equilibrium concentration of the adsorptive-solute solution after exposure to a known amount of adsorbent, and n is a dimensionless parameter defined as the *adsorption intensity*, which is reflected in the shape of the adsorption curve.

A plot of q_e versus C_e gives the adsorption curve. When $1/n$ is less than 1, and n is greater than 1, the adsorption curve is concave to the x-axis (C_e). This isotherm is similar to the Type I curves. As discussed earlier, this indicates *favorable adsorption*. When n equals 1, the Freundlich equation reduces to Henry's equation. When $1/n$ is greater than 1, and n is less than one, the adsorption curve is convex as observed from the x-axis. This curve is similar to isotherm Type III curves and implies *unfavorable adsorption*.

Equation (8.3) can be linearized by taking the log of both sides of the equation to give:

$$\log\ q_e = \log\ K_F + 1/n\ \log\ C_e \tag{8.4}$$

It is generally accepted that adsorption data that fits the Freundlich equation can accommodate monolayer and multilayer results from adsorption on heterogeneous adsorbent sites. A short coming of the Freundlich equation is that it does not predict adsorption saturation as seen with Type I, IV, and V isotherms. A key drawback is that the Freundlich equation does not accurately predict adsorption behavior at high pressures.

8.3.2 Langmuir Adsorption Isotherm Equation

Irving Langmuir first reported his well-known isotherm model in 1916 and 1918. The research was published in 1918. In 1932, he was awarded the Nobel Prize in chemistry for his research in surface chemistry. The Langmuir equation is applicable for gas/vapor-solid adsorption processes. As stated earlier, a different derivation is required for the adsorption-from-solution case. The thermodynamically derived equation is referred to as the Langmuir-like equation. The derivation for both systems follows.

The question is often asked: "Why derive these equations?" The short answer is deriving them shows us why they work. As will be highlighted in the derivations, every equation has limitations and assumptions. Derivations help identify these limitation and assumptions and explain their context. Being aware of these helps us choose when equations are valid and when they might lead to inaccurate results. This is crucial to making informed decisions. Finally, by deriving these equations, the underlying principles and assumptions leads to deeper discernment and understanding of the thermodynamic processes at play and how to apply them to real-world problems.

8.3.2.1 Langmuir Equation for Vapor/Gas Adsorption onto an Adsorbent

Langmuir imagined a solid surface to be comprised of high-vapor adsorptive–adsorbent interaction sites. Derivation of the Langmuir equation provides insight into the physical meaning of the equation and the individual parameters. The derivation follows that of Connors and Mecozzi (2010). The derivation assumes the adsorption and desorption processes occur at a constant temperature and are described by first-order rate processes shown in Eqs. (8.5) and (8.6).

$$\text{Adsorption rate} = k_a \cdot P/P^o{}_{\textit{adsorptive-vapor}} \cdot A_{\textit{unbound sites}} \tag{8.5}$$

$$\text{Desorption rate} = k_d \cdot A_{\textit{adsorbate bound}} \tag{8.6}$$

where k_a is the adsorption rate constant, $P/P^o{}_{adsorptive\text{-}vapor}$ is the relative pressure of the adsorptive vapor where P is the equilibrium vapor pressure and P^o is the saturation vapor pressure at the experimental temperature, $A_{unbound\ sites}$ is the solid adsorbent surface area per unit mass in units of $cm^2/g_{adsorbent}$ that is unbound, k_d is the desorption rate constant, and $A_{adsorate\ bound}$ is the solid adsorbent surface area per mass adsorbent that has *bound* adsorbate. $A_{adsorbate\ bound}$ is frequently given in units of $cm^2/g_{adsorbent}$. The adsorption rate equation suggests that the adsorption rate is directly proportional to the adsorptive vapor relative vapor pressure and the adsorbent surface area of the unbound sites. On the other hand, the desorption rate is directly proportional to the surface area occupied by adsorbate molecules.

At equilibrium, the adsorption and desorption rates are equal (see Section 5.1.1.3)

$$k_a \cdot P/P^o{}_{\textit{adsorptive}-\textit{vapor}} \cdot A_{\textit{unbound sites}} = k_d \cdot A_{\textit{adsorbate bound}} \tag{8.7}$$

The total surface area of available absorption sites is equal to the surface area of the unoccupied or unbound sites and adsorbate-bound sites, which is given by Eq. (8.8).

$$A_{\textit{total}} = A_{\textit{unbound sites}} + A_{\textit{adsorate bound}} \tag{8.8}$$

and

$$A_{\textit{unbound sites}} = A_{\textit{total}} - A_{\textit{adsorbate bound}} \tag{8.9}$$

where A_{total} is the total surface area per unit mass of the adsorbent.

At equilibrium, the equilibrium constant for the adsorption/desorption process is K_L, the Langmuir constant. K_L is equal to k_a/k_d,(see Section 5.1.1.3). K_L is related to the adsorption capacity of the adsorbent. Rearranging Eq. (8.7), it follows that:

$$k_a/k_d = K_L = A_{adsorbate\ bound}/P/P^o{}_{adsorptive\text{-}vapor} \cdot A_{unboundsites} \tag{8.10}$$

Substituting for $A_{unbound\ sites}$ from Eq. (8.9) gives:

$$K_L = A_{adsorbate\ bound}/P/P^o{}_{adsorptive\text{-}vapor} \cdot (A_{total} - A_{adsorbate\ bound}) \tag{8.11}$$

Multiplying both sides by $P/P^o{}_{adsorptive\text{-}vapor}$ gives:

$$K_L \cdot P/P^o{}_{adsorptive\text{-}vapor} = A_{adsorbate\ bound}/(A_{total} - A_{adsorbate\ bound}) \tag{8.12}$$

A new term θ, *degree of saturation,* is defined as

$$\theta = A_{adsorbate\ bound}/A_{total} \tag{8.13}$$

θ is the fraction of the total available adsorption sites that are bound.

It follows that:

$$A_{adsorbate\ bound} = \theta \cdot A_{total} \tag{8.14}$$

Substituting into Eq. (8.12) for $A_{adsorbate\ bound}$ from Eq. (8.14) gives:

$$K_L \cdot P/P^o{}_{adsorptive\text{-}vapor} = \theta \cdot A_{total}/(A_{total} - \theta \cdot A_{total}) \tag{8.15}$$

On the right-hand side of Eq. (8.15), the numerator and denominator are divided by A_{total} and results in Eq. (8.16).

$$K_L \cdot P/P^o{}_{adsorptive\text{-}vapor} = \theta/(1-\theta) \tag{8.16}$$

Equation (8.16) can be rearranged so *degree of saturation* is expressed as the dependent variable. This is accomplished in five steps:

$$\begin{aligned}
&\textit{Step 1.}\ \theta = K_L \cdot P/P^o{}_{adsorptive\text{-}vapor} \cdot (1-\theta)\\
&\textit{Step 2.}\ \theta = K_L \cdot P/P^o{}_{adsorptive\text{-}vapor} - K_L \cdot P/P^o{}_{adsorptive\text{-}vapor}\,(\theta)\\
&\textit{Step 3.}\ \theta = K_L \cdot P/P^o{}_{adsorptive\text{-}vapor}/\left(1 + K_L \cdot P/P^o{}_{adsorptive\text{-}vapor}\right)\\
&\textit{Step 4.}\ \theta\left(1 + K_L \cdot P/P^o{}_{adsorptive\text{-}vapor}\right) = K_L \cdot P/P^o{}_{adsorptive\text{-}vapor}\\
&\textit{Step 5.}\ \theta + K_L \cdot P/P^o{}_{adsorptive\text{-}vapor}(\theta) = K_L \cdot P/P^o{}_{adsorptive\text{-}vapor}
\end{aligned} \tag{8.17}$$

Now, Eqs. (8.13) and (8.17) can be set equal to each other to give the Langmuir equation at constant temperature (8.18).

$$\theta = A_{adsorbate\ bound}/A_{total} = K_L \cdot P/P^o{}_{adsorptive\text{-}vapor}/\left(1 + K_L \cdot P/P^o{}_{adsorptive\text{-}vapor}\right) \tag{8.18}$$

θ, degree of saturation, can be determined by using sensitive modern gas adsorption instrument. By knowing the surface area of the adsorbent and the volume of gas adsorbed, θ can be calculated.

The general shape of the Langmuir equation at constant temperature is shown in Figure 8.2.

At very low relative pressures, $K_L \cdot P/P^o << 1$, the denominator approaches 1 and $\theta = K_L \cdot P/P^o{}_{adsorptive\text{-}vapor}$. The Langmuir equation is linear in this portion of the curve and is often referred to as the Henry's law region. At very high relative pressures, $K_L \cdot P/P^o >> 1$, the

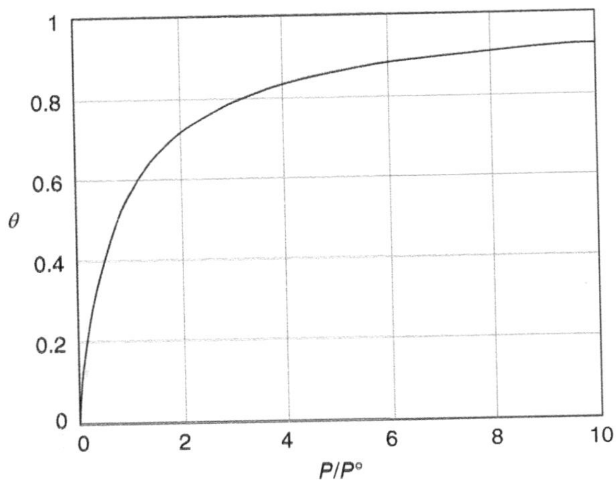

Figure 8.2 Generic concave plot of the Langmuir equation (Eq. 8.18), where θ is the fraction of the total available adsorption sites that are bound at a relative vapor (P/P^o).

denominator approaches $K_L \cdot P/P^o$, and the fraction of the total available adsorption sites that are bound increases asymptotically to 1. This indicates that adsorption becomes increasingly more difficult.

As seen from Figure 8.2, the Langmuir equation is not a linear equation. The next series for algebraic steps are performed to linearize the equation. Equation (8.19) can be derived from Eq. (8.18)

$$A_{adsorbate\ bound} = A_{total} \cdot K_L \cdot P/P^o/(1 + K_L \cdot P/P^o) \tag{8.19}$$

It is generally accepted that the volume of a gas that adsorbs onto an adsorbent area is proportional to the volume of the gas adsorbed. A typical vapor adsorption instrument measures the volume of adsorptive gas/vapor that is adsorbed at a constant temperature and relative pressure. In this case, the fraction of adsorptive gas bound can be given in terms of volume:

$$\theta = V/V_m \tag{8.20}$$

where V is the volume of adsorbed gas and V_m is the volume of gas required to form a monolayer. Substituting V for $A_{adsorbate\ bound}$ and V_m for A_{total} in Eq. (8.19) results in the Langmuir equation that is based directly on gas volume measurements taken at constant pressure.

$$V = V_m \cdot K_L \cdot P/P^o/(1 + K_L \cdot P/P^o) \tag{8.21}$$

Equation (8.21) can be linearized by making some algebraic rearrangements. First, multiple both sides of Eq. (8.21) by $(1 + K_L \cdot P/P^o)$ to give:

$$V(1 + K_L \cdot P/P^o) = V_m \cdot K_L \cdot P/P^o \tag{8.22}$$

Multiply V through the $(1 + K_L \cdot P/P^o)$ expression:

$$V + V \cdot K_L \cdot P/P^o = V_m \cdot K_L \cdot P/P^o \tag{8.23}$$

Divide both sides of Eq. (8.23) by P/P^o to give:

$$V/\ (P/P^o) + V \cdot K_L = V_m \cdot K_L \tag{8.24}$$

Now, subtract $V \cdot K_L$ from both sides to get the linear form of the Langmuir equation in terms of readily measurable quantities:

$$V/(P/P^o) = V_m \cdot K_L - V \cdot K_L \tag{8.25}$$

where a plot of $V/(P/P^o)$ versus V results in a slope of $-K_L$ and an intercept of $V_m \cdot K_L$. Once the slope is determined, V_m can be calculated from the intercept of $V_m \cdot K_L$. As a reminder, K_L is equal to k_a/k_d, the adsorption and desorption rate constants, and is related to the adsorption capacity of the adsorbent.

The operating assumptions for the kinetically derived Langmuir equation are:

1) The surface consists of a number of adsorption sites.
2) All of the sites are energetically equal.
3) The adsorbed phase is confined to a monolayer.
4) Lateral interactions between adsorbate molecules do not occur.

8.3.2.2 Langmuir-Like Equation for Adsorptive-Solute Solution Adsorption onto an Adsorbent

As mentioned earlier, an adsorptive solute – adsorbent system has a minimum of three-components, consisting of solvent, solute, and solid surface. The surface of the solid is never bare; all of the solid surface must be covered by either solvent, solute, or a mixture of both. The derivation of the Langmuir-like equation follows that of Wurster (2024).

The Langmuir-like equation model assumptions are:

1) A given number of sites are available on the surface for adsorption.
2) The adsorption sites are energetically equal.
3) The adsorbed phase is confined to a monolayer.
4) There are no lateral interactions between adsorbed molecules. That is, an ideal surface solution is assumed in which lateral interactions are independent of surface concentration.
5) Molecules of component 1 and component 2 occupy the same area on the solid surface. This assumption is seldom true, but the lack of obeyance is seldom troublesome.
6) The solution from which adsorption is occurring is dilute.
7) There is no mixed film formation at maximum surface coverage of the solute.

The adsorptive solute – adsorbent system can be described by the following equilibrium scheme:

$$\text{Adsorbed solvent} + \text{solute in solution} \rightleftharpoons \text{adsorbed solute} + \text{solvent in solution} \tag{8.26}$$

The Langmuir-like equilibrium expression is:

$$K_{Langmuir\text{-}like} = \left({a^{surface}}_2 \cdot {a^{solution}}_1\right) / \left({a^{surface}}_1 \cdot {a^{solution}}_2\right) \tag{8.27}$$

where a is the activity coefficient; subscripts 1 and 2 represent the solvent and solute, respectively; and the superscripts *surface* and *solution* indicate surface and bulk constituents, respectively. If the two-dimensional surface solution is assumed to be ideal, the activity coefficients can be replaced by mole fraction at the surface, resulting in Eq. (8.28).

$$K_{Langmuir\text{-}like} = \left({X^{surface}}_2 \cdot {a^{solution}}_1\right) / \left({X^{surface}}_1 \cdot {a^{solution}}_2\right) \tag{8.28}$$

The adsorbed surface phase contains only two components, the adsorptive solute and the solvent. It follows:

$$X^{surface}{}_1 + X^{surface}{}_2 = 1 \tag{8.29}$$

and Eq. (8.28) can be rewritten:

$$K_{Langmuir\text{-}like} = (X^{surface}{}_2 \cdot a^{solution}{}_1)/[(1 - X^{surface}{}_2) \cdot a^{solution}{}_2] \tag{8.30}$$

Equation. (8.30) is multiplied by $[(1 - X^{surface}{}_2) \cdot a^{solution}{}_2]$ on both sides to give:

$$K_{Langmuir\text{-}like}[(1 - X^{surface}{}_2) \cdot a^{solution}{}_2] = (X^{surface}{}_2 \cdot a^{solution}{}_1) \tag{8.31}$$

Now, both sides of Eq. (8.31) are divided by $a^{solution}{}_1$:

$$X^{surface}{}_2 = K_{Langmuir\text{-}like}[(1 - X^{surface}{}_2) \cdot a^{solution}{}_2] \; /a^{solution}{}_1 \tag{8.32}$$

Then multiply $[(1 - X^{surface}{}_2) \cdot a^{solution}{}_2]$ by $K_{Langmuir\text{-}like}$ to give:

$$X^{surface}{}_2 = [K_{Langmuir\text{-}like} \cdot a^{solution}{}_2 - K_{Langmuir\text{-}like} \cdot a^{solution}{}_2 \cdot X^{surface}{}_2] \; /a^{solution}{}_1 \tag{8.33}$$

Divide the right side of Eq. (8.33) by $a^{solution}{}_1$.

$$X^{surface}{}_2 = K_{Langmuir\text{-}like} \cdot a^{solution}{}_2/a^{solution}{}_1 - K_{Langmuir\text{-}like} \cdot (a^{solution}{}_2/a^{solution}{}_1) \cdot X^{surface}{}_2 \tag{8.34}$$

Add $K_{Langmuir\text{-}like} \cdot (a^{solution}{}_2/a^{solution}{}_1) \cdot X^{surface}{}_2$ to both sides of Eq. (8.34).

$$X^{surface}{}_2 + K_{Langmuir\text{-}like} \cdot (a^{solution}{}_2/a^{solution}{}_1) \cdot X^{surface}{}_2 = K_{Langmuir\text{-}like} \cdot a^{solution}{}_2/a^{solution}{}_1 \tag{8.35}$$

Collect common terms on the left side of the Eq. (8.35).

$$X^{surface}{}_2 \, (1 + K_{Langmuir\text{-}like} \cdot (a^{solution}{}_2/a^{solution}{}_1) = K_{Langmuir\text{-}like} \cdot a^{solution}{}_2/a^{solution}{}_1 \tag{8.36}$$

Divide both sides for Eq. (8.36) by $(1 + K_{Langmuir\text{-}like} \cdot (a^{solution}{}_2/a^{solution}{}_1)$.

$$X^{surface}{}_2 = (K_{Langmuir\text{-}like} \cdot a^{solution}{}_2/a^{solution}{}_1)/(1 + K_{Langmuir\text{-}like} \cdot (a^{solution}{}_2/a^{solution}{}_1) \tag{8.37}$$

In dilute solutions where $X^{solution}{}_1 \geq 0.99$, the activity of the solvent in solution remains almost constant. Now we assign a new equilibrium constant $K'_{Langmuir\text{-}like} = K_{Langmuir\text{-}like/}a^{solution}{}_1$. Now substitute $K'_{Langmuir\text{-}like}$ for $K_{Langmuir\text{-}like/}a^{solution}{}_1$ in Eq. (8.37) to give:

$$X^{surface}{}_2 = (K'_{Langmuir\text{-}like} \cdot a^{solution}{}_2/(1 + K'_{Langmuir\text{-}like} \cdot (a^{solution}{}_2) \tag{8.38}$$

Mole fraction, $X^{surface}{}_2$ is converted to moles to get Eq. (8.38) in terms of measurable quantities, where:

$$X^{surface}{}_2 = n^{surface}{}_2/n^{surface}{}_{total} \tag{8.39}$$

$n^{surface}{}_2/n^{surface}{}_{total}$ can be substituted into Eq. (8.38) to give:

$$n^{surface}{}_2/n^{surface}{}_{total} = \left(K'_{Langmuir\text{-}like} \cdot a^{solution}{}_2/\left(1 + K'_{Langmuir\text{-}like} \cdot \left(a^{solution}{}_2\right)\right.\right. \tag{8.40}$$

Equation (8.40) is multiplied by $n^{surface}{}_{total}$ resulting in:

$$n^{surface}{}_2 = \left(n^{surface}{}_{total} \cdot K'_{Langmuir\text{-}like} \cdot a^{solution}{}_2\right)/\left(1 + K'_{Langmuir\text{-}like} \cdot \left(a^{solution}{}_2\right)\right. \tag{8.41}$$

In dilute solution, the activity coefficient is 1, and $a^{solution}{}_2$ becomes $C^{solution}{}_2$, which is the adsorptive solute concentration in solution. Equation (8.42) becomes the Langmuir-like equation used for adsorption from solution compared to the Langmuir equation used for adsorption from vapor, $V = V_m \cdot K_L \cdot P/P^o/(1 + K_L \cdot P/P^o)$:

$$n^{surface}{}_2 = \left(n^{surface}{}_{total} \cdot K'_{Langmuir\text{-}like} \cdot C^{solution}{}_2\right)/\left(1 + K'_{Langmuir\text{-}like} \cdot \left(C^{solution}{}_2\right)\right. \tag{8.42}$$

Equation (8.42) is the Langmuir-like equation for adsorptive solute adsorption onto an adsorbent. The primary difference between the Langmuir equation, which describes a vapor/gas adsorption process, and the Langmuir-like adsorptive solute adsorption process equation is the meaning of Langmuir constants. In the case of the Langmuir vapor/gas adsorption process, the Langmuir constant is $K_L = k_a/k_d$. k_a is the adsorption rate constant and k_d is the desorption rate constant. This system consists of a certain fraction of the solid surface that is covered by the adsorptive gas/vapor and a certain surface fraction that remains bare or unbound. The Langmuir-like constant is related to the equilibrium constant of an adsorptive-solute adsorption process. This system can be described by the following equilibrium scheme: adsorbed solvent + solute in solution ⇆ adsorbed solute + solvent in solution. The solid surface of this system is never bare; all of the solid surface must be covered by either solvent, solute, or a mixture of both.

8.3.3 Brunauer, Emmett, and Teller (BET) Adsorption Isotherm Equation

This theory was published by BET in 1938. The BET theory applies to multilayer adsorption systems that describe Type II isotherms. As shown earlier, Type II isotherms are sigmoidal in shape and result from adsorption onto nonporous or microporous solid adsorbents to form a mono-adsorbate layer and then multi-adsorbate layers. The BET equation is arguably one of the most used equations in physical chemistry, having some 20,000 citations (Osterrieth et al., 2022). It is used routinely in many different industries to measure material surface area.

The original vapor/solid Langmuir equation (kinetic derivation) is the foundation for the BET theory. The BET theory assumes that the Langmuir equation applies to each layer of adsorbate. As the Type II isotherm shows, the adsorption process can result in infinite adsorption compared to Langmuir's theory, which posits finite monolayer adsorption as seen in the Type I isotherms. The assumptions necessary to derive the BET equation include the Langmuir assumptions in addition to specific additional assumptions.

Langmuir kinetically-derived equation assumptions:

1) The surface consists of some number of adsorption sites.
2) All of the sites are energetically equal.
3) The adsorbed phase is confined to a monolayer.

4) Lateral interactions between adsorbate molecules do not occur.
   Additional assumptions necessary to derive the BET equation
5) Each adsorbed molecule is a possible site for subsequent adsorption.
6) The differential heat of adsorption of the first layer of adsorbate is assumed to have a unique value. The differential heat of adsorption of all subsequent layers is assumed to be equal to the heat of condensation of the liquid adsorbate.
7) Adsorption and desorption can only occur onto or from exposed sites.

The BET equation is used by many industries to determine adsorbent surface area. The form of the BET equation for surface area analysis using nitrogen gas adsorption is:

$$1/q_e[(P^o/P)-1] = 1/(q_m \cdot K_{BET}) + [(K_{BET}-1)/q_m \cdot \mathrm{K}_{BET}](P/P^o) \tag{8.43}$$

where q_e is the mass of nitrogen adsorbed per gram of adsorbent, P^o is the saturation nitrogen pressure at the experimental temperature, P is the equilibrium nitrogen pressure, q_m is the monolayer capacity in mass of nitrogen per gram adsorbent, and K_{BET} is the BET constant, which is related to the heats of adsorption and vaporization. P/P^o is the relative nitrogen gas pressure. The K_{BET} is:

$$K_{BET} = e^{[(\Delta Hadsorption - \Delta Hvaporization)/RT]} \tag{8.44}$$

where $\Delta H_{adsorption}$ is the heat of adsorption and $\Delta H_{vaporization}$ is the heat of vaporization from the layer.From Eq. (8.43), a plot of $1/q_e$ $[(P^o/P)-1]$ versus P/P^o gives a straight line over a relative pressure range up to 0.25. A typical five-point BET experiment uses P/P^o values of 0.05, 0.10, 0.15, 0.20, and 0.25. Capillary condensation is a concern above 0.4 to 0.5 relative pressure. The slope is:

$$\text{Slope} = (K_{BET}-1)/(q_m \cdot \mathrm{K}_{BET}) \tag{8.45}$$

and the intercept is:

$$\text{Intercept} = 1/(q_m \cdot K_{BET}) \tag{8.46}$$

Dimensional algebraic analysis can be used to show that q_m, the amount in the monolayer is equal to 1/(slope + intercept). The derivation starts with Eq. (8.47), where Eq. (8.45) is substituted for the slope and Eq. (8.46) is substituted for the intercept in Eq. (8.47).

$$\text{Slope + intercept} = [(K_{BET}-1)/(q_m \cdot \mathrm{K}_{BET})] + 1/(q_m \cdot K_{BET}) \tag{8.47}$$

Both terms on the right-hand side of Eq. (8.47) have the same denominator, $(q_m \cdot K_{BET})$. When the two terms on the right-hand side of Eq. (8.47) are added, this gives Eq. (8.48).

$$\text{Slope + intercept} = [(K_{BET}-1)+1)/(q_m \cdot K_{BET}) \tag{8.48}$$

The −1 and + 1 cancel each other, and cancelling the K_{BET} terms in the numerator and denominator gives Eq. (8.49):

$$\text{Slope + intercept} = [(\cancel{K_{BET}} - \cancel{1}) + \cancel{1})/(q_m \cdot \cancel{K_{BET}}) = 1/q_m \tag{8.49}$$

and solving for q_m gives Eq. (8.50).

$$q_m = 1/(\text{slope + intercept}) \tag{8.50}$$

To confirm Eq. (8.50)'s equality, the right side of Eq. (8.45) is substituted for the slope in Eq. (8.50), and the right side of Eq. (8.46) is substituted for the intercept in Eq. (8.50). This gives Eq. (8.51).

$$q_m = 1/\left[(K_{BET}-1)(q_m \cdot \mathrm{K}_{BET})^{-1} + 1(q_m \cdot K_{BET})^{-1}\right] \tag{8.51}$$

Using a series of algebraic steps, it can be shown that the right side of Eq. (8.50) equals q_m. The algebraic steps are provided in Eqs. (8.52) through (8.54). First, on the right-hand side of Eq. 8.51, multiply Eq. (8.51)'s numerator and denominator by $q_m \cdot K_{BET}$ and cancel common terms. Simple algebra leads to Eq. (8.54), where $q_m = q_m$, which confirms Eq. (8.50) and $q_m = 1/(\text{slope} + \text{intercept})$.

$$q_m = q_m \cdot K_{BET}/[(K_{BET}-1)(\cancel{q_m \cdot K_{BET}})^{\cancel{-1}} + (\cancel{q_m \cdot K_{BET}})^{\cancel{-1}}]\ \cancel{q_m \cdot K_{BET}} \tag{8.52}$$

$$q_m = q_m \cdot K_{BET}/[(K_{BET}\cancel{-1}) + \cancel{1}] \tag{8.53}$$

$$q_m = q_m \cdot \cancel{K_{BET}/K_{BET}} \tag{8.54}$$

Equations (8.45) and (8.46) can be used to arrive at Eq. (8.55). As mentioned earlier, K_{BET} is the BET constant, which is related to the opposing processes of adsorption and vaporization (desorption). The first step in determining K_{BET} and Eq. (8.55) is to solve Eqs. (8.45) and (8.46) for q_m, the amount on the monolayer.

$$K_{BET} = (\text{slope}/\text{intercept}) + 1 \tag{8.55}$$

Solving Eq. (8.45) for q_m.

$$\text{Slope} = (K_{BET}-1)/q_m \cdot K_{BET} \tag{8.45}$$

$$q_m = (K_{BET}-1)/(K_{BET} \cdot \text{slope}) \tag{8.56}$$

Solving Eq. (8.46) for q_m.

$$\text{Intercept} = 1/q_m \cdot K_{BET} \tag{8.46}$$

$$q_m = 1/(K_{BET} \cdot \text{intercept}) \tag{8.57}$$

Now Eqs. (8.56) and (8.57) can be set equal to each other and solved for K_{BET}.

$$(K_{BET}-1)/\ (K_{BET} \cdot \text{slope}) = 1/(K_{BET} \cdot \text{intercept}) \tag{8.58}$$

Multiply both sides of Eq. (8.58) by K_{BET} to give Eq. (8.59) followed by several algebraic steps:

$$(K_{BET}-1)/\text{slope} = 1/\text{intercept} \tag{8.59}$$

$$(K_{BET}-1) = \text{slope}/\text{intercept} \tag{8.60}$$

$$\underline{K_{BET} = \text{slope}/\text{intercept} + 1} \tag{8.55}$$

Note: once q_m is known, the surface area of the adsorbate can be calculated.

Example 8.1 Using the BET Equation to Determine the Surface Area of an Adsorbent
Use the BET equation and the BET data from the table to determine the surface area of an adsorbent using nitrogen gas as the adsorptive.

P/P^o	Volume adsorbed (mL N_2 g^{-1} adsorbent)	$1/[q_e\ (P^o/P - 1)]$
9.9767E-2	59.6360	1.487
2.0441E-1	67.4405	3.048
3.0508E-1	74.4430	4.718

Given

The BET equation is $1/q_e\ [(P^o/P)-1] = 1/q_m \cdot K_{BET} + [(K_{BET}-1)/q_m \cdot K_{BET}](P/P^o)$
The cross-section area of N_2 is 0.162 nm^2
1 nm^2 = 1E-18 m^2
Avogadro's number is 6.023E23 molecules per mole
Nitrogen's molecular weight is 28.02 g mol^{-1}
A plot of $1/q_e[(P^o/P)-1]$ versus P/P^o gives:

- the slope is 15.7316 g adsorbent/g adsorbed N_2
- the intercept of −0.1105 g adsorbent/g adsorbed N_2

$q_m = 1/(\text{slope} + \text{intercept})$ (8.50)

Solution

1) First, q_m needs to be calculated using Eq. (8.50) and slope and intercept values obtained from the plot of $1/q_e[(P^o/P)-1]$ versus P/P^o.

$$q_m = 1/(\text{slope} + \text{intercept})$$
$$q_m = 1/(15.7316\ \text{g adsorbent/g adsorbed}\ N_2 - 0.1105\ \text{g adsorbent/g adsorbed}\ N_2)$$
$$q_m = 1/15.6205\ \text{g adsorbent/g adsorbed}\ N_2 = 0.06402\ \text{g}\ N_2/\text{g adsorbent}$$

2) The adsorbent surface area can now be calculated using dimensional analysis and the givens.

$$\text{Surface area} = (0.06402\ \text{g}\ N_2/\text{g adsorbent}) \cdot (1\ \text{mol}\ N_2/28.02\ \text{g}\ N_2) \cdot (6.023\text{E}23\ \text{molecules/mol}) \cdot (0.162\ \text{nm}^2/\ N_2\ \text{molecule})\ (1\text{E-18 m}^2\ \text{nm}^{-2}) = 222.9\ \text{m}^2\ \text{g}^{-1}\ \text{adsorbent}$$

<u>Surface area = 223 $m^2\ g^{-1}$ adsorbent</u> with 3 significant figures.

This is a high surface area. To provide a perspective, the typical surface areas for lactose, magnesium stearate, and fumed colloidal silica are 0.1 $m^2\ g^{-1}$, 5–20 $m^2\ g^{-1}$, and 90–400 $m^2\ g^{-1}$, respectively.

8.4 Other Adsorption Isotherm Equations

Table 8.3 summarizes other isotherm equations that have been used to model adsorption data. As mentioned earlier, two, three, four, and five-parameter adsorption isotherm equations have been used to model the adsorption process. Typically, the coefficient of determination, R^2, is used to measure the *goodness of fit* of various model isotherms. If one compares the fit of various models that have a different number of descriptor variables, it is recommended to compare the models' *adjusted* R^2 values. Like R^2, *adjusted* R^2 is an indication of the quality of a model and how well it fits the data. The benefit of *adjusted* R^2 is that the *adjusted* R^2 "accounts" for the number of terms in the model. It is well known that adding terms to a model will improve the fit of the data set. However, one needs to be cognizant of overfitting the data by potentially using too many variables. In the case of overfitting, the R^2 values are typically high, indicating a good data fit, but the predictability of the model adequately fitting the data for other data sets is often poor. So, the *adjusted* R^2 is used to fairly compare the fit of one model with 2 variables and another model with 4 variables.

Table 8.3 Other Reported Adsorption Isotherms[a].

Adsorption isotherm model	Mathematical expression	Comments
Bauder isotherm	$q_e = \dfrac{q_m b_0 C_e{}^{1+x+y}}{1 + b_0 C_e{}^{1+x}}$ [b]Where q_e is the equilibrium amount of adsorptive solute adsorbed by the weight of the adsorbent. [c]Where q_m is the maximum amount of adsorptive solute adsorbed or the adsorbed amount, also maximum adsorption capacity ($mg\ g^{-1}$), b_0 is the equilibrium constant, and x and y are Bauder parameters. [d]C_e ($mg_{adsorptive\text{-}(solute)}/L_{solution}$) is the equilibrium concentration of the adsorptive solute solution after exposure to a known amount of adsorbent. A nonlinear fit is preferable in order to obtain the necessary terms from the equation.	• For lower surface coverage, this model acts similarly to the Freundlich isotherm.
Dubinin-Radushkevich isotherm	$\ln q_e = \ln q_m - \beta\varepsilon^2$ $\epsilon = RT\ln\left(1 + \dfrac{1}{C_e}\right)$ $E = \dfrac{1}{\sqrt{2\beta}}$ A plot of $\ln q_e$ and ε^2 produces a straight line with a slope equal to $(-\beta)$. Where β is the Dubinin-Radushkevich constant, which is a measure of the activity coefficient related to mean sorption energy ($mol^2\ kJ^{-2}$), ε is known as the Polanyi potential, and E is mean adsorption energy.	• Equation describes adsorption onto heterogeneous surfaces. • It is only applicable for intermediate range adsorbate concentrations. • This semi-empirical model for adsorption is used to describe physical adsorption processes with multilayer van der Waal's forces.
Elovich isotherm	$\ln \dfrac{q_e}{C_e} = \ln(K_E q_m) - \dfrac{1}{q_m} q_e$ Where K_E is the model's constant. Values of this constant and q_m can be obtained from the linear expression $\ln (q_e/C_e)$ versus q_e.	• The model assumes multilayer formation, and the number of adsorption sites increases exponentially with adsorption. • Works well with chemisorption.
Flory-Huggins isotherm	$\ln \dfrac{\theta}{C_0} = \ln K_{FH} + n\ln(1-\theta)$ Where K_{FH} is the Flory-Higgins equilibrium constant ($L\ mol^{-1}$) and n represents the number of adsorbates occupying the adsorption sites. The equilibrium constant is related to Gibbs free energy: $\Delta G^0 = RT\ln K_{FH}$	• This equation allows the determination of adsorption spontaneity. Adsorption is spontaneous when ΔG^0 is negative.
Fowler-Guggenheim model	$\ln \dfrac{C_e(1-\theta)}{\theta} = -\ln K_{FG} - \dfrac{2w\theta}{RT}$ Where K_{FG} is the Fowler-Guggenheim equilibrium constant ($L\ mg^{-1}$) and w is the interaction energy between adsorbed molecules ($kJ\ mol^{-1}$)	• Applicable where absorbate molecules interact laterally. • A linear fit of the data yields the values for K_{FG} and w. • This model is restricted to situations where $\theta < 60\%$. • Values for w can be negative, indicating the adsorbed molecular interaction is repulsive. Zero indicates there is no interaction between the adsorbed molecules. A positive w value indicates attractive interactions occur between adsorbed molecules.

(*Continued*)

Table 8.3 (Continued)

Adsorption isotherm model	Mathematical expression	Comments
Freundlich isotherm	$\log q_e = \log K_F + n \log C_e$ Where K_F is the Freundlich constant and reflects adsorption capacity and n is a constant that relates to adsorption intensity.	• The model assumes a monolayer formation of the adsorbate on the surface of a heterogeneous adsorbent.
Fritz and Schlunder Five-Parameter isotherm	$q_e = \frac{q_{mFS5} K_1 C_e^{m_1}}{1 + K_2 C_e^{m_2}}$ Where q_{mFS5} is the model maximum adsorption capacity (mg g^{-1}) (5 in the subscript was added to signify the "five-parameters" in the model). The parameters describe a nonlinear fit of the data to the model.	• The model resembles the Langmuir isotherm as the exponent values approach unity (m_1 and $m_2 \rightarrow 1$). • At higher adsorbate concentrations, the model is similar to the Freundlich isotherm.
Fritz-Schlunder isotherm	$q_e = \frac{q_{mFS} K_{FS} C_e}{1 + q_{mFS} C_e^{MFS}}$ Where q_{MFS} is the Fritz-Schlunder maximum adsorption capacity (mg g^{-1}), K_{FS} is an equilibrium constant for the model, and the exponent MFS is obtained by non-linear regression analysis. As the value of MFS approaches unity (MFS $\rightarrow$ 1), the model becomes more and more similar to the Langmuir isotherm.	• It is a versatile model. • At high concentrations of the adsorbate, the model resembles the Freundlich isotherm.
Halsey isotherm	$q_e = \frac{1}{n_H} \ln K_H - \frac{1}{n_H} \ln C_e$ Both n_H and K_H are constants and can be obtained from the slope and the y-intercept of the q_e versus ln C_e line.	• This isotherm assumes multilayer adsorption, where some of the layers are relatively far from the surface where the adsorption is taking place.
Harkin-Jura isotherm	$\frac{1}{q_e^2} = \frac{B}{A} - \left(\frac{1}{A}\right) \log C_e$ The straight-line equation resulting from plotting ($1/q_e^2$) versus log C_e produces the values of the constants, A and B.	• The module assumes multilayer adsorption on a heterogeneous pore surface.
Henry's isotherms	$q_e = K_{HE} C_e$ Where q_e is the amount of the absorbate per unit weight of adsorbent (mg g^{-1}), K_{HE} is Henry's constant, and C_e is the equilibrium adsorptive solute concentration.	• A simple model • Works well with low concentration of the adsorbate (usually a gas).
Hill isotherm	$\log \frac{q_e}{q_H - q_e} = n_H \log C_e - \log K_D$ The model has three constants: q_H, n_H, and K_D.	• Hill isotherm is used to assess the influence of different adsorbates on the binding sites. • It assumes a homogeneous surface for the adsorbent.
Hill-DeBoer model	$\ln \frac{C_e(1-\theta)}{\theta} - \frac{\theta}{(1-\theta)} = -\ln K_1 - \frac{K_2 \theta}{RT}$ Where θ is the fraction of the adsorbent surface occupied by the adsorbate, K_1 and K_2 are the Hill-DeBoer constants K_1 (L mg^{-1}), and K_2 is the energetic constant of the interaction between adsorbed molecules (kJ mol^{-1}), respectively. R is the gas constant (0.00831 kJ mol^{-1} K^{-1}) and T is the temperature in kelvin.	• A linear plot of the equation yields the values of the constants. • Applicable where absorbate molecules interact laterally and there is a mobile adsorption process.

Table 8.3 (Continued)

Adsorption isotherm model	Mathematical expression	Comments
Jossens isotherm	$\ln\left(\frac{C_e}{q_e}\right) = -\ln H + F{q_e}^p$ Where H, F, and p are constants and can be obtained by linear regression analysis. The exponent p is characteristic of the adsorbent, and it is independent of the temperature.	• It is based on the energy of adsorbate–adsorbent interaction at the adsorption sites. • It assumes a heterogeneous surface. • At low concentration of the adsorbate, the model resembles Henry's isotherm.
Jovanovic isotherm	$\ln q_e = \ln q_m - K_J C_e$ K_J is a constant obtained from the linear expression.	• It assumes that some "mechanical" interaction occurs between the adsorbate and the surface of the adsorbent.
Kahn isotherm	$q_e = \frac{q_m b_K C_e}{(1 + b_K C_e) a_K}$ Where a_K is the Kahn isotherm exponent and b_K is its constant.	• Applies to the adsorption of two adsorbates, both present at low concentration in the solution.
Kiselev isotherm	$\frac{1}{C_e(1-\theta)} = \frac{K_1}{\theta} + K_1 K_n$ Where K_1 is the Kiselev equilibrium constant ($L\ mg^{-1}$) and K_n is the equilibrium constant of the formation of a complex between adsorbed molecules. The data is fitted to the linear plot between $[1/C_e\ (1-\theta)]$ versus $(1/\theta)$.	• The model is only applicable for surface coverage $\theta > 0.68$. • It is a localized monomolecular layer model.
Koble-Carrigan isotherm	$\frac{1}{q_e} = \frac{1}{A_K {C_e}^p} + \frac{B_K}{A_K}$ Where A_K, B_K, and p are the isotherm constants.	• A hybrid isotherm for Langmuir and Freundlich isotherms. • The model is applicable when the exponent $p \geq 1$.
Langmuir isotherm	$\frac{C_e}{q_e} = \frac{1}{q_m K_L} + \frac{C_e}{q_m}$ Where q_e is the amount of the adsorbate on the surface of the adsorbent ($mg\ g^{-1}$) and q_m is the value of q_e when the surface reaches a saturation point. K_L is the Langmuir constant that reflects adsorption capacity. K_L is related to a parameter known as the separation factor (R_L): $R_L = \frac{1}{1 + K_L C_0}$ Where C_0 is the initial concentration of the adsorbate ($mg\ g^{-1}$). R_L is a dimensionless parameter. The value of R_L can be greater than 1, indicating unfavorable adsorption conditions. An R_L value of zero suggests the adsorption process is irreversible. An R_L of 1 is a linear adsorption process. Favorable adsorption occurs when R_L values are between 0 and 1.	• The model assumes a dynamic adsorption and desorption process. • Useful for monolayer formation.
Langmuir–Freundlich isotherm	$q_e = \frac{q_{MLF}(K_{LF} C_e)^{MLF}}{1 + (K_{LF} C_e)^{MLF}}$ Where q_{MLF} is the Langmuir–Freundlich maximum adsorption capacity ($mg\ g^{-1}$), K_{LF} is the equilibrium constant for heterogeneous solid, and the MLF exponent is a number between 0 and 1 obtained by regression analysis.	• At low adsorbate concentration, this model follows the Freundlich isotherm model, while at high adsorbate concentration, it becomes a better fit to the Langmuir isotherm.

(Continued)

Table 8.3 (Continued)

Adsorption isotherm model	Mathematical expression	Comments
Marczewski-Jaroniec isotherm	$q_e = q_{M_{MJ}} \left[\frac{(K_{MJ}C_e)^{n_{MJ}}}{1+(K_{MJ}C_e)^{n_{MJ}}}\right]^{\frac{M_{MJ}}{n_{MJ}}}$ Where K_{MJ} is a constant. The parameters n_{MJ} and M_{MJ} describe the spreading of distribution in the path of lesser and higher adsorption energy, respectively.	• When n_{MJ} and $M_{MJ} = 1$, the model produces similar results as those obtained from the Langmuir isotherm. • If $n_{MJ} = M_{MJ}$, then the model is similar to the Langmuir–Freundlich isotherm. • The four-parameter general Langmuir equation.
Radke-Prausniiz isotherm	$q_e = \frac{q_{MPR}K_{RP}C_e}{(1+K_{RP}C_e)^{MRP}}$ Where q_{MRP} is the Radke-Prausnitz maximum adsorption capacity (mg g^{-1}), K_{RP} is the Radke-Prausnitz equilibrium constant, and MRP is the Radke-Prausnitz model exponent.	• Works well at low adsorbate concentrations.
Redlich-Peterson isotherm	$\ln \frac{C_e}{q_e} = \beta \ln C_e - \ln A$ Where β and A are constants. The constant β can take on values between 0 and 1.	• A hybrid model that can convert to Langmuir or Freundlich isotherms.
Sips isotherm	$\beta_S \ln C_e = -\ln\left(\frac{K_S}{q_e}\right) + \ln a_S$ β_S is the Sips exponent and a_S (L g^{-1}) and K_S (L g^{-1}) are Sips constants.	• The model assumes a multilayer formation on heterogeneous surfaces. • It is a hybrid model for Langmuir and Freundlich isotherms. • Sips isotherm is concentration-dependent and is a function of the pH and temperature of the system.
Temkin isotherm	$q_e = \frac{RT}{b}\ln K_T + \frac{RT}{b}\ln C_e$ Where b is the Temkin constant, which is related to the heat of sorption (J mol^{-1}), and K_T is the Temkin isotherm constant (L g^{-1}). A plot between q_e and ln C_e allows calculations of the constants b and K_T.	• The model assumes that the heat of adsorption (ΔH_{ads}) of all molecules in the layer decreases linearly as a result of increased surface coverage.
Tóth isotherm	$\ln \frac{q_e{}^n}{q_m{}^n - q_e{}^n} = n \ln K_L + n \ln C_e$ Both constants n and K_L have units of (mg g^{-1}). When $n = 1$, this model is similar to the Langmuir isotherm. If n deviates appreciably from unity, then the system describes a heterogeneous surface for adsorption.	• A similar model to the Langmuir-like isotherm can describe a heterogeneous surface adsorption.
Weber-Van Vliet isotherm	$C_e = p_1 q_e{}^{(p_2 q_e{}^{p_3} + p_4)}$ Where p_1, p_2, p_3, and p_4 are the model parameters. Nonlinear fit is required for this model.	• Widely applicable to adsorption data.

[a] From Ayawei et al., 2017, and adapted from Al-Achi et al. 2023.
[b] Where q_e is the equilibrium amount of adsorptive solute adsorbed by the weight of the adsorbent.
[c] Where q_m is the maximum amount of adsorptive solute adsorbed or the adsorbed amount, also maximum adsorption capacity (mg g^{-1}).
[d] C_e ($mg_{adsorptive\text{-}(solute)}/L_{soltuion}$) is the equilibrium concentration of the adsorptive solute solution after exposure to a known amount of adsorbent.

8.5 Factors Influencing Adsorption

The nature of the adsorbent and adsorptive, adsorptive pressure, concentration and solubility, temperature, and pH are important factors that affect the adsorption process.

8.5.1 Adsorbent Nature

The adsorbent surface nature can significantly affect the adsorption process. The Freundlich equation (Eq. 8.3) states that the adsorption extent is directly proportional to the Freundlich constant, K_F, which is a measure of the adsorption capacity of the adsorbent. The Langmuir derivation assumed that the adsorption rate is directly proportional to the number of unbound adsorbent sites (Eq. 8.6). As mentioned earlier, adsorption is a surface phenomenon. It follows that the higher the surface area, the greater the extent of binding, q_e, Eq. (8.23). The level of adsorbent porosity also affects the surface area of the material and the potential number of adsorption sites. Surface roughness can also increase surface area and the opportunity for binding. Generally, the higher the number of adsorption sites per surface, the greater the potential extent of binding.

Chemisorption may occur if the surface has specific sites for a chemical reaction. Common suspending viscosity-enhancing clays like bentonite, hectorite, kaolin, and veegum are charged materials and can form electrostatic interactions with adsorptives of the opposite charge. This form of adsorption is driven by electrostatic interactions, and can be considered a type of ionic bonding. This reaction is often referred to as an ionic exchange process.

8.5.2 Adsorptive Nature

Considering gas adsorption, gases with higher enthalpies of condensation like SO_2, Cl_2, NH_3, and O_2 are generally better adsorptives than gases with lower heats of condensation such as N_2, Ar, and Kr. However, gases like SO_2, Cl_2, and NH_3 can exhibit significant chemisorption on many surfaces, especially those with potential active sites. Also, these gases can react with the surface of the material being analyzed, altering its properties and leading to inaccurate surface area estimations. Additionally, gases with high enthalpies of condensation can form multiple layers of adsorbate on the surface, even at relatively low pressures. All these properties make SO_2, Cl_2, NH_3, and O_2 poor choices for an adsorptive for surface area measurements.

N_2, Ar, and Kr are considered inert gases and are better adsorptives. Nitrogen is the most commonly used adsorptive. Nitrogen is relatively inert, meaning it doesn't readily undergo chemical reactions with most surfaces. This is crucial because surface area measurements rely on physisorption not chemisorption. Nitrogen is readily available in high purity, making it a practical and cost-effective choice for laboratory use. The adsorption behavior of nitrogen at condensation temperatures is well characterized. Overall, nitrogen provides a balance of availability, purity, and well-defined behavior making it a reliable and convenient adsorptive for determining surface areas.

Although nitrogen has been the standard adsorptive agent for BET surface area and porosity measurements, its physiochemical properties have disadvantages arising from its diatomic structure and quadrupole moment. Interactions may occur between nitrogen's higher electron density ends and an adsorbent surface compared to monoatomic Ar or Kr. Because N_2 is diatomic, it may also have different orientations on the adsorbent surface. These issues have resulted in up to 20% error in surface area data.

Nongaseous adsorptives can demonstrate various dipole and hydrogen bonding interactions with a surface. Nonpolar adsorptives tend to favor adsorption onto nonpolar adsorbents, while more polar adsorptives adsorb more to polar adsorbents. As discussed earlier under adsorbent nature, nongaseous ionic adsorptives can undergo electrostatic interactions with adsorbents. Polymers and proteins are well known to undergo adsorption onto a variety of adsorbent surfaces. Adsorption between proteins and glass and plastic surfaces has been commonly reported. Globular adsorptive proteins frequently unfold at the adsorption sites. The protein adsorption process is often slow because the binding occurs onto a large number of adjacent adsorption sites, forming *'trains."* The high protein molecular weight also leads to a slower adsorption.

8.5.3 Adsorptive Pressure, Concentration, and Solubility

The Langmuir-like Eq. (8.42) shows that $n^{surface}{}_2$, the equilibrium extent of adsorption, is proportional to $C^{solution}{}_2$, the concentration of adsorptive solute in solution.

Based on the Lundelius Rule (Lundelius, 1920), the adsorption extent is generally inversely proportional to the adsorptive solute's solubility. Greater adsorption occurs for adsorptive solutes for a given value of S/S^0 that have lower solubilities. This concept is important if one desires to increase the extent of adsorption. The idea is based on the principles discussed in Sections 7.5 and 7.6. In this case, the adhesive solute–solvent and solute–solute interactions need to be less than the adsorptive solute-adsorbent forces so that solute–solvent adhesive forces can be disrupted in favor of the adsorptive solute-adsorbent interactions.

8.5.4 Temperature

Adsorption is an equilibrium process that has a negative differential heat of adsorption or isosteric heat of adsorption ($\Delta istH_{absorption}$). Adsorption decreases with increasing temperature.

In addition to the differential enthalpy of adsorption, the differential entropy of adsorption needs to be considered. When an adsorptive molecule adsorbs to the adsorbent, the adsorptive molecule becomes more ordered. This increase in order results in a decrease in the differential adsorption entropy, and $\Delta istS_{adsorption}$ is negative. When the temperature is increased, the adsorbate's kinetic energy is increased, causing it to dissociate from the adsorbent. Once dissociated, the adsorbate is no longer bound and becomes more disordered. The differential entropy is then increased, and $\Delta istS_{desorption}$ is positive. As such, an increase in temperature favors disorder and desorption.

Therefore, as the temperature of an adsorbed system is increased, enthalpy of adsorption is exothermic and adsorption is discouraged. Increase temperature increases the entropy of desorption and desorption is favor.

8.5.5 pH

In general, ionized species are not as well adsorbed as nonionic species. As such, the percent adsorbed mirrors the nonionized species percent determined from the Henderson-Hasselbalch equation (see Section 5.9.3). This superimposition of the percent adsorbed drug and percent unionized drug was observed by Richards and Meaken (1974). Thus, the solubility of the unionized form will generally determine the extent of adsorption.

8.6 Adsorption–Desorption Hysteresis

Adsorption and desorption profiles were introduced earlier in Sections 8.1.3 and 8.1.4. ADH was also briefly discussed in Section 8.2.5, Reversible Type IV Isotherm. As a reminder, *porosity* is the ratio of the total pore volume to the volume of the material. Porous adsorbents are divided into

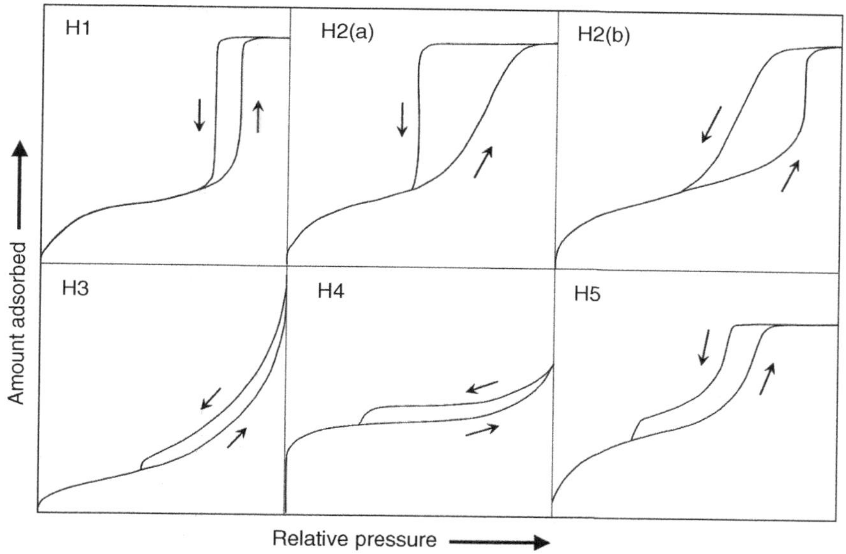

Figure 8.3 Hysteresis classification types. *Source:* Thommes et al. (2015)/Walter de Gruyter GmbH.

external and *internal surfaces. Internal surfaces* contain *macropores*, *mesopores*, and *micropores* having pores and widths of greater than 50 nm, between 2 nm and 50 nm, and not exceeding 2 nm, respectively. The external surface is noted as a nonmicroporous surface and may feature surface roughness. It is not uncommon to encounter ADH, as shown in Figure 8.3. The recent IUPAC Technical Report (Thommes et al., 2015) added two ADH isotherms, H2(b) and H5.

Hysteresis or "loop" formation occurs when mesopores are present in the adsorbent. In this case, the mechanisms of gas filling and evacuation are different, as was discussed in Section 8.2.5. Mesopores fill with multiple gas/vapor layers, which is followed by capillary condensation. A closed-ended mesopore capillary is shown in Figure 8.4. Filling occurs by the gas forming an adsorbed

Figure 8.4 Illustrates the H1 adsorption-desorption hysteresis isotherm for adsorbents having narrowly distributed cylindrical pore size and relatively ordered 3D pore networks. *Source:* Adapted from Cychosz and Thommes (2018).

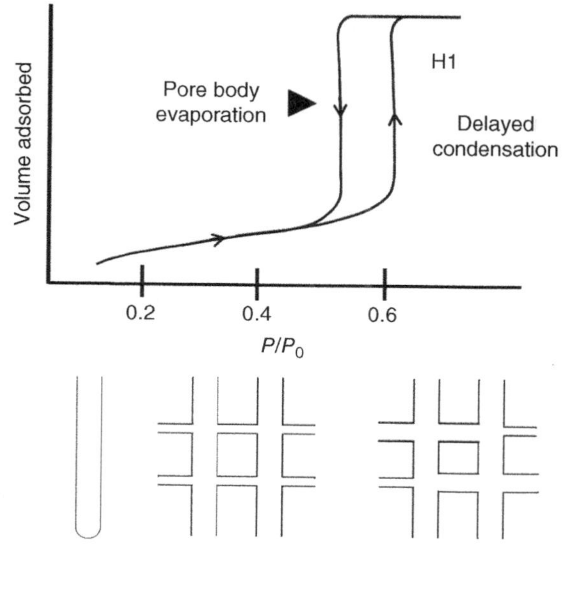

monolayer on the sides and bottom of the closed-ended capillary. In the highly confined 2–50 nm capillary space, van der Waal interactions become high enough to cause vapor condensation. Gas condensation tends to accumulate at the bottom of the capillary. This condensation occurs below the saturation gas vapor pressure. This unique phenomenon is described by the use of the Kelvin equation and the Young-Laplace equation, which are closely related and interconnected. Essentially, the Kelvin equation explains how the curvature-induced pressure difference that is described by the Young-Laplace equation influences the equilibrium vapor pressure (Eq. 8.61).

$$\ln p_{equilibrium}/p_{saturation} = H\gamma V_{liqiud\ molar\ volume}/RT \tag{8.61}$$

where $p_{equilibium}$ is the equilibrium vapor pressure at the liquid–vapor interface inside the pore, $p_{saturation}$ is the saturation vapor pressure outside the pore, H is the mean surface curvature of the meniscus at the liquid–vapor interface, and it originates from the Young-Laplace equation, γ is the liquid–vapor surface tension, R is the gas constant, and T is the temperature in kelvin.

$$H = (1/R_1 + 1/R_2) \tag{8.62}$$

where R_1 is the radius of a circle tangent to the meniscus face and R_2 is the radius of a second circle tangent to the meniscus face that is perpendicular to the circle created by R_1. H replaces the radius of a particle used in the Kelvin equation.

For a hemispherical meniscus, $R_1 = R_2$ and

$$H = 2/R_{lv} \tag{8.63}$$

In this case, R_{lv} equals the pore radius, r_{pore}.

Equation (8.61) becomes Eq. (8.64)

$$\ln p_{equilibrium}/p_{saturation} = -2\gamma V_{liqiud\ molar\ volume}/r_{pore}\ RT \tag{8.64}$$

An explanation of the negative sign follows. The adsorbate adhesion causes a concave meniscus, similar to water adhering to a glass capillary. The concave surface is curved inward when viewed from the top of the meniscus. H is negative because the center of the tangent circle to the meniscus is outside the meniscus' face or surface.

The ADH profile is the result of the mesopore size and shape. H1, H2(a), and H2(b) are discussed in more detail to illustrate this point (Cychosz and Thommes, 2018). Figure 8.4 illustrates the H1 ADH isotherm for adsorbents having narrowly distributed cylindrical pore size and relatively ordered 3D pore networks. The H1 adsorption arm exhibits a Type IV(a) isotherm. It demonstrates delayed pore condensation, leading to a sharp increase in the amount adsorbed at higher relative pressures. During the post-adsorption phase, as the relative pressure is decreased, the gas–liquid equilibrium is disturbed. This desorption process is referred to as *equilibrium pore evaporation or pore evaporation*.

Figure 8.5a displays a subtype for the H2(a) ADH profile and is indicated in Figure 8.5a as H2(a1). Here, the adsorbent pores are ink-bottle shaped with a closed end or an open-ended ink-bottle-shaped capillary. Here, the pore cavity size distribution is wider than the neck size distribution. When the neck size is larger than a *critical size*, cavitation is unlikely. Compared to the H1 ADH isotherm shown in Figure 8.3, the Figure 8.5a H2(a) ADH profile exhibits a slower gas adsorption rate and a delayed condensation. The characteristic H2(a) isotherm is the result of the substrate having a wide range of pore sizes. In this case, equilibrium pore evaporation does not occur. The wider pore body remains filled with liquid until a lower relative pressure is reached that supports evaporation from the smaller, more restrictive neck. The neck's smaller size and narrower size

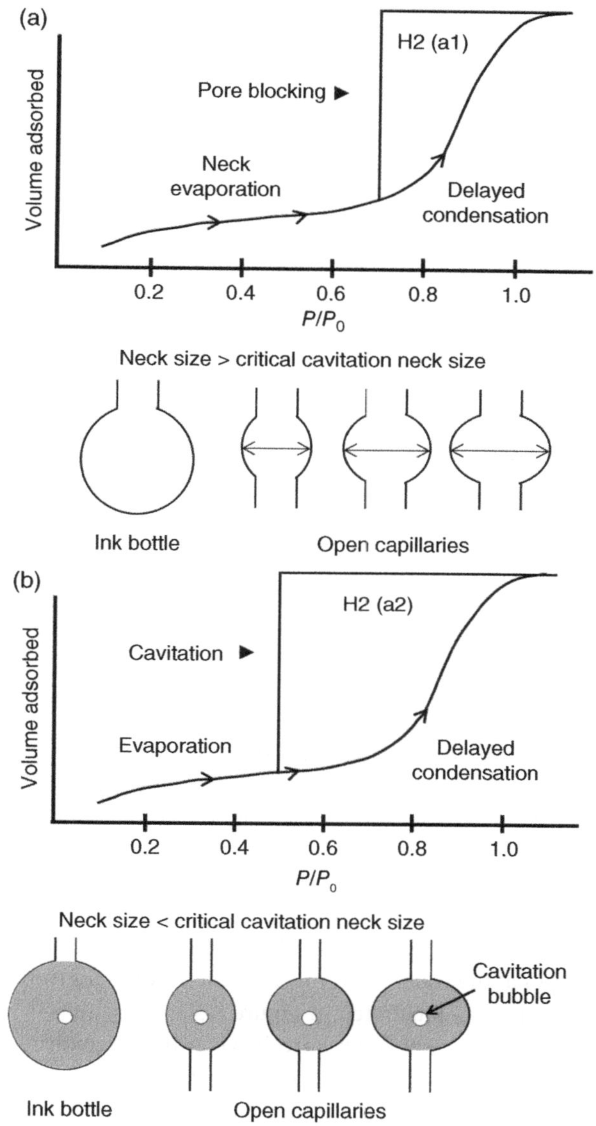

Figure 8.5 (a) The pore cavity size distribution is wider than the neck size distribution. When the neck size is larger than a *critical size*, cavitation is unlikely. The desorption process is the result of *pore blocking*. (b) Cavitation evaporation.

distribution control when evaporation can occur. Once the critical P/P^o is reached, neck evaporation occurs. The evaporation from the pores occurs almost simultaneously, resulting in a steep desorption step with respect to the amount adsorbed. This desorption process is called *pore blocking*. Other H2(a1)-type adsorbent materials will exhibit a similar ADH profile.

H2(a2), *cavitation evaporation* materials have a similar pore size distribution as H2(a1) materials. However, the H2(a2) neck size distribution is smaller than a *critical size*, $w_{neck\ size} < w_{critical\ neck\ size}$ (see Figure 8.5b). This neck size creates a highly restrictive pore opening, which can lead to

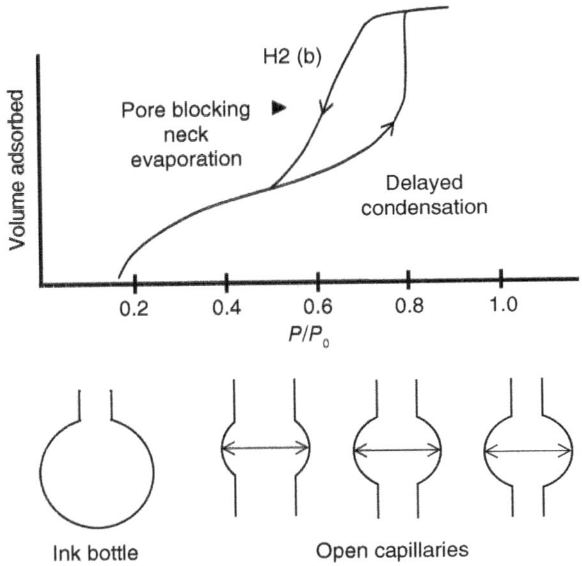

Figure 8.6 ADH isotherm resulting from a narrow pore size distribution and relatively wider neck size distribution. Neck evaporation occurs at a slower rate than seen with Figure 8.5a,b.

cavitation. Cavitation bubble formation results when the condensed phase evaporates within the liquid phase. Evaporation occurs at given temperature and pressure conditions when the liquid pore pressure becomes smaller than the liquid's vapor pressure. When the liquid evaporates, it forms a vapor bubble within the liquid. The pressure–temperature phase diagram for the material of interest is the basis for this evaporation process (see section 5.3). In an open system like a pore system, the bubbles that reach the surface burst. They burst because the Laplace pressure inside the curved bubble surface is greater than the ambient pressure on the outside of the bubble (also see Sections 7.10). This vapor originating from the bubble is then released from the pore, much like boiling water bubbles burst into steam water vapor at the water's surface. This release of vapor from the pore leads to *desorption* resulting from cavitation as illustrated in Figure 8.5b.

ADH isotherms that exhibit a narrow pore size distribution and relatively wider neck size distribution are shown in Figure 8.6. Adsorption and delayed condensation occur at a faster rate at generally lower P/P^0 than H2(a1) and H2(a2) isotherms because of the uniform pore size. H2(b) neck evaporation desorption is a slower process compared to H2(a1) and H2(a2) isotherms. The relatively wide neck size distribution is responsible for the slower desorption process. This desorption process is called *neck evaporation*.

ADH isotherms H3 are associated with agglomerated microporous plate-like materials for which the pores are not completely filled with condensate. The adsorption portion of the H3 curve is of Type II. The abrupt closure of the loop is characteristic of H3, H4, and H5 isotherms. The abrupt closure is due to a sudden cavitation that typically occurs around 0.4 to 0.5 P/P^o.

Type H4 adsorption arm is a hybrid curve of Type I and Type II isotherms. The relatively fast and high adsorption at low P/P^o is due to the filling of micropores. The Type II portion of the composite adsorption curve is the result of filling meso- and macropores. The desorption curve is similar to the desorption phase of H3.

Type H5 isotherm profile exhibits a unique adsorption and desorption phase compared to H3 and H4. The adsorption segment is consistent with a Type IV profile. Materials associated with H5 ADH consist of open and restricted pores or network geometries.

Example 8.2 Application: Water Vapor Adsorption–Desorption Characterization of Two Different Grades of Magnesium Stearate from the Same Manufacturer
Kosher vegetable magnesium stearate (KV MgSt) and food-grade vegetable MgSt (FGV MgSt) were characterized by dynamic water vapor adsorption and desorption (Haware et al., 2016).

The sorption–desorption isotherms are shown in Figure 8.7.

Differential scanning calorimetry, powder X-ray diffraction (PXRD), and solid-state nuclear magnetic resonance confirmed that KV MgSt was more disordered than FGV MgSt. PXRD data

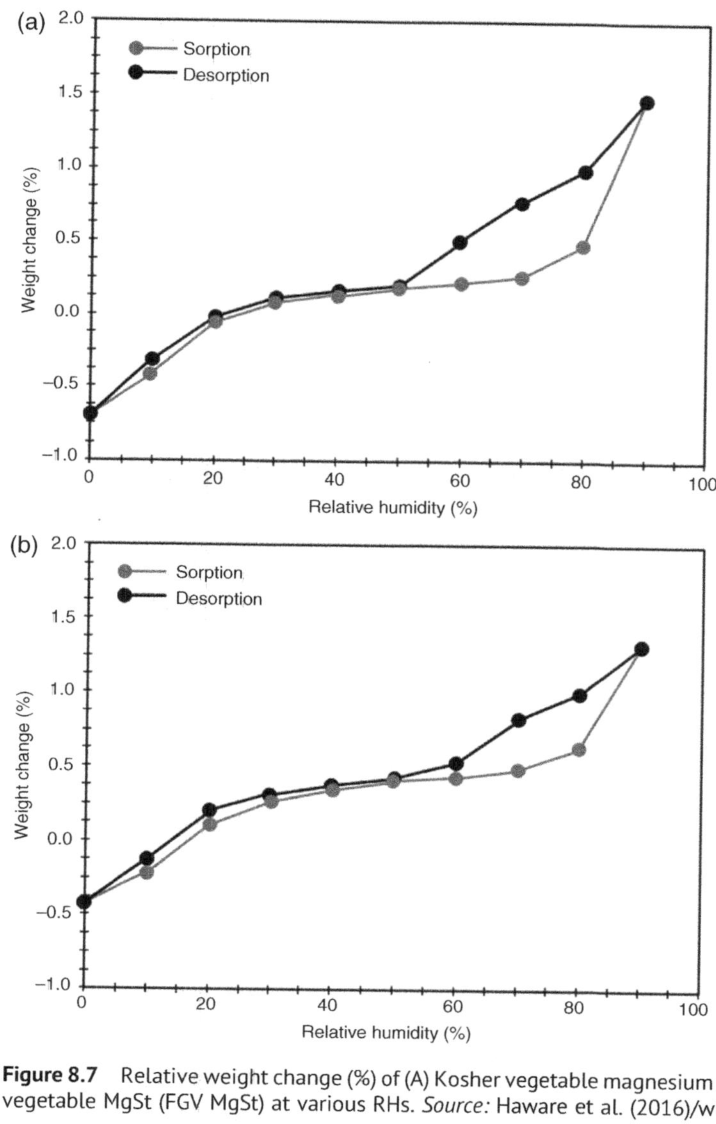

Figure 8.7 Relative weight change (%) of (A) Kosher vegetable magnesium stearate (KV MgSt) and food-grade vegetable MgSt (FGV MgSt) at various RHs. *Source:* Haware et al. (2016)/with permission of Elsevier.

analysis indicated that KV MgSt was 27.5% disordered compared to FGV MgSt disorder content of 15.6%.

The ADH profiles of KV MgSt and FGV MgSt are consistent with H2(b) isotherms. These isotherms are associated with wider pore size distribution and desorption by pore evaporation. KV MgSt shows more hysteresis than FGV MgSt. It was concluded that the water vapor adheres to KV MgSt to a higher degree than FGV MgSt during both sorption and desorption because of its higher level of disorder and higher energy sites.

8.7 Solid-Water Vapor Adsorption Systems

Many drugs and excipients are known to adsorb and absorb moisture/water vapor. Adsorbed and absorbed water can affect material stability and function. It is surprising very few pharmaceutical textbooks discuss hygroscopicity, since moisture content can cause significant negative performance and stability consequences. For purposes of this discussion, *hygroscopicity is defined as the ability of a material to attract and hold environmental water vapor by adsorption and absorption mechanisms*. Since the mechanism of water uptake can often be unclear, the term *sorption* can be used. Materials unaffected by RH are designated as nonhygroscopic materials. Those substances that attract and hold environmental water vapor are called *hygroscopic* materials. A common question is "how hygroscopic is the material?" Does it contain 1% moisture, which is most likely adsorbed moisture or 20% to 30% moisture, which is most likely absorbed water vapor? Material holding a significant amount of water can lead to inaccurate quantities of the desired substance if the water content is not accounted for.

Callahan et al. (1982) proposed a hygroscopicity classification system to help address the question, "how hygroscopic is the material?" The hygroscopicity classification system is shown in Table 8.4. It is useful in describing a material's *relative degree of hygroscopicity*. It should be noted that the relatively high percentages of moisture content presented in Table 8.4 suggest that absorption is the most prominent mechanism of attracting and holding water for many pharmaceutical materials.

Figure 8.8 is a Type III adsorption–desorption isotherm of a nonhygroscopic, highly crystalline drug substance. This crystalline drug meets the HCS *Class I nonhygroscopic* classification since there is less than 0.1% moisture increase below 90% RH.

Table 8.4 Hygroscopicity Classification System (HCS)[a].

Class I: Nonhygroscopic. Essentially no moisture increases occur at RHs below 90%. Furthermore, the increase in moisture content after storage for 1 week above 90% RH is less than 20%.
Class II: Slightly hygroscopic. Essentially no moisture increases occur at RHs below 80%. The increase in moisture content after storage for 1 week above 80% RH is less than 40%.
Class III: Moderately hygroscopic. The moisture content does not increase above 5% after storage at RHs below 60%. The increase in moisture content after storage for 1 week above 80% RH is less than 50%.
Class IV: Very hygroscopic. Moisture increase may occur at RHs as low as 40 to 50%. The increase in moisture content after storage for 1 week above 90% RH may exceed 30%.

[a] Adapted from Callahan et al. (1982), with permission from Informa Healthcare.

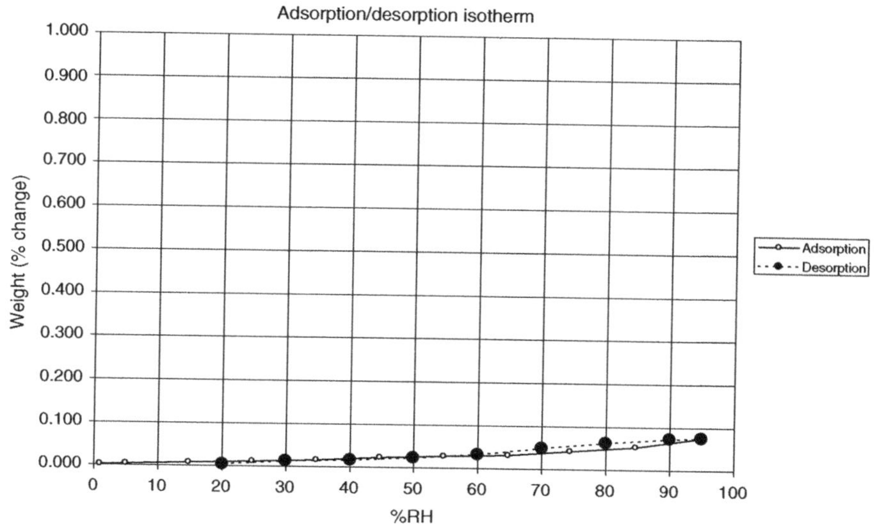

Figure 8.8 Type III adsorption–desorption isotherm of a highly crystalline drug substance.

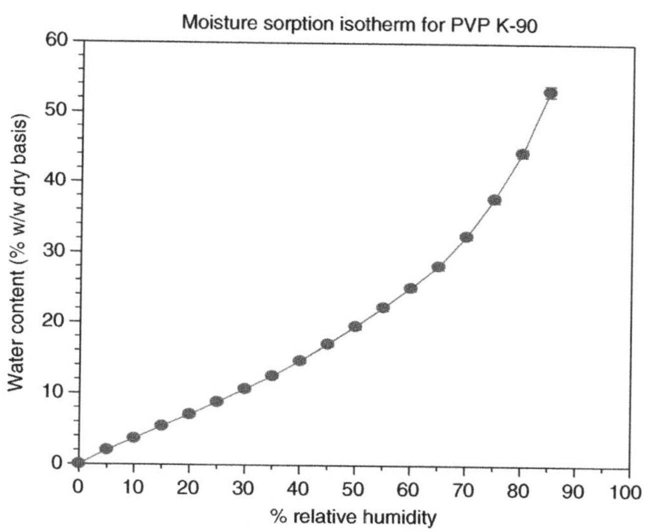

Figure 8.9 Moisture sorption isotherm of PVP-K90. The standard deviation bars are barely visible (*N* = 3).

Figure 8.9 is a sorption isotherm of PVP-K90 (polyvinylpyrrolidone polymer). This profile is a typical example of amorphous materials. PVP-K90 meets the *very hygroscopic* Class IV hygroscopicity classification. A significant increase in moisture of ~10% w/w is observed at 25% RH. Even at common manufacturing-controlled RHs ranging from 45% to 65% RH, there is a 20% to 30% increase in weight when using predried PVP-K90 excipient.

Figures 8.10 and 8.11 exemplify a *deliquescent* material. A *deliquescent* material has the ability to absorb sufficient water vapor that the solid dissolves in the attracted moisture. Sorbitol exhibits

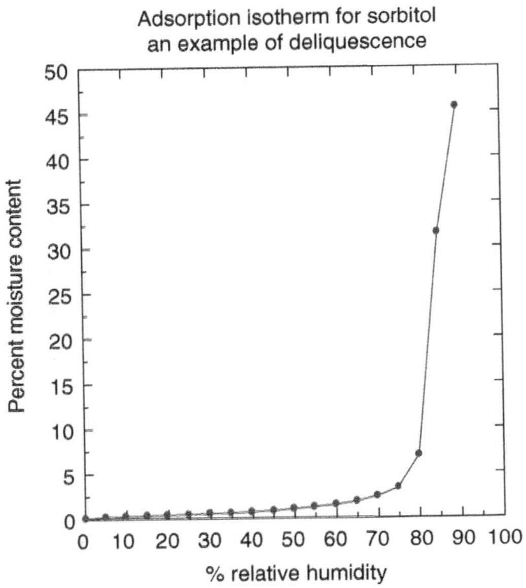

Figure 8.10 Sorbitol deliquescence at high percent RH.

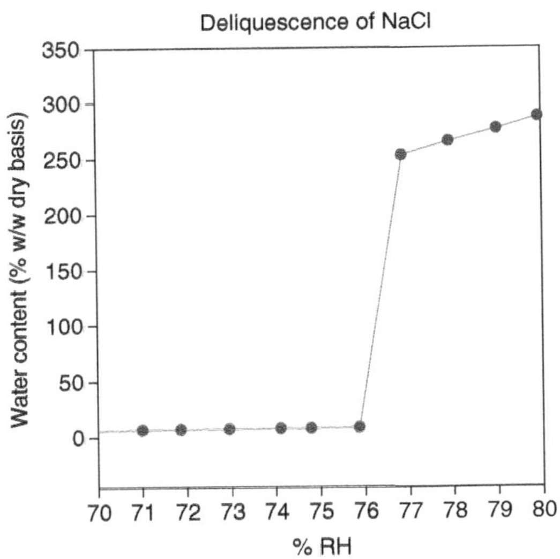

Figure 8.11 Sodium chloride deliquescence.

marginal weight gain until 65% to 75% RH. Above 75% RH, there is a steep increase in the moisture content of the sorbitol. 75% RH is called the *deliquescent relative humidity*.

Figure 8.11 shows the deliquescence of sodium chloride. Sodium chloride does not take up moisture until about 76% RH and then undergoes deliquescence. There is a remarkable 250% weight increase due to moisture uptake in a one unit increase in %RH. As an aside, this may explain why potato chips quickly become soggy at a picnic when the humidity is high. If one considers the moisture sorption isotherms for starch and NaCl, the deliquescence of NaCl is most likely the culprit.

Figure 8.12 has an interesting sorption and desorption isotherm profile that is not similar to physisorption types presented earlier. Figure 8.12 illustrates a chemisorption-like reaction between

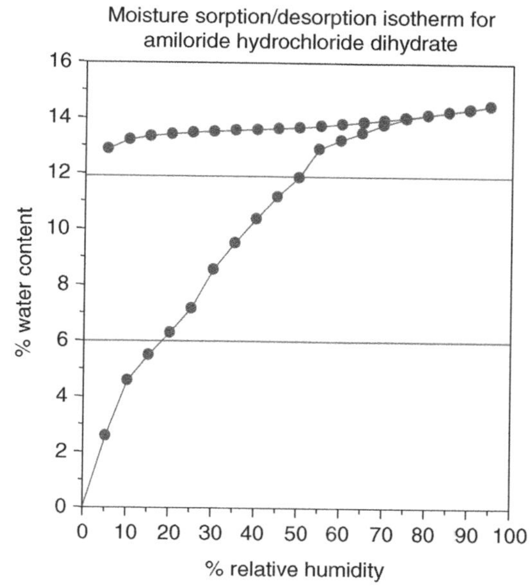

Figure 8.12 Chemisorption-like reaction of water vapor sorptive and amiloride hydrochloride anhydrate adsorbent, which forms a stable amiloride hydrochloride dihydrate.

anhydrous amiloride hydrochloride and water vapor to form a stable amiloride hydrochloride dihydrate. In this case, the chemisorption-like reaction does not involve the typical chemisorption characteristic covalent bond formation and monolayer uptake, but the chemical formula for anhydrous material changes to account for the dihydrate formation, and the hydrate is not reversibly changed to the anhydrate form at low %RH.

8.7.1 Author's Anecdote- Failure of a Commercial Controlled-Release Product Caused by Sodium Chloride Deliquescence

A controlled-release tablet was developed using a controlled-release polymer and sodium chloride (NaCl) as a pore former. The NaCl pore former was used to create "holes" or pores in the polymer to fine-tune the drug-release profile. The product was FDA approved, commercialized, and on the market for a number of years. One day, I was contacted by manufacturing and told that a number of product batches had failed dissolution testing. Dissolution testing showed a "burst" release causing the failure. After extensive review of the manufacturing batch records and exhaustive review of quality acceptance records for all of the materials, there was no clear explanation for the dissolution failure. Microscopy analysis of the film coat was undertaken. It indicated that the pore structure was inconsistent when compared to approved batches. FTIR microscopy of the failed batches suggested that some of the pores contained water, which was not observed in the passed batches.

It was hypothesized that NaCl was sorbing water and in some cases deliquescing. This was surprising since Tang et al. (1977) reported a widely accepted NaCl deliquescent RH value of 75.5% RH at 25 °C.

As background, the manufacturing area, including dispensing, weighing, film-coating, in-process storage, and packaging, had %RH controls that kept the percent RH below 75.5%. The *target operating* RH range was 45% to 65% RH. The *alert control limits* were 40% and 70% RH, and the *action limits* were 35% and 75% RH. The product batch records were once again exhaustively reviewed for RH excursions. Interestingly, there was a period that corresponded to the dissolution failures where the humidity consistently ran at the alert upper limit % RH range of 70%, but no reported values were noted above 70% RH.

More accurate NaCl water vapor sorption studies were performed, which demonstrated an onset of moisture uptake at % RH values as low as 70%. The definitive isotherms were similar to Figure 8.9 shown for sorbitol. There was a small %RH range over which NaCl began to gradually sorb moisture prior to ensemble NaCl deliquescence. The manufacturing percent RH control limits were changed to address this unexpected dissolution problem caused by NaCl moisture uptake at 70% RH values.

Post the time of this anecdote, research by Wise et al. (2008) reported the presence of aqueous solutions of NaCl and water at RHs as low as 70% RH. They made this observation using an environmental electron transmission microscope. This study challenged the conventional thinking that deliquescence is a phase transition and occurs at specific temperature, pressure, and RH.

8.8 Solid-Solution Adsorption Systems

Sections 8.3.1, Freundlich Adsorption Isotherm Equation and 8.3.2, Langmuir Adsorption Isotherm Equation discuss the concepts and theory of adsorption. One of the most common applications of solid-solution adsorption is the adsorption of an adsorptive solute onto filters. Determination of drug adsorption onto filters should be part of every analytical method development and validation project if filters are used to clarify a solution for analysis. Similarly, the drug adsorption onto filters should also be studied for every product design and manufacturing process that uses filters.

Example 8.3 Sucrose Octaacetate (SOA) Drug-Filter Adsorption Study
SOA is nontoxic and has a number of uses based on its bitter taste. It is used in tonics for its bitter taste. It has been used as a surrogate for bitter-tasting drugs for taste-masking product design studies.

A drug-filter adsorption study was undertaken using a validated stability indicating assay. An SOA solution sample was passed through a filter and was collected at 1 mL intervals for a total of 10 mL. The results are shown in Figure 8.13.

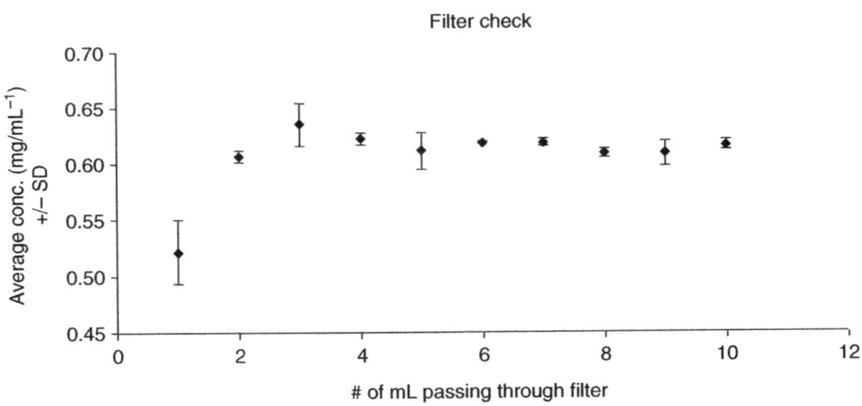

Figure 8.13 Sucrose octaacetate (SOA) drug-filter adsorption study. Ten mL of an SOA solution was passed through a 0.2 um Acrodisc filter. One-mL sample was collected and assayed for SOA. The data represents three independent adsorption trials. The standard deviation for each sample point from each trial is represented by the bars.

There was a 20% loss of SOA for the first 1 mL sample due to SOA being adsorbed onto the filter. An equilibrium SOA concentration was achieved after ~5 mL of solution was passed through the filter. Figure 8.13 clearly highlights large assay variability in the first 1 mL sampled. It also shows that assay precision is improved after 5mL of solution are pushed through the filter. To perform accurate and precise SOA analysis, this data suggests that at least the first 5 mL of SOA solution should be discarded before performing any analysis.

References

Adamson, A.W. and Gast, A.P. (1997). *Physical Chemistry of Surfaces*, 6e, 390–393. New Delhi: John Wiley & Sons.

Al-Achi, A., Gupta, M.R., and Stagner, W.C. (2023). *Integrated Pharmaceutics: Applied Preformulation, Product Design, and Regulatory Science*, 2nde, 231–232. Hoboken: John Wiley & Sons.

Ayawei, N., Ebelegi, A.N., and Wankasi, D. (2017). Modelling and interpretation of adsorption isotherms. *J. Chem.* 30398717. https://doi.org.10.1155/2017/3039817.

Brunauer, W., Emmett, P.H., and Teller, E. (1938). Adsorption of gases in multimolecular layers. *J. Am. Chem. Soc.* 60: 309–319. https://doi.org/10.1021/ja01269a013.

Callahan, J.C., Cleary, G.W., Eefant, M. et al. (1982). Equilibrium moisture content of pharmaceutical excipients. *Drug Dev. Ind. Pharm.* 8: 355–369.

Connors, K.A. and Mecozzi, S. (2010). *Thermodynamics of Pharmaceutical Systems: An Introduction to Theory and Applications*, 2e. Hoboken, NJ: John Wiley & Sons.

Cychosz, K.A. and Thommes, M. (2018). Progress in the physisorption characterization of nanoporous gas storage materials. *Engineering* 4: 559–566. https://doi.org/10.1016/j.eng.201806.001.

Florence, A.T. and Attwood, D. (2011). *Physiochemical principles of Pharmacy*, 5e, 205, 207. London: Pharmaceutical Press.

Freundlich, H. (1907). Über die Adsorption in Lösungen. *Z. Phys. Chem.* 57: 385–470.

Haware, R.V., Dave, V.S., Kakarala, B. et al. (2016). Vegetable-derived magnesium stearate functional evaluation by DM^3 approach. *Eur. J. Pharm. Sci.* 89: 115–124. https://doi.org/10.1016/j.ejps.2016.04.019.

Langmuir, L. (1916a). Part 1: The research library of the general electric company. 2221.

Langmuir, L. (1916b). Part 2: The research library of the general electric company. 1848.

Langmuir, I. (1918). The adsorption of gases on plane surface of glass, mica and platinum. *J. Am. Chem. Soc.* 40: 1361–1402.

Lennard-Jones, J.E. (1931). Cohesion. *Proc. Phys. Soc.* 43 (5): 461–482. https://doi.org/10.1088/0959-5309/43/5/301.

Lundelius, E.F. (1920). Adsorption and solubility. *Kolloid Z.* 26: 145.

Nambu, N., Sakaurai, S., and Nagai, T. (1975). Adsorption from solution, permeation through cellulose membrane, partition coefficient and surface tension of several antidepressants and phenothiazines. *Chem. Pharm. Bull.* 23: 1404.

Osterrieth, J.W.M., Rampersad, J., Madden, D. et al. (2022). How reproducible are surface areas calculated from the BET equation? *Adv. Mater.* 34 (27): https://onlinelibrary.wiley.com/doi/10.1002/adma.202201502.

Richards, N.E. and Meaken, B.J. (1974). The sorption of benzocaine from aqueous solution by nylon 6 powder. *J. Pharm. Pharmacol.* 26: 166. https://doi.org/10.1111/j.2042-7158.1974.tb09249.x.

Tang, I.N., Munkelwitz, H.R., and Davis, J.G. (1977). Aerosol growth studies-II. Preparation and growth measurements of monodisperse salt aerosols. *J. Aerosol Sci.* 8: 149–159.

Thommes, M., Kaneko, K., Neimark, A.V. et al. (2015). Physisorption of gases, with special reference to the evaluation of surface area and pore size distribution (IUPAC technical report). *Pure Appl. Chem.* https://doi.org/10.1515/pac-2014-1117.

Wise, M.E., Martin, S.T., Russell, L.M., and Buseck, P.R. (2008). Water uptake by NaCl particles prior to deliquescence and the phase rule. *Aerosol Sci. Technol.* 42: 281–294. https://doi.org/10.1080/02786820820802047115.

Wurster, D.E. (April 2024). Personal communication.

Chapter 8 Problem set

8.1 Define adsorbate, adsorbent, and adsorptive.

8.2 Fill out the table to identify the differences between physisorption and chemisorption.

Properties	Physisorption	Chemisorption
Nature		
Bonding interaction		
Adsorption enthalpies		
Saturation uptake		
Surface specificity		
Temperature range		
Adsorption kinetics		

8.3 Fill out the table to identify the differences between adsorption and absorption.

Properties	Adsorption (physical)	Absorption
Phenomenon		
Enthalpy		
Temperature		
Reaction rate		
Concentration		

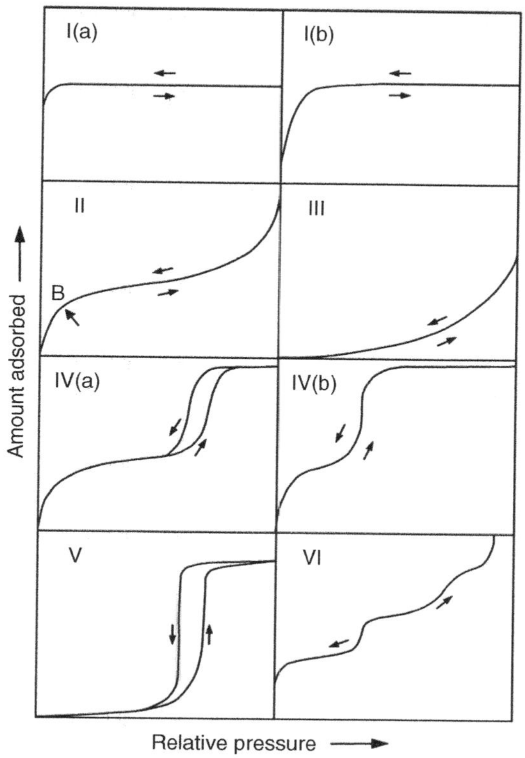

8.4 Adsorption–desorption isotherms.

8.4a Describe the pertinent features of the Type I isotherm.

8.4b Describe the pertinent features of the Type II isotherm.

8.4c Describe the pertinent features of the Type III isotherm.

8.4d Describe the pertinent features of the Type IV isotherm.

8.4e Describe the pertinent features of the Type V isotherm.

8.4f Describe the pertinent features of the Type VI isotherm.

8.5a What is the Freundlich equation?

8.5b What is the significance of K_F *and* n?

8.5c Using the Freundlich equation and the data table below, determine K_F and n for the adsorption of tetracaine onto activated charcoal at 25 °C.

Adsorption of Tetracaine onto Activated Charcoal (from Florence and Attwood, 2011)

Equilibrium concentration, C_e, mg L^{-1}	0.155	0.468	1.259	2.510	5.370
Amount adsorbed/adsorbent, q_e, mg g^{-1}	202.8	217.7	232.0	243.2	254.7

8.7 Using the Langmuir isotherm shown below and the Langmuir equation:

$$\theta = A_{adsorbate\ bound}/A_{total} = K_L \cdot P/P^o{}_{adsorptive\text{-}vapor}/\left(1 + K_L \cdot P/P^o{}_{adsorptive\text{-}vapor}\right)$$

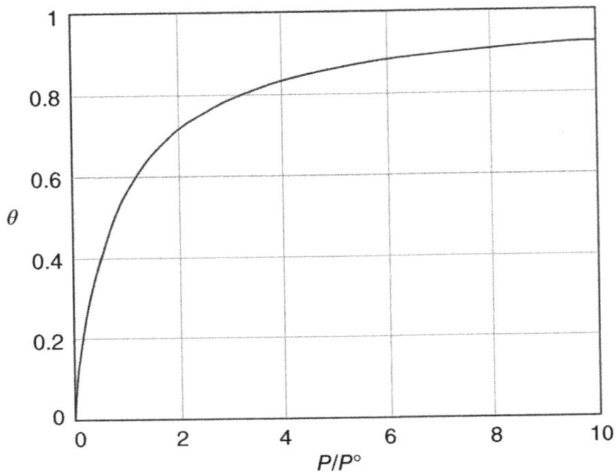

8.7a Where is the Henry's law portion of the curve?

8.7b What does it mean when the line asymptotically approaches 1.

8.8 Provide the linear form of the Langmuir equation and state the meaning of each term.

8.9 What are the four assumptions related to the Langmuir equation?

8.10 The following data is taken from Florence and Attwood (2011), which was extracted from Nambu et al. 1975) for the adsorption of amitriptyline onto carbon black.

8.10a Using a modified Langmuir-like equation, $C_{adsorptive\ solute}/q_e = 1/(q_{max} \cdot K_L) + C_{adsorptive\ solute}/q_{max}$ and the data given in the table below, determine K_L and q_{max}.

8.10b What physical meaning does K_L and q_{max}?

Amount adsorbed/adsorbent, q_e (mol g^{-1})	0.75	0.95	1.10	1.25	1.40	1.55	1.65
Equilibrium concentration, C_e, mg L^{-1}	0.25	0.40	0.60	0.70	1.10	1.35	1.95
C_e/q_e, g L^{-1}	3.33	4.21	5.45	5.60	7.86	8.71	11.82

8.11 True (T) or False (F)

- The Elovich isotherm is a good model for chemisorption.
- At higher adsorbate concentrations, the Fritz and Schlunder five-parameter equation gives a similar curve to the Freundlich equation.
- The Hill isotherm is used to assess the influence of different adsorptives upon adsorbent binding sites.

8.12 List seven factors that can affect the rate and extent of adsorption.

8.13 True or False Capillary condensation occurs below the saturation gas vapor pressure. This unique phenomenon is described by the Kelvin-Laplace equation.

8.14 Describe the pertinent features of an H1 isotherm.

8.15 Define deliquescence.

8.16 Calculate the surface area of a drug given the following information.
Givens:

- The cross-section area of N_2 is 0.162 nm^2
- 1 nm^2 = 1E-18 m^2
- Avogadro's number is 6.023E23 molecules per mole
- Nitrogen's molecular weight is 28.02 g mol^{-1}
- $q_m = 140.124$ cm^3 N_2/1 g drug
- $K_{BET} = 163.5479$

Solved Problem Set

Chapter 1 Solved Problem Set

1.1 List the seven fundamental physical quantities.

Moles or particles, time, length, mass, temperature, electric current, and luminous intensity.

1.2 What are the fundamental dimensions associated with the fundamental physical quantities?

Moles	[n]
Time	[t]
Length	[L]
Mass	[M]
Temperature	[T]
Electric current	[I]
Luminous intensity	[I_v]

1.3 What are the formulas for the derived physical quantities: area, volume, velocity, concentration, and density?

Area	Length × width	[L^2]
Volume	Length × width × height	[L^3]
Velocity	Distance per second	[Lt^{-1}]
Concentration	Amount per volume	[nL^{-3}]
Density	Mass per volume	[ML^{-3}]

1.4 Define dimensional analysis and list three examples where dimensional analysis is useful.

Definition of dimensional analysis: Dimensional analysis of a physical quantity shows the association and interrelationships between or among two or more fundamental or derived physical quantities.

Thermodynamics for Drug Product Design: A Guide for Pharmaceutical Scientists, First Edition. William Craig Stagner.

Uses of dimensional analysis:

1) DA can be used to check for the dimensional consistency of an equation and associated property values.
2) Physical properties of interest can be reduced to its fundamental dimensions of amount, time, length, mass, temperature, electrical current, and luminous intensity by employing DA.
3) Dimensional analysis plays a key role in converting units of one system to another.
4) Dimensions can be used to determine unknown relationships and equations by utilizing fundamental dimensions for known quantities.
5) Dimensional consistency requires that the units and dimensions on the left-hand and right-hand sides be the same.

1.5 Explain dimensional quantity algebra.

A physical quantity value can be expressed as a *product* of a *numerical value* and a *unit*;

$$\text{Physical quantity} = \text{numerical value} \times \text{unit}$$

By convention for convenience and efficiency, the "times sign" is not used. The commonly accepted formatting without the "times sign" for temperature is temperature = 298 K.

1.6 Explain conversion ratio. What is another name for conversion ratio?

Using quantity algebra, all conversion factors can be rearranged as quotients that are equal to 1. These quotients are called unitary conversion quotients or conversion ratios. Given the conversion factor 1 erg = 1E-7 J; the unitary conversion quotients are 1 erg/1E-7 J = 1 or 1E-7 J/1 erg = 1.

1.7 Seven Using dimensional analysis, determine the units of the drug complexation equilibrium constant, K_{11}, for the equation given below when S_t, S_o, and L_t have units of mol L^{-1}.

$$S_t = S_o + (K_{11}S_oL_t)/(1 + K_{11}S_o)$$

S_t is the total molar concentration of the drug after complexation, S_o is the solubility of the drug before a complexing agent (ligand) is added, L_t is the total ligand molar concentration added to the drug liquid, and K_{11} is the 1 : 1 drug: ligand equilibrium constant. Note: to be dimensionally correct, both sides of the equation need to have the same units and dimensionality.

$$S_t = S_o + (K_{11}S_oL_t)/(1 + K_{11}S_o)$$

Express all terms using their units.

$$\text{mol L}^{-1} = \text{mol L}^{-1} + \left(K_{11} \times \text{mol L}^{-1} \times \text{mol L}^{-1}\right)/\left(1 + K_{11} \times \text{mol L}^{-1}\right)$$

In the denominator of the second term on the right-hand side of the equation, "1" is a dimensionless number, so the $K_{11} \times$ mol L^{-1} term also needs to be dimensionless. This term becomes dimensionless when K_{11} units are L mol^{-1}.

Likewise, the right-hand side numerator term ($K_{11} \times$ mol $L^{-1} \times$ mol L^{-1}) must have units mol L^{-1} to be consistent with the left-hand side of the equation. Therefore, K_{11} must have units of L mol^{-1}. This can be confirmed by substituting for K_{11}, giving L $\text{mol}^{-1} \times$ mol $L^{-1} \times$ mol L^{-1}, which equals mol L^{-1}. These units are consistent with the left-hand side of the equation.

Substituting mol L^{-1} for K_{11} into the complete equation shows that both sides of the equation have units of mol L^{-1}.

1.8 Write the following numbers in E scientific notation:

$$-200.12, 0.00347, \text{ and } 69201748.$$

-200.12 is -2.0012E2, 0.00347 is 3.47E-3, and 69 201 748 is 6.9201748E7

1.9 Discuss accuracy and precision and how they relate to a valid measurement.

Accuracy of a measurement system is the difference between the true value or target value of a quantity and the measured value or mean of measured values. *Precision* deals with the closeness or difference of measured values independent of the measured values accuracy. Precision is related to the variability of the measured quantities. Precision is also related to the *repeatability* and *reproducibility or robustness* of the analytical method. A reported measurement is considered *valid* when it is both accurate and precise.

1.10 How does uncertainty or measurement error relate to how data is reported?

The level of uncertainty determines the number of significant figures in the recorded measurement. There are several rules in reporting uncertainty.

Rule 1- Stating Uncertainties: experimental uncertainties should be expressed to only one significant figure because the uncertainty is an estimate and carries no more significance than one digit.

Rule 2- Stating measurements or values: the last significant figure in a reported quantity should have the same place or decimal place value as the uncertainty.

1.11 How many significant figures (SFs) are in

a) 109.04 5 SFs
b) 0.01903 4 SFs
c) 1090 ambiguous 3 or 4 SFs

1.12 Report to the correct number of significant figures?

a) 707
b) 6.31E10

1.13 Report to the correct number of significant figures.

$$21.1 + 2.037 + 6.13 = 29.267 = \underline{29.3}$$

1.14 What is the rounded number?

If the final number is reported to three significant figures, and the final number is 89.87, then this value would be rounded off to 89.9

If the final number is reported to three significant figures, 89.83 is rounded to 89.8.

1.15 Using exponent rules, calculate the buffer capacity for a buffer having a total buffer concentration of 1 M, $[H_3O^+] = 1.0 \times 10^{-4}$ M, and $K_a = 1.47 \times 10^{-4}$, using the general buffer capacity equation: buffer capacity $= 2.303C \dfrac{K_a[H_3O^+]}{\{K_a + [H_3O^+]\}^2}$ (Martin A, Swarbrick J, Cammarata

A. Solubility and distribution phenomena. In: *Physical Pharmacy: Physical Chemical Principles in the Pharmaceutical Sciences*, 3rd ed. Philadelphia: Lea & Febiger; 1983. pp. 272–313.)

$$\text{Buffer capacity} = 2.303\,C\frac{K_a[\mathrm{H_3O^+}]}{\{K_a + [\mathrm{H_3O^+}]\}^2}$$
$$= (2.303)(1)\frac{(1.47\times10^{-4})(1.010^{-4})}{[(1.47\times10^{-4}) + (1.010^{-4})]^2} = \underline{0.56\ \text{2SF}}$$

1.16 Convert the nature log form of ln(concentration) = 1.85 − 0.45 × time (h) to its base 10 log form.

ln(concentration) = 1.85 − 0.45 × time (h) becomes

2.303 log(concentration) = ln (concentration)

Substituting gives

2.303 log(concentration) = 1.85 − 0.45 × time (h)
log(concentration) = [1.85 − 0.45 × time (h)]/2.303
<u>log(concentration) = 0.803 − 0.20 × time (h) Check SFs.</u>

1.17 What is the derivative of the zero-order reaction, $[A] = [A_0] - kt$ with respect to t?

$d[A]/dt = d[A_0]dt - (-kt)dt$ gives

$d[A]/dt = 0 - k = -k$

1.18 The diffusion coefficient (D) is a function of temperature (T), the solution viscosity (η), and particle radius (r). A is Avogadro's number and R is the gas constant

$$D = \frac{RT}{\pi\eta r A}$$

When all variables change simultaneously, what is the equation for the diffusion coefficient?

$$dD = \left(\frac{\partial D}{\partial T}\right)_{\eta,r} dT + \left(\frac{\partial D}{\partial \eta}\right)_{T,r} d\eta + \left(\frac{\partial D}{\partial r}\right)_{T,\eta} dr$$

At constant temperature, D is a function of viscosity and the particle radius, $D = f(\eta, r)$.

$$D = \frac{RT}{\pi A}\eta^{-1}r^{-1}$$
$$D = (\text{constant})\,\eta^{-1}\ r^{-1}$$
$$\left(\frac{\partial D}{\partial \eta}\right)_{T,r} = (\text{constant}\,r^{-1})\,\eta^{-2}$$
$$\left(\frac{\partial D}{\partial r}\right)_{T,\eta} = (\text{constant}\,\eta^{-1})\,r^{-2}$$

Therefore, the diffusion coefficient, D, at constant temperature and changes in viscosity and particle radius is the sum of the partial derivatives of D at constant temperature and particle radius and the partial derivative of D at constant temperature and constant viscosity.

$$\left(\left(\frac{\partial D}{\partial \eta}\right)\frac{\partial D}{\partial r}\right)_T = D = \left(\frac{RT}{\pi A}r^{-1}\right)\eta^{-2} + \left(\frac{RT}{\pi A}\eta^{-1}\right)r^{-2}$$

1.19 What is the definite integral of zero-order rate expression $\frac{dA}{dt} = -k_0A_0$?

$$dA = -k_0 dt$$

$$\int_0^A dA = -k_0\int_0^t dt$$

$$A - A_0 = -k_0(t - t_0)$$

1.20 Find the range, mean, median, mode, standard deviation, and % RSD for this data set.

90, 76, 53, 78, 88, 80, 81, 91, 99, 68, 62, 78, 67, 82, 88, 89, 78, 72, 77, 96, 93, 88, 88

Range = 99 − 53 = 46; mean = 81; median = 81; and mode = 88

Standard deviation, s: **11.271356540402**

Count, N:	23
Sum, Σx:	1862
Mean, $\bar{x}$:	80.95652173913
Variance, s^2:	127.04347826087

Steps

$$s = \sqrt{\frac{1}{N-}\sum_{i=1}^{N}(x_i - \bar{x})^2},$$

$$s^2 = \frac{\sum(x_i - \bar{x})^2}{N-1}$$

$$= \frac{(90 - 80.95652173913)^2 + \dots + (88 - 80.95652173913)^2}{23 - 1}$$

$$= 127.04347826087$$

$$s = \sqrt{127.04347826087}$$

$$= 11.271356540402$$

$$\% \text{RSD} = (11.271356540402/81)_* 100\% = 13.9\,\%\,\text{RSD}$$

Chapter 2 Solved Problem Set

2.1 What is the average kinetic energy of helium, neon, argon, krypton, xenon, and radon at 25.0 °C?

$$KE_{ave} = 3/2(RT)$$

$= 3/2(8.314\ \text{J mol}^{-1}\ \text{K}^{-1})\ (298.15\ \text{K}) = \underline{3718.2\ \text{J mol}^{-1}}$ or $\underline{3.7182\text{E}3\ \text{kJ mol}^{-1}}$ for all the gases. Kinetic energy is only dependent on temperature for the monatomic gases.

2.2 Work is defined as force times distance. Work is a form of energy and has units of joules. Using dimension algebra, show that force times distance equals joules.

Force is in Newtons. Newton has fundamental units of m kg s^{-2}.
w = m kg s^{-2} (m) = $\underline{m^2}$ $\underline{kg\ s^{-2}}$ and the fundamental units of joule are $\underline{m^2}$ $\underline{kg\ s^{-2}}$ (see Table 1.2)

2.3 Define isolated, closed, and open systems.

Neither matter nor energy can flow into or out of an *isolated* system.
Energy can flow into or out of a *closed* system, but matter cannot cross the boundary between the system and its surroundings.
Both energy and matter can flow in and out of an open system.

2.4 List four state variables.

Temperature, pressure, volume, and composition.

2.5 Define intensive and extensive variables. Name several intensive and extensive variables.

An intrinsic physical property or variable that is independent of the size of the system is called an *intensive* property or variable. Temperature, pressure, surface tension, density, melting point, viscosity, concentration, and average kinetic energy are intensive variables.

Physical properties or variables that are additive and are quantity- or capacity-dependent are known as *extensive* properties or variables. Mass, moles, and volume are extensive variables.

2.6 Forms of energy are products of intensive and extensive variables. What are the intensive and extensive variables for thermal energy, work of expansion, surface energy, and chemical energy?

Energy form	Intensive (I) variable	Extensive (E) variable	Product I · E[1]
Thermal energy (heat)	Temperature (K)	ΔS (J/K)	J
Work of expansion	Pressure (N m^{-2})	ΔV (m^3)	N m = J
Surface energy	Surface tension (N m^{-1})	ΔA (m^2)	N m = J
Chemical energy	Chemical potential (J mol^{-1})	Mol (m)	J

2.7 Calculate the enthalpy of solution in joules per mole for sodium hydroxide when 50.0 g of NaOH is completely dissolved in 1000 g of water at 20.0 °C. The final temperature is 33.2 °C. Assume there is no heat loss in the calorimeter and the heat capacity of the solution is the same as water, 4.184 J^{-1} $°C^{-1}$.

$$Q_{reaction} = -\left(m_{solution} \cdot C_{solution\ heat\ capacity} \cdot \Delta T\right) = \Delta H_{soluiton}$$

$$= -\left[1050\,\cancel{g} \cdot 4.184\,J\,\cancel{g}^{-1\circ}\cancel{C}^{-1} \times (33.2 - 20.0)^{\circ}\cancel{C}\right]$$

$$\Delta H_{soluiton} = -5.799E10^{4}\,J$$

The molar mass of NaOH is 40.00 g mol^{-1}.

$$\text{Mole of NaOH added to the water} = 50.0\,\cancel{g}/\left(40.00\,\cancel{g}\,mol^{-1}\right) = 1.250\,mol\,NaOH$$

$$\Delta H_{soluiton} = \left[-5.799E10^{4}\,\cancel{J}/(1.25\,mol\,NaOH)\right] \cdot \left(1\,kJ \cdot 1000\,\cancel{J}^{-1}\right) = \underline{-46.4\,kJ\,mol^{-1}}$$

2.8 In Section 2.7.1, Pharmaceutical Application: Using an Adiabatic Calorimeter to Determine the Enthalpy of Solution of a Drug or Material of Interest, identify the boundary between the system and surroundings.

The boundary is the surface of the solid ammonium nitrate.

2.9 Describe a thermodynamic state function. Name several thermodynamic state functions.

A *state function* is a thermodynamic property whose value does not depend on the path taken to reach the *final state.*

Temperature (T), pressure (P), volume (V), composition or mass, density (ρ), specific volume (ν), and energy [internal energy (U), enthalpy (H), Gibb's free energy (G), Helmholtz energy (A), and entropy (S)] are thermodynamic state functions.

2.10 Describe a thermodynamic path function. Name several thermodynamic state functions.

Path functions depend on the path taken to get to the final state.

Heat (q) and work (w) are path functions,

2.11 Define a thermodynamic spontaneous process. Give several examples of spontaneous processes.

A spontaneous process occurs naturally without help or intervention.

As the volume of a container is expanded, the gas in the container will spontaneously fill the new volume. Heat always spontaneously flows from a material at a higher temperature to a material at a lower temperature. A thermodynamically unstable polymorphic form of a drug spontaneously converts to the more stable crystalline drug form.

Chapter 3 Solved Problem Set

3.1 What is the work of reversible isothermal expansion of an ideal gas when carried out under constant temperature?

At constant temperature, $P = nRT/V$. From Eq. (3.6), $-w = P\Delta V$.

$$-w_{rev} = nRT \int_{initial}^{final} dV/V$$

$$-w_{rev} = nRT \ \ln \ V_2/V_1$$

3.2 What is the reversible work of expansion of one mole of ideal gas at 25.00 °C when the gas expands from 5.00 L to 25.00 L?

$$-w_{rev} = nRT \ \ln \ V_2/V_1$$

$$-w_{rev} = (1 \ \cancel{\text{mol}}) \ (8.314 \text{ J } \cancel{\text{K}}^{-1} \cancel{\text{mol}}^{-1}) \ (298.15 \ \cancel{\text{K}}) \ \ln \ (25 \ \cancel{\text{L}}/5 \cancel{\text{L}})$$

$$-w_{rev} = 3.990\text{E}3 \text{ J}$$

$$\underline{w_{rev} = -3.99\text{E}3 \text{ J}.}$$

The reversible work is negative because the work is done by the system, and its energy is given up doing the work.

3.3 Hand warmers can provide comfort during winter outdoor activities. These reusable hand warmers can be made using a 50% w/w % sodium acetate supersaturated water solution. A thin metal disc inside the solution plastic bag container can be clicked to initiate precipitation of the supersaturated solution.

Given:

a) Assume 82.03 g NaAcetate (one mole) is dissolved in 82 g water.
b) The crystallization reaction at constant pressure is

$$CH_3CO_2Na + 3\,H_2O\,(l) \rightarrow CH_3CO_2Na \cdot 3\,H_2O\,(\text{solid})$$

c) $CH_3CO_2Na \cdot 3\,H_2O \ \Delta H_{crystallization} = -30.744 \text{ kJ mol}^{-1}$

$$= -30,744 \text{ J } \cancel{\text{mol}}^{-1} \cdot \left(82.03 \text{ g} \cdot \cancel{\text{mol}}^{-1}\right)^{-1} = -374.8 \text{ J g}^{-1}$$

d) At the final temperature:
- 56.44 g $CH_3CO_2Na \cdot 3\,H_2O$ solid in the liquid mixture
- The solution portion of the liquid mixture contains 59.59 g water and 47.97 g dissolved CH_3CO_2Na (aq)

e) Heat capacities are

$CH_3CO_2Na \cdot 3\,H_2O$ (solid) $2.40 \text{ J g}^{-1} \text{K}^{-1}$;
water $4.184 \text{ J g}^{-1} \text{K}^{-1}$; CH_3CO_2Na
$2.58 \text{ J g}^{-1} \text{K}^{-1}$

A) Is crystallization an exothermic or endothermic process?
Crystallization is an exothermic process. The system releases heat.

B) What is the final temperature if the hand warmer is activated at 25.0 °C?
From Eq. (2.22)
$q_{reaction} = -(m \cdot C_p \cdot \Delta T)$, where $q_{reaction}$ is the heat of the reaction, m is the mass of the materials having their respective heat capacities, C_ps, and ΔT is the temperature change that occurs during the crystallization reaction. Since the crystallization takes place under constant pressure, $q_{reaction} = \Delta H$, Eq. (3.31).

A) Is crystallization an exothermic or endothermic process?

Crystallization is an exothermic process. The system releases heat.

B) What is the final temperature if the hand warmer is activated at 25.0 °C?

We solve for ΔT knowing that the initial temperature was 25.0 °C. Therefore,

$$\Delta T = -q_{reaction} = -\Delta H_{crystallization} / \sum m \cdot C_p$$

$$\Delta T = -(-374.8\,\text{J}\,\text{g}^{-1}) \cdot 56.44\,\text{g} / [(56.44\,\text{g} \cdot 2.40\,\text{J}\,\text{g}^{-1}\,\text{K}^{-1}) + (59.59\,\text{g} \cdot 4.184\,\text{J}\,\text{g}^{-1}\,\text{K}^{-1}) + 47.97\,\text{g} \cdot 2.58\,\text{J}\,\text{g}^{-1}\,\text{K}^{-1})]$$

$$\Delta T = 2.115\text{E}4\,\text{J}/(135.5 + 249.3 + 123.8)\,\text{J}\,\text{K}^{-1}$$

$$\Delta T = 2.115\text{E}4\,\text{J}/(508.6\,\text{J}\,\text{K}^{-1})$$

$$\Delta T = 41.6\,\text{K or } 41.6\,°\text{C}$$

The equilibrium temperature would be

Final Temp − 25.0°C = 41.6°C

Final Temp = 41.6°C + 25.0°C = 66.6°C or 152°F.

3.4 Explain the different features of the modulated differential scanning calorimetry thermogram.

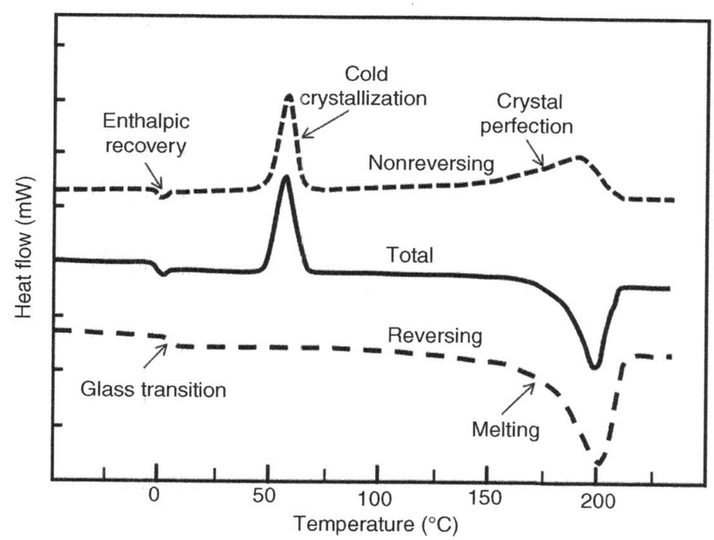

The total heat flow curve exhibits three thermal events.

The first event around 0 °C shows an initial change in heat flow indicative of a heat capacity change associated with a glass/amorphous solid phase transition. On closer inspection, there is a very slight endothermic peak at the end of the glass transition. The reversing and nonreversing heat flows deconvolute the total heat flow of the thermal event. The reversing heat flow shows the typical *Cp* glass transition phase change. The nonreversing heat curve separates the heat capacity change from the slight endothermic peak seen in the total heat flow curve. The nonreversing curve shows a small separate endothermic peak, which is attributed to the amorphous powder's *enthalpic recovery* at the glass transition. Enthalpic recovery is the absorption of energy that the sample lost as it relaxed toward its more preferred equilibrium state. Enthalpic recovery is an endothermic process since heat/energy is absorbed by the sample.

The second total heat flow transition is an exotherm that occurs around 45 to 50 °C. This thermal event is totally accounted for as a nonreversing heat capacity behavior. This kinetic heat flow is credited to *cold crystallization*. Cold crystallization refers to a crystallization process that happens during heating a glass or amorphous powder above its glass transition temperature. Cold crystallization occurs above the drug's glass transition point and below its crystalline melting point.

The third total heat flow peak also has hidden thermal events that cannot be resolved by DSC. The MDSC nonreversing heat flow uncovers an additional thermal behavior associated with the drug crystal lattice imperfections rearranging to form a more perfect crystal. This process is like *heat annealing* material to obtain a more stable material form.

3.5 How much energy is required to melt 25 g of ice.

Given:

a) The heat required for a phase transition is

$q = m\,\Delta H_f$, where q is heat/energy, m is mass, and ΔH_f is the enthalpy or heat of fusion.

b) ΔH_f of ice is 334 J g^{-1}

Substituting values for the variables indicated above gives

$$q = 25\,\text{g} \cdot \left(334\,\text{J}\,\text{g}^{-1}\right) = \underline{8.4\text{E}3\,\text{J}}$$

It takes 8.4E3 J to melt 25 g of ice.

Note:

1) There is no temperature term in the heat equation shown above. This is because the temperature remains constant during a phase transition. A phase transition is an equilibrium process.
2) The number of significant figures is 2 since 25 g has only two significant figures.

3.6 When 25 g of ice is melted, what is the entropy of the system?

Given

a) ΔH_f of ice is 334 J g^{-1}

b) $S_{system} = \Delta H_f / T_{transition\ temperature}$

$$S_{system} = 25\,\text{g} \cdot \left(334\,\text{J}\,\text{g}^{-1}\right)/273.15\,\text{K} = \underline{31\,\text{J}\,\text{K}^{-1}}$$

The entropy of 25 g of melted ice is 31 J K^{-1}

3.7 Which molecule has the greatest entropy? Explain.

a) 1 mol SO_2 or 1 mol SO_3	SO_3: For similar molecules, larger mass ↑ S
b) water at 5 °C or water at 25 °C	Water at 25 °C: Increase in temperature ↑ S
c) He or Ne	Ne: For similar molecules, larger mass ↑ S
d) water or ethanol	Ethanol: weaker hydrogen bonds ↑ S

3.8 Is entropy positive or negative for each reaction? Explain.

a) $N_{2\ (g)} + 3H_{2\ (g)} \rightarrow 2NH_{3\ (g)}$ — Negative; The product has fewer gas molecules

b) $H_{2\ (g)} + I_{2\ (s)} \rightarrow 2HI_{\ (g)}$ — Positive; The product has more gas molecules

c) $Ag^{+}_{\ (aq)} + Cl^{-}_{\ (aq)} \rightarrow AgCl_{(s)}$ — Negative; The product has fewer particles or ions

d) Ice $\rightarrow$ water vapor — Positive; Sublimation results in gas which has $\uparrow S$

Chapter 4 Solved Problem Set

4.1 Fill in the table shown below.

ΔH_{system} sign	ΔS_{system} sign	$-T\Delta S_{system}$ term sign	ΔG_{system} sign[a]	Spontaneity description
− Exothermic	+	−	−	ΔG^{b} is always negative. Process proceeds spontaneously at ALL temperatures.
− Exothermic	−	+	(+/−)[c]	ΔG is negative at low temperatures and proceeds spontaneously. ΔG is positive at high temperatures, and the forward reaction is not spontaneous. The reverse reaction is spontaneous at high temperatures.
+ Endothermic	+	−	(+/−)[c]	ΔG is negative at high temperatures and is spontaneous. ΔG is positive at low temperatures. The reverse reaction is spontaneous at low temperatures.
+ Endothermic	−	+	+	ΔG is always positive, and the forward reaction is never spontaneous at any temperature. The reverse reaction is spontaneous at ALL temperatures.

[a] $\Delta G_{system} = \Delta H_{system} - T\Delta S_{system}$.
[b] ΔG represents the system Gibbs free energy change.
[c] Sign and spontaneity is temperature dependent.

4.2 Fill in the figure when $\Delta H_{system} < 0$ and $\Delta S_{system} > 0$.

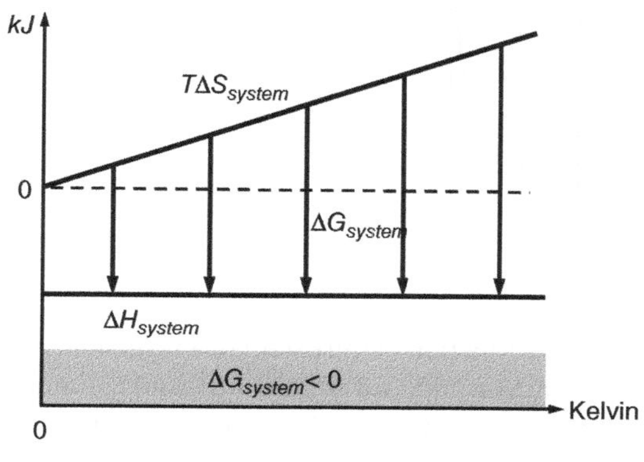

4.3 Explain the above figure.

ΔH_{system} is negative, indicating an exothermic reaction. The $T\Delta S$ line is positive. The ΔG_{system} arrows point down, indicating ΔG_{system} is negative and spontaneous over all temperatures. The figure also shows that as the temperature increases, the magnitude of ΔG_{system} becomes more negative.

4.4 Fill in the figure when $\Delta H_{system} > 0$ and $\Delta S_{system} < 0$.

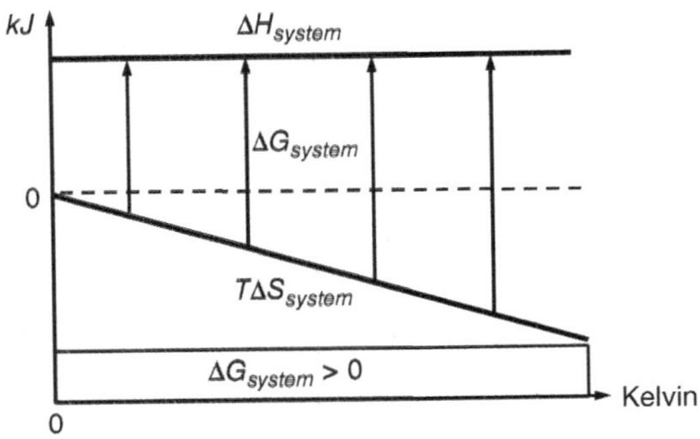

4.5 Explain the above figure.

In this case, $\Delta S_{system} < 0$ and the slope of the $T\Delta S_{system}$ line is negative. ΔH_{system} is positive, indicating an endothermic reaction. The ΔG_{system} arrows point up, indicating ΔG_{system} is positive, and the reaction is nonspontaneous over all temperatures. The figure also shows that as the temperature increases, the magnitude of ΔG_{system} becomes more positive.

4.6 Fill in the figure.

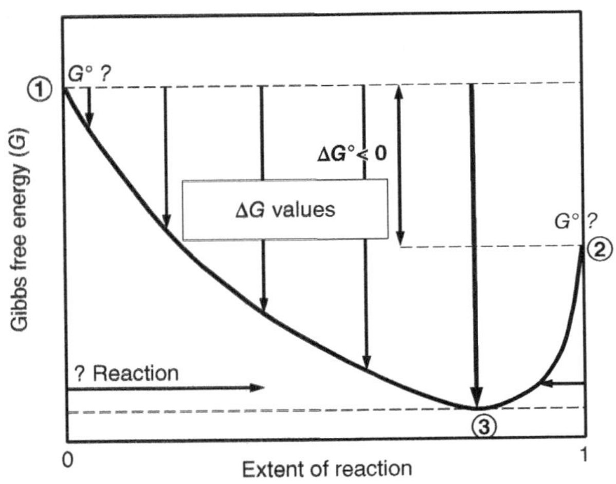

4.7 Explain the above figure

This figure shows a Gibbs free energy curve as a function of the extent of a reversible reaction. When the extent of the reaction is 0, the reaction mixture is 100% reactants A and B (see Eq. [4.30], $aA + bB \leftrightharpoons cC + dD$). When the extent of the reaction is 1, the reaction mixture is 100% products C and D. As indicated in Figure 4.2a, $G^o{}_{products} < G^o{}_{reactants}$. In this case, the standard Gibbs free energy of the reactants (1) needs to be greater than the $G^o{}_{products}$ (2), and $\Delta G^o{}_{reaction} = (G^o{}_{products} - G^o{}_{reactants})$ needs to be less than 0. This is shown in Figure 4.2a.

Depending on the position of the reaction along the extent axis, the reaction will move either toward the reactants or products to achieve equilibrium. The equilibrium position is shown as the lowest point on the Gibbs free energy axis (3).

4.8 Calculate the final mixture volume when 100 mL of 1-propanol is added to 100 mL of water at 20 °C.

Given

1-propanol density at 20 °C is 0.803 g mL^{-1}
Water density at 20 °C is 0.998 g mL^{-1}
1-propanol MW is 60.09 g mol^{-1}
Water MW 18.02 g mol^{-1}
At a mole fraction of 0.194, 1-propanol, the molar volume of 1-propanol is 72 mL mol^{-1} and water molar volume is 18 mL mol^{-1}

Solution:

Step 1. Calculate the mole fraction of the 100 mL 1-propanol and 100 mL water mixture.

$n_{1\text{-}propanoll} = 100\ \cancel{mL}\ (0.0.803\ \text{g}\,\cancel{mL^{-1}})/(60.09\ \text{g}\,\text{mol}^{-1}) = 1.336\ \text{mol}$

$n_{water} = 100\ \cancel{mL}\ (0.998\ \cancel{\text{g}mL^{-1}})/(18.02\ \text{g}\,\text{mol}^{-1}) = 5.538\ \text{mol}$

Mole fraction 1-propanol = (1.336 mol)/(1.336 mol + 5.538 mol) = 0.1943

Step 2. Substitute the appropriate values into Eq. (4.65)

$Vfinal_{T,P} = n_{1\text{-}propanol}\, V_{M,1\text{-}propanol} + n_{water}\, V_{M,water}$

$\cancel{1.336\ mol}_{1\text{-}propanol}(72\ \text{mL}\cancel{mol^{-1}\ 1\text{-}propanol}) + 5.538\ \cancel{mol\ H_2O}\ (18.02\ \text{mL}\ \cancel{mol^{-1}\ H_2O})$

= 96.2 mL + 99.8 mL = 196 mL mixture; 2.0E2 mL of mixture with 2 significant figures

4.9 Show that $dG = -dw_{nonPV}$

It can be shown that $dG = -dw_{nonPV}$ by employing earlier equations that are given below.
Given

1) Equation (3.4) $dU = dq - dw$ (where the minus sign indicates that the work is done by the system)
2) Equation (3.75) $dq = Tds$ (for reversible process at constant pressure)
3) Rearrangement and derivatization of Eq. (3.30), $H = U + PV$ gives $dU = dH - PdV$
4) Equation (3.6) $dw_{PV} = PdV$ (work of expansion for a closed system and when work is done on the system)
5) Equation (4.2) $\Delta G = \Delta H - T\Delta S$ infinitesimal form $dG = dH - TdS$

Starting with Eq. (3.4), $dU = dq - dw$, substitute Eq. (3.75) for dq into Eq. (3.4) to give

$$dU = Tds - dw \tag{4.79}$$

Substitute $dU = dH - PdV$ for dU in Eq. (4.79).

$$dH - PdV = Tds - dw \tag{4.80}$$

Substitute $dw_{PV} + dw_{nonPV}$ for dw Eq. (4.80).

$$dH - PdV = Tds - dw_{PV} - dw_{nonPV} \tag{4.81}$$

From Eq. (3.6) above, substitute dw_{PV} for PdV in Eq. (4.81).

$$dH - \cancel{dw_{PV}} = Tds - \cancel{dw_{PV}} - dw_{nonPV} \tag{4.82}$$

Rearrange Eq. (4.82) to give

$$dH - Tds = -dw_{nonPV} \tag{4.83}$$

From number 5 Eq. (4.2), substitute dG for dH–TdS in Eq. (4.83).

$$dG = -dw_{nonPV} \tag{4.84}$$

Chapter 5 Solved Problem Set

5.1 You routinely have to prepare a phosphate buffer for dissolution studies. Preparing large quantities takes a long time. You decide to heat the phosphate to increase its solubility and so it will dissolve faster. The company chemical store has two sodium phosphates to choose from: disodium phosphate and disodium phosphate dodecahydrate.

a) Does it matter which disodium phosphate you choose? yes
b) If it matters, what is your rationale?

Disodium phosphate has an exothermic −5.6 ΔH of solution, and disodium phosphate dodecahydrate has an endothermic +22.8 ΔH of solution.

The general generic reaction expressions for the two compounds solution processes are

disodium phosphate ⇆ disodium phosphate + 5.6 Kcal mol^{-1}.
disodium phosphate dodecahydrate +22.8 Kcal mol^{-1} ⇆ disodium phosphate dodecahydrate.

Note: Section 5.2.3, Effect of Temperature Change Stress on Equilibrium states that in an endothermic reaction, the absorbed heat is considered a reactant. For exothermic reactions, the heat given up is considered a product.

According to Le Châtelier's principle, adding heat to the disodium phosphate system will result in driving the solution reaction from right to left, resulting in a slower dissolution process, which is not the desired outcome. On the other hand, adding heat to the disodium phosphate dodecahydrate drives the equilibrium to the right which, will speed up the phosphate dodecahydrate solution process.

5.2a Explain the sodium sulfate curve shown in Figure 5.6

From 0 °C to ~30 °C, the slope to the $Na_2SO_4 \cdot 10\,H_2O$ is positive, which means its solubility is increasing with increasing temperature. SEE Problem 1. Applying Le Châtelier's principle, adding heat to an endothermic reaction causes the reaction to proceed to the right, producing more products. This is observed in Figure 5.6. $Na_2SO_4 \cdot 10\,H_2O$ solubility increases with an increase in temperature.

Then a sodium sulfate phase change occurs at ~30 °C. The transforms from the decahydrate to an anhydrous form. It dehydrates in this transition temperature in water. In addition to this phase transition, the slope of the solubility line changes from a positive slope to a negative slope. This tells us the heat of solution for Na_2SO_4 exothermic and gives up heat at this point. Any addition of heat above ~30 °C will result in a decrease in the solubility of Na_2SO_4.

5.2b From 0 °C to ~30 °C, what sodium sulfate species would you find as the excess solid? At ~30 °C and higher, what sodium sulfate species would you find as the excess solid?

First, it is important to note that solubility is an equilibrium system. Solubility is defined as the amount of a species in solution that is in *equilibrium with the excess solid* determined at a specified temperature and pressure. As part of a solubility experiment, it is best practice to carefully characterize the excess solid associated with the species in solution. The solid material added to the solvent at the beginning of the solubility study may have undergone a transition. If a transition occurs, the measured solubility is species in equilibrium with the transformed solid NOT the original solid that was added initially. Characterization of the excess solid can be done with infrared spectroscopy, DSC, X-ray, NMR, and so on.

Now to the answer. From 0 °C to ~30 °C, the excess solid would be $Na_2SO_4 \cdot 10\ H_2O$ as labeled in Figure 5.6. The excess solid from ~30 °C would be Na_2SO_4 as labeled.

5.3 The vapor pressure of water changed 27.17 mmHg from 99 °C to 100 °C. The corresponding volume change from water to vapor over this 1 °C was 3.0114E-2 m^3 mol^{-1}. Calculate the water enthalpy of vaporization at 373.15 K. (adapted from https://carnotcycle.wordpress.com/2013/04/01/the-clausius-clapeyron-equation/).

The Clapeyron equation can be used to calculate the heat of vaporization.

$$dP/dT = \Delta H_{vap}\ /T\Delta V \ (5.13)$$

where dP/dT is 27.17 mmHg K^{-1}, T is 373.15 K, and ΔV is 3.0114E-2 m^3 mol^{-1}.

Solving for ΔH_{vap} using Eq. (5.13) is

$$\Delta H_{vap} = (dP/dT)(\ T\Delta V)$$

The terms are now substituted with the known quantities.

$$\Delta H_{vap} = \left(27.17 \text{ mm Hg K}^{-1}\right)(\ 373.15 \text{ K})\ 3.0114\text{E-2 m}^3 \text{ mol}^{-1})$$

All the units need to be in SI units. The following conversion ratios (see Section 1.2.5 Dimensional Quantity Algebra) can be employed where 1 atm/760 mmHg = 1; 1.013E5 Nm^{-2}/1 atm = 1 and 1 J/1 Nm = 1.

$$\Delta H_{vap} = \big(27.17 \cancel{\text{mm Hg}}\ \cancel{\text{K}^{-1}} \cdot 1\ \cancel{\text{atm}}/760\ \cancel{\text{mmHg}} \cdot 1.013\text{E5}\ \cancel{\text{N}}\text{m}^{-2}/1\ \cancel{\text{atm}} \cdot 1\text{J}/1\cancel{\text{N}}\text{m}\big)\\ (373.15\ \cancel{\text{K}})\ 3.0114\text{E-2 m}^3 \text{ mol}^{-1})$$

and the cancelling the meter units gives

$$\Delta H_{vap} = \big(27.17 \cancel{\text{mm Hg}}\ \cancel{\text{K}^{-1}} \cdot 1\ \cancel{\text{atm}}/760\ \cancel{\text{mmHg}} \cdot 1.013\text{E5}\ \cancel{\text{Nm}^{-2}}/1\ \cancel{\text{atm}} \cdot \text{J}/1\cancel{\text{Nm}}\big)\\ (373.15\ \cancel{\text{K}})\ 3.0114\text{E-2}\ \cancel{\text{m}}\text{ mol}^{-1})$$

$$\Delta H_{vap} = \underline{4.017\text{E4 J mol}^{-1}} \text{ four significant figures.}$$

5.4 Using the table provided (data determined from Table 5.3), calculate water's autoionization-dissociation enthalpy

$1/T$	$\ln K_w$
0.00353	−33.467
0.00341	−33.032
0.0033	−31.851
0.00319	−31.165
0.00309	−30.593
0.00300	−29.972

The Van't Hoff equation can be used to calculate water's enthalpy of autoionization-dissociation.

$$\ln K = -\Delta H/RT + \Delta S/R \; (5.88)$$

A plot of $\ln K$ versus $1/T$, T is in kelvin gives a slope of $-\Delta H/R$, where ΔH is the heat of autoionization-dissociation.

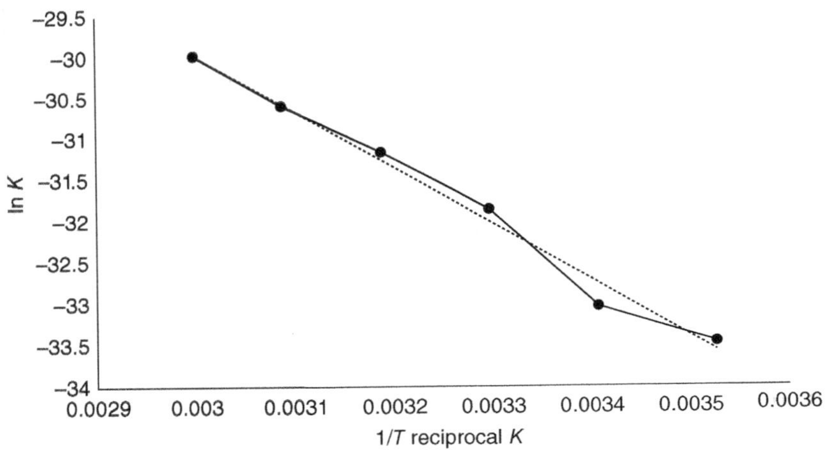

The slope of this graph is

Slope $= -\Delta H/R = -6849.18$ K (keeping all the digits until final selection of significant figures).

$+\Delta H = +6849.18 \text{ K } (8.314\text{E-3 kJ mol}^{-1}\text{ K}) = +\underline{56.9 \text{ kJ mol}^{-1}}$ $\underline{\text{or } 13.6 \text{ kcal mol}^{-1}}$ (3 sig. fig.)

Using dimensional analysis:

$$(56.9\,\cancel{\text{kJ}}\cdot\text{mol}^{-1})(10^3\,\cancel{\text{J}}\cdot\cancel{\text{kJ}^{-1}})\left[(1\,\text{kcal}\cdot\cancel{\text{mol}^{-1}})\left(4.184\text{E}-3\,\cancel{\text{J}}^{-1}\cdot\cancel{\text{mol}^{1}}\right)\right] = \underline{13.6\,\text{kcal mol}^{-1}}.$$

The autoionization-dissociation enthalpy is endothermic. This tells us that it takes energy for water to undergo autoionization-dissociation.

5.5 Using the table below, calculate the enthalpy and entropy of protein unfolding. The equilibrium constant for each temperature is K = *concentration unfolded/concentration folded*. Discuss the meaning of the enthalpy and entropy terms.

1/*T*	ln *K*
0.00345	−2.12
0.00339	−1.38
0.00333	−0.67
0.00328	0.017
0.00323	0.68
0.00317	1.33
0.00313	1.95

The Van't Hoff equation can be used to calculate the protein's unfolding enthalpy and entropy.

$$\ln K = -\Delta H/RT + \Delta S/R \text{ (5.88)}$$

A plot of ln $K_{unfolding}$ versus $1/T$, where T is in kelvin gives a slope of $-\Delta H/R$ and intercept $\Delta S/R$. From the slope and intercept, the enthalpy and entropy of protein unfolding can be calculated.

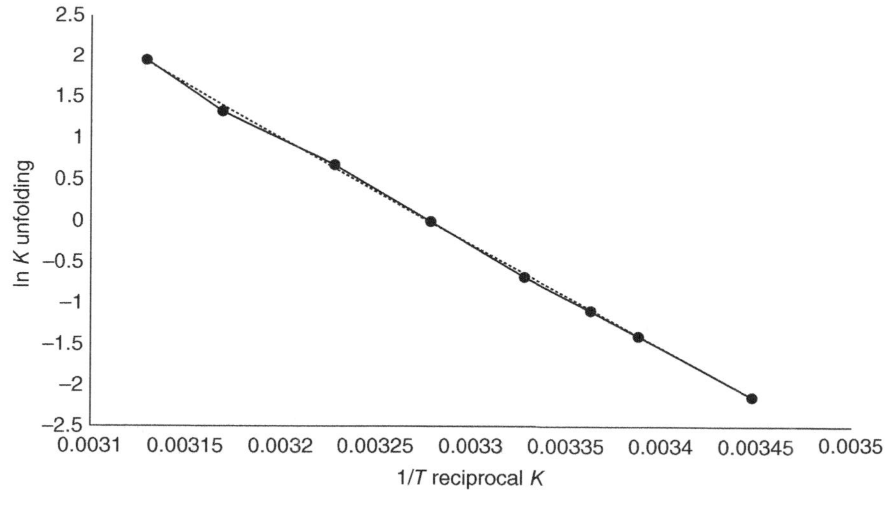

The slope is of the line is −1.264E4 K (Note: slope = $\Delta y/\Delta x$. Considering the units for the slope; no units/(1/T) = T or K). The intercept is 4.146E1 (Note: the intercept has no unites because ln K has no units). The enthalpy and entropy calculations are given later.

Unfolding Enthalpy: −1.264E4 K = $-\Delta H/RT$ and ΔH = 1.264E4 K (1.987E-3 Kcal mol^{-1} K) where dimensional algebra gives

ΔH = <u>+25 Kcal mol^{-1}</u>. There are 2 significant figures because the mantissa of ln K has 2 significant figures.

Unfolding Entropy: 4.146E1 = $\Delta S/R$ and ΔS = 4.146E1 (1.987E-3 Kcal mol^{-1} K)

ΔS = <u>0.083 Kcal mol^{-1} K</u>

The positive enthalpy indicates that energy is required for protein unfolding to occur. The positive entropy means that entropy or protein disorder increases when the protein unfolds.

$$\Delta S = \underline{0.824\,\text{kcal} \cdot \text{mol}^{-1} \cdot \text{K}^{-1}}$$

Chapter 6 Solved Problem Set

6.1 Give two different but equivalent definitions of *solubility*.

Solubility is defined as the maximum amount of solute that can dissolve in a given quantity of solvent at a specified set of conditions.

Solubility is defined as the concentration of solute in a *saturated solution* at a specified set of conditions when the solute is in equilibrium with the solute's undissolved solid phase.

6.2 Define *unsaturated* and *supersaturated* solutions.

An *unsaturated* solution contains less solute in solution than the solvent has capacity to dissolve.

A *supersaturated* solution accommodates more solute in solution than the solvent is capable of dissolving. Supersaturated solutions are thermodynamically unstable, and the solute will spontaneously come out of solution. The timing of this precipitation or crystallization is unpredictable.

6.3 What is the mole fraction, percent w/w, molarity, molality, and percent w/v for lidocaine when 2.000 g lidocaine is dissolved in 100.0 g ethanol?

Given

Lidocaine (L) MW 234.3 g mol^{-1}

Ethanol (E) MW 46.07 g mol^{-1}; density 0.7892 g mL^{-1}

Using dimensional analysis, see Section 1.2, Physical Quantity Dimensions and Dimensional Analysis

a) Mole fraction lidocaine, mole component/total moles

$$\text{Moles L} = 2\,\text{g}\,(1\,\text{mol}/234.3\,\text{g}) = 8.536\text{E}-3\,\text{mol L}$$

$$\text{Moles E} = 100\,\text{g}\,(1\,\text{mol}/0.7892\,\text{g}) = 1.267\text{E}2\,\text{mol E}$$

$$\underline{\text{Mole fraction}} = 8.536\text{E}-3\,\text{mol L}/(8.536\text{E}-3\,\text{mol L} + 1.267\text{E}2\,\text{moles/E}) = \underline{6.737\text{E}}$$

b) Percent w/w

$$\underline{\text{Percent w/w}} = (2.000\,\text{g L})/(2.000\,\text{g L} + 100.0\,\text{g E}) \cdot 100\% = \underline{1.961\%\text{w/w L in E solution}}$$

c) Molarity (assuming lidocaine makes a negligible contribution to the solution density)

$$100.0\ \text{g E}\ (1\ \text{mL}/0.7892\ \text{g}) = 126.7\ \text{mL E and solution volume}$$

$$\underline{\text{Molarity}} = (8.536\text{E}-3\ \text{mol L}/126.7\ \text{mL of solution}) \cdot \left(\cancel{1000\ \text{mL}}\ \text{L}^{-1}\ \text{solution}\right) = \underline{6.737\text{E}-2\ \text{M}}$$

d) Molality

$$\underline{\text{Molality}} = (8.536\text{E}-3\ \text{mol L}/\cancel{100.0\ \text{g E}}) \cdot (\cancel{1000\ \text{g E}}/1\ \text{kg E}) = \underline{8.530\text{E}-2\ \text{m}}$$

e) Percent w/v (assuming lidocaine makes a negligible contribution to the solution density)

$$\underline{\text{Percent w/v}} = \left(2.000\ \text{g L}/\cancel{100.0\ \text{g E}}\right) \cdot (\cancel{0.7892\ \text{g E}}/\text{mL E}) = \underline{1.578\text{E}-2\%\text{w/v}}$$

6.4 Name the six main interaction forces and briefly describe each one and their relative strength.

Bond type	**Dissociation energy (kcal mol^{-1})**
Ionic lattice (primary valence bonds)[1]	100–200
Covalent (primary valence bonds)[1]	50–150
Hydrogen bonds[2]	1–12
Keeson dipole–dipole[2]	0.5–2
Debye dipole-induced dipole[3]	Up to 0.5
London dispersion force[2]	<1–15

An ionic bond (valence bond) arises from electrostatic forces of attraction between two oppositely charged ions. Ionic bonds are formed between metals and nonmetals.

A covalent bond, also a valence bond, occurs as the result of sharing electrons. Covalent bonds join nonmetals to nonmetals.

Hydrogen bonds generally form strong electrostatic dipole–dipole interactions. *Dipole–dipole* forces are attractive forces between the positive end of one polar molecule and the negative end of another polar molecule.

van der Waals forces are a group of relatively weak attraction forces. These forces include *Keesom* forces (weak dipole–dipole interactions), *Debye* forces (dipole-induced dipole), and *London* or *dispersion* forces (induced dipole-induced dipole). A dipolar substance can be thought of as having a pair of equal and oppositely charged areas or *poles* that are separated by a distance. *Keesom* forces can only occur between polar molecules, which exhibit permanent dipole moments. The interaction is electrostatic in nature and based on Coulomb's law. *Debye* forces occur from a permanent dipole-induced dipole interaction. The permanent dipole induces a polarization of a second molecule that leads to mutual attraction of the induced dipolar molecule. *London or dispersive* forces is temporary dipole-induced dipole interactions. London forces are caused by random fluctuations of electron density. London forces exist for all substances, including nonpolar, polar, or ionic materials.

6.5a $\Delta H_{solution} = \Delta H_1 + \Delta H_2 + \Delta H_3$ is an example of what law? Define that law.

Hess's Law

Hess's Law states that the total enthalpy change of a reaction is independent of whether the reaction takes place in one step or any number of steps because enthalpy is a state function.

6.5b For the enthalpy of solution model, what are $\Delta H_1 + \Delta H_2 + \Delta H_3$?

ΔH_1 is the enthalpy of fusion of the crystalline solute (ΔH_{fusion}), ΔH_2 is the energy required to form a cavity in the solvent, and ΔH_3 is the solute-solvent interaction or enthalpy of mixing.

6.6 What is the heat of sublimation of naphthalene if the enthalpy of fusion is 19.01 kJ/mol and the enthalpy of vaporization is 47.44 kJ mol^{-1}?

$$\Delta H_{sublimation} = \Delta H_{fusion} + \Delta H_{vaporization}$$

$$\Delta H_{sublimation} = 19.01 \text{ kJ mol}^{-1} + 47.44 \text{ kJ mol}^{-1} = 66.45 \text{ kJ mol}^{-1}$$

6.7 When does $\Delta H_{solution} = \Delta H_{fusion}$?

When the solution is ideal and there is no enthalpy resulting from the solute-solvent interaction.

6.8 Does the Figure 6.1a below represent an endothermic or exothermic heat of solution process? Explain the steps

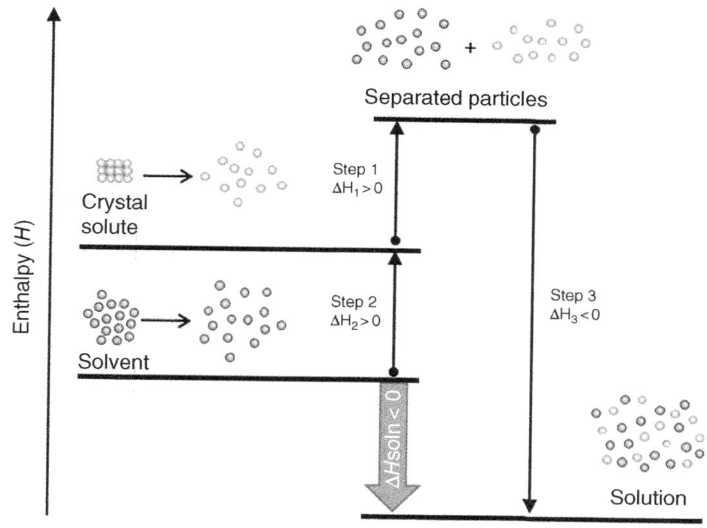

(a) Exothermic solution formation

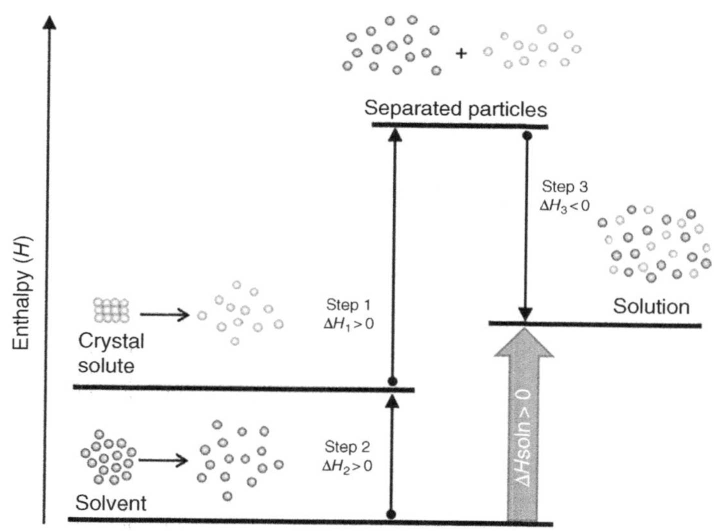

(b) Endothermic solution formation

Exothermic solution process.

Here, Step 1 involves the melting of the crystal solute and forming a supercooled liquid. The heat of fusion is endothermic since heat/energy needs to be absorbed to overcome the crystal lattice structure. Step 2 is another endothermic reaction that requires energy to break up the solvent-solvent attraction forces so a cavity can be formed in the solvent. The enthalpy of solvation is negative ($-\Delta H_{solvation}$), resulting from intermolecular solute-solvent attraction forces. For an exothermic heat of solution ($-\Delta H_{solution}$) to occur, the negative enthalpy of solvation must be greater than the sum of the endothermic reactions in Steps 1 and 2.

6.9 Does the Figure 6.1b below represent an endothermic or exothermic heat of solution process? Explain the steps.

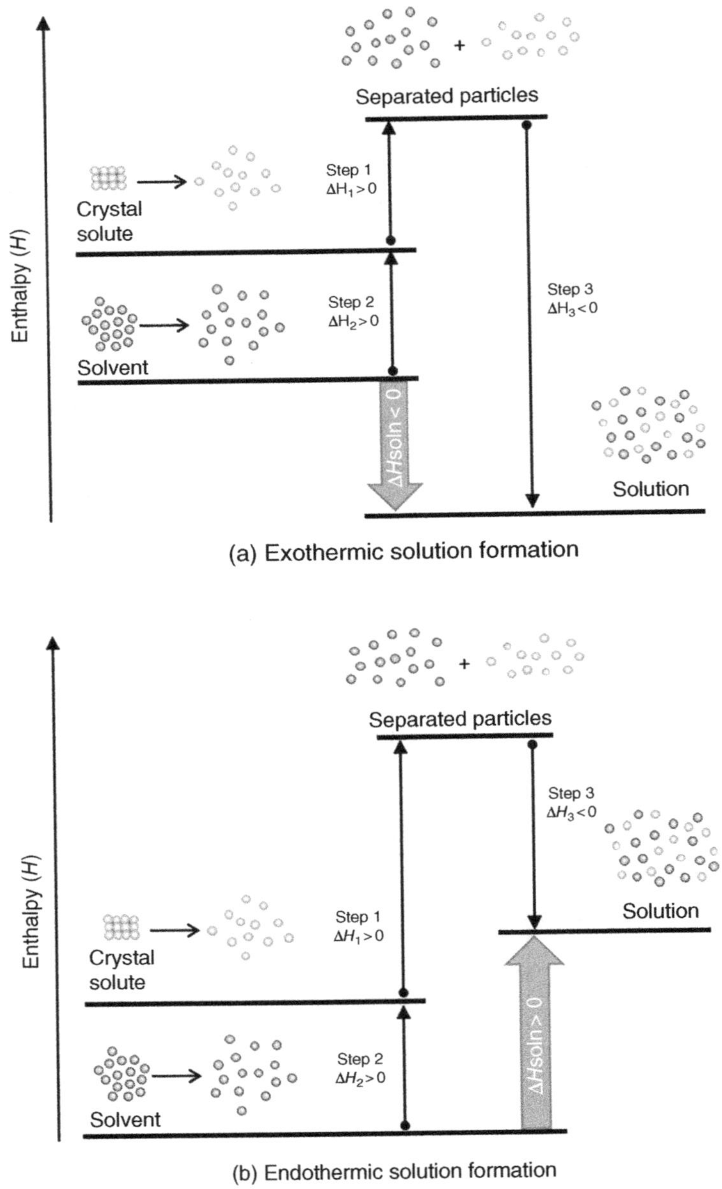

Explains the enthalpy changes that lead to an endothermic heat of solution. In this context, Steps 1 and 2 are endothermic in nature, like the exothermic solution formation example in Figure 6.1a. Similarly, the enthalpy of Step 3 of Figure 6.1b is negative, as noted for Step 3 of Figure 1.6a. The difference in this case is the negative enthalpy of solvation is smaller than the sum of the endothermic contributions associated with Steps 1 and 2. This results in an overall endothermic enthalpy of solution ($+\Delta H_{solution}$).

6.10 Determination of $\Delta G^{o}{}_{solution}$ for $MgCl_2$ at 25 °C

Given

$$\Delta H^{o}{}_{solution} = -38.24\,\text{kcal mol}^{-1}$$

$$\Delta S^{o}{}_{solution} = -0.0274\,\text{kcal K}^{-1}\cdot\text{mol}$$

$$\Delta G^{o}{}_{solution} = -38.24\,\text{kcal mol}^{-1} - (298.15\,\cancel{\text{K}})\left(-0.0274\,\text{kcal}\,\cancel{\text{K}}^{-1}\cdot\text{mol}\right)$$

$$\underline{\Delta G^{o}{}_{solution}} = -38.24\,\text{kcal mol}^{-1} + 8.17\,\text{kcal mol}^{-1} = \underline{-30.1\,\text{kcal mol}}^{-1}$$

$\underline{\Delta G^{o}{}_{solution}}$ is negative and solution formation occurs spontaneously. The $MgCl_2$ solution process is enthalpic driven. The entropy of solution is negative resulting from strong $MgCl_2$-water interactions that lead to increased order. In the Gibbs free energy equation, the entropy term becomes positive and opposes a spontaneous solution.

6.11 Calculate naphthalene's solubility at 25.0 °C using the idea solubility equation (ΔH_{fusion} = 19.01 kJ mol^{-1} and melting point 80.0 °C)

$$\ln X_{2,i} = -\left(\Delta H_{fusion}/R\right)\,[(T_m - T)/T\cdot T_m] \qquad (6.43)$$

$$\ln X_{2,i} = -(19010\,\text{J/mol}/8.314\,\text{J/mol/K})[(353.15\,\text{K}- 298.15\,\text{K})/353.15\,\text{K}\times 298.15\,\text{K}]$$
$$= -1.1943$$
$$X_{2,i} = \underline{0.303}$$

6.12 Using the Van't Hoff plot from the solubility data, calculate nitrofurantoin enthalpy of solution. Also calculate the entropy at 24 °C.

Given

$$\text{Van't Hoff equation for solution: } \ln X_2 = -\Delta H^{\circ}_{solution}/RT + \Delta S^{\circ}_{solution}/R \qquad (6.38)$$

Nitrofurantoin solubility data

Temperature (°C)[a]	**Temperature (*K*)**	**Solubility, mole fraction (X_2)**
24	297.15	6.01E-6
30	303.15	8,57E-6
37	310.15	1.32E-5
45	318.15	1.90E-5

[a] Adapted from Chen et al. (1976) and Connors and Mecozzi (2010).

A plot of ln X_2 versus $1/T$ gives a slope of $-\Delta H^{\circ}{}_{solution}/R$. $\Delta S^{\circ}{}_{solution}$ is $-\Delta H^{\circ}{}_{solution}/T$ where T is in Kelvin.

Conversion of the solubility data to ln X_2 temperature data to $1/K$

Ln X_2	**1/*K***
−12.02	3.3653E-3
−11.66	3.2987E-3
−11.24	3.2242E-3
−10.87	3.1432E-3

The slope of the plotted line is −2.300E3 K.
From the Given equation: $-\Delta H^{\circ}_{solution}/R = \text{slope}$
Therefore,

$$-\Delta H^{\circ}_{solution}/R = -2.300\text{E}3\ \text{K, and}$$

$$\cancel{\Delta H^{\circ}_{solution}} = 2.300\text{E}3\ \cancel{\text{K}} \cdot 8.314\ \text{J mol}^{-1}\ \cancel{\text{K}^{-1}} = \underline{1.9\text{E}4\ \text{J mol}^{-1}} \text{ or } \underline{1.9\text{E}1\ \text{kJ mol}^{-1}}$$

(note two significant figures because there are only two significant figures for the centigrade temperatures)

$$\underline{\Delta S^{\circ}_{solution,24oC}} = \Delta H^{\circ}_{solution}/T_{solution\ temperature\ in\ K} = 1.9\text{E}1\ \text{kJ mol}^{-1}/297.15\ \text{K}$$

$$= \underline{6.8\text{E}-2\ \text{kJ mol}^{-1}\ \text{K}^{-1}}$$

6.13 Provide the reference state and standard state conditions for solutions.

Reference State Conditions:

$$\text{Solution activity}, a = \text{solution concentration } (X, M, m) \tag{6.54}$$

$$\text{Activity coefficient}, \gamma = 1 \text{ where solution activity/solution concentration} = 1 \tag{6.55}$$

Standard State Conditions:

$$a_i = 1 \tag{6.56}$$

where a_i is the activity of the ith component. Using Eq. (6.52), a *solution component standard state also occurs when the chemical potential of the component equals its standard chemical potential.*

$$\mu_i = \mu_i^{\circ} + RT\ \ln\ (a_i = 1) \tag{6.52}$$

$$\mu_i = \mu_i^{\circ} \tag{6.57}$$

6.14 List the reference and standard states for solutions.

Reference States:

1) Solvent- The reference state for a solvent using a mole fraction scale is the pure solvent. In infinitely dilute solutions, the solvent behaves ideally, and $a = X = 1$.
2) Miscible Liquid Solute- The liquid solute reference state is the pure liquid solute. Concentration is expressed in mole fraction.
3) Partially Soluble Liquid or Solid Solute- The reference state is an infinitely dilute solution, where Eqs. (6.54) and (6.55) apply.

Standard States:

1) Solvent- The solvent standard state is set as the *pure liquid at 1 atm, where* $a = 1$.
2) Miscible Liquid Solute-The liquid solute standard state is *the pure liquid at 1 atm and* $a = 1$

Partially Soluble Liquid or Solid Solute- The definition of the standard state of a partially soluble liquid or solid solute involves the definition of a hypothetical solution using X, M, or m concentration scales. The partially soluble liquid or solid solute standard state *is a hypothetical solution of* unit *concentration at infinitely dilute concentration that acts as an ideal solution.*

Chapter 7 Solved Problem Set

7.1 Describe the interfacial surface region

An interfacial surface at equilibrium does not have a distinct boundary or boundaries. The phases mix at the interface and exhibit a gradual change in a species' density, concentration, and other extensive properties.

The interfacial surface region can be thought of as a film-like phase region. This film-like region behaves differently than the bulk phases. [Note: Once again, I don't like the use of the word "elastic."]

7.2 Define interfacial tension

Interfacial tension results from unbalanced forces acting at the interface. It has units of $N\ m^{-1}$ or force per unit length.

7.3 Define *interfacial energy*

Interfacial energy is the energy or work necessary to overcome intermolecular attractive forces to increase the surface area of an interface. Interfacial energy has units of $J\ m^{-2}$.

7.4 Show why interfacial tension and interfacial energy values can be interchanged

Interfacial tension and interfacial energy have the same fundamental units, kg/s^2 and therefore have the same values.

$$\text{Interfacial tension units : } \mathrm{N\,m^{-1}} = (\cancel{\mathrm{m}} \cdot \mathrm{kg\,s^{-2}})/\cancel{\mathrm{m}} = \underline{\mathrm{kg\,s^{-2}}} \tag{7.5}$$

$$\text{Interfacial energy units : } \mathrm{J\,m^{-2}} = \mathrm{N \cdot m\,m^{-2}} = \{[(\cancel{\mathrm{m}} \cdot \mathrm{kg})\,\mathrm{s^{-2}}] \cdot \cancel{\mathrm{m}}\}\,\cancel{\mathrm{m}}^{-\cancel{2}} = \underline{\mathrm{kg\,s^{-2}}} \tag{7.6}$$

7.5 What is the difference between surface tension and interfacial tension?

Surface tension is a subcategory of *interfacial tension*. When a gas or vapor is the phase in contact with a liquid or solid, the interfacial tension is designated as *surface tension*. By convention the liquid–vapor and solid–vapor interface is called a *surface*. This subcategorization is primarily based on the relatively large liquid–liquid and solid–solid cohesive forces and comparatively weak liquid–vapor and solid–vapor adhesive forces operating at the surface compared to the relatively stronger adhesive forces of liquid–liquid, solid–liquid, and solid–solid interfaces.

7.6 Name four techniques that can be used to measure the surface tension of liquid–vapor interfaces and briefly describe them.

Capillary Method-

If a glass capillary is placed in a bulk liquid, the liquid in the tube may rise above or fall below the bulk liquid level. This spontaneous movement is known as *capillarity* and depends on the liquid's relative cohesion and adhesion forces. Capillarity is the result of the liquid's surface tension and the liquid-glass interfacial contact angle.

$$\text{Upward force} = (2\,\pi r) \cdot \gamma \cdot \cos\,\theta$$

$$\text{Downward force} = \left(\pi r^2 h\right) \cdot g \cdot d$$

$$\gamma = \left[\left(\pi r^2 h\right) \cdot g \cdot d\right] / [(2\,\pi r) \cdot \cos\,\theta]$$

Du Noüy Ring Method-
The du Noüy ring tensiometer method is used to measure a liquid's surface tension and the interfacial tension between two liquids. It is a dynamic technique and does not measure an equilibrium interfacial tension. It is a relatively simple and fast way to measure surface tension. The main advantage of the du Noüy method is its ability to measure liquid–liquid interfacial tensions.

An ultraclean platinum ring (typically) is submerged below the interface. After immersion, the ring is pulled up through the interface. The maximum upward force is given by;

$$\gamma = \left(F - w_{ring}\right) / (4\,\pi R)$$

where γ is the interfacial tension, F is the maximum force required to pull the ring through the interface, w_{ring} is the weight of the ring minus the liquid buoyant force of the lower phase, R is the average of the inner and outer radius of the ring, and $4\pi R$ is twice the radius of the ring. The buoyant force is a function of the liquid density the ring is immersed in, the volume displaced by the ring, and the gravity constant.

Wilhelmy Plate Method-
The plate method can be performed in a static or dynamic mode. The plate is typical roughened platinum. Platinum has high surface energy and generally leads to complete wetting and a contact angle (θ) of 0°. Unlike the du Noüy method, no correction is required for the weight of the ring or ring buoyancy. In the static mode, the plate is lowered to "bridge" the two phases (liquid–vapor or liquid–liquid phases) and is not moved during testing. This allows direct measurements of surface kinetics over a wide range of timescales.

$\gamma = F/(L \cos\,\theta)$, where F is the tension acting on the plate, L is the *wetted length* (plate perimeter) at the level of the liquid surface, θ is the contact angle between the plate and the liquid, and γ is the surface/interfacial tension.

Maximum Bubble Pressure Method-
The maximum bubble pressure method is a dynamic method used for surface tension measurements. This method can only be used for determining a liquid–vapor surface tension.

$$\gamma = \left[\left(P_{max} - P_{hydrostatic\ pressure}\right) \cdot r_{capillary}\right] / 2$$

The test liquid's surface tension is automatically reported by the newer maximum-bubble-pressure instruments. Additional corrections for gravity and viscosity effects can improve the accuracy of the reported surface tension.

7.7 How is surface tension effected by temperature?

Surface tension decreases with an increase in temperature because the increase in thermal energy increases molecular motion and decreases cohesion.

7.8 Using surface tension and interfacial tension terms, what are the expressions for cohesive and adhesive work?

$$\underline{w_c} = 2\gamma$$

$$w_a = \gamma_1 + \gamma_2 - \gamma_{12}$$

7.9 What is the equation that describes the spreading of two immiscible liquids: one test liquid on the surface of a substrate liquid?

$$S_{test\ liquid/substrate\ liquid} = \gamma_{substrate\ liquid} - \left(\gamma_{test\ liquid} + \gamma_{test\ liquid/substrate\ liquid}\right)$$

7.10 What is the spreading coefficient when a drop of water is placed on mineral oil?

Given:

Surface tension of water is 71.8 mN m^{-1}
Surface tension of mineral oil is 28 mN m^{-1}
Interfacial tension of water-mineral oil is 4.0 mN m^{-1}

$$S_{test\ liquid/substrate\ liquid} = \gamma_{\text{substrate liquid}} - \left(\gamma_{\text{test liquid}} + \gamma_{\text{test liquid/substrate liquid}}\right)$$

$S = 28$ mN m^{-1} – (71.8 mN m^{-1} + 4 mN m^{-1}) = –47.8 mN m^{-1} = <u>–48 mN m^{-1}</u> rounded to the correct number of significant figures. The spreading coefficient is negative indicating the water drop will not spread on the mineral oil.

7.11 What is the spreading coefficient of a film-coating solution consisting of 9% hydroxypropyl methylcellulose (HPMC) and 1% polyethylene glycol (PEG) 400 on a compressed tablet consisting of 75.2% microcrystalline cellulose (MCC), 24.2% lactose, 0.4% magnesium stearate, and 0.2% silicon dioxide (all w/w %).

Given:

Film coating surface tension is 45.6 mN m^{-1}
Contact angle is 43°

$$S_{test\ liquid/substrate\ solid} = S = \gamma_{lv}\ (\cos\ \theta - 1)$$

$S = 45.6$ mN m^{-1} (0.7314–1) = 45.6 mN m^{-1} (–0.2686) = –12.25 mN m^{-1} = <u>–12.3 mN m^{-1}</u> rounded to the correct significant figures.

7.12 What is the adhesive work equation?

$$w_a = w_{sl} = \gamma_{lv}\ (\cos\theta + 1)$$

7.13 Using the OWRK model equation, calculate a polymer's polar surface energy effects.

Given:

Water

$$\gamma_{Wlv}{}^{Tot} = 72.8\ \text{mN m}^{-1}$$

$$\gamma_{Wlv}{}^{d} = 21.8\ \text{mN m}^{-1}$$

$$\gamma_{Wlv}{}^{p} = 51.0\ \text{mN m}^{-1}$$

$$\theta = 83.0°$$

Glycerin

$$\gamma_{Wlv}{}^{Tot} = 64.0\ \text{mN m}^{-1}$$

$$\gamma_{Wlv}{}^{d} = 34.0\ \text{mN m}^{-1}$$

$$\gamma_{Wlv}{}^{p} = 30.0\ \text{mN m}^{-1}$$

Polymer

$$\gamma P_{sv}{}^{d} = 40.1\ \text{mN m}^{-1}$$

OWRK Equation

$$0.5\,\gamma_{lv}\,(\cos\,\theta + 1) = \left(\gamma_{sv}{}^{d}\,\gamma_{lv}{}^{d}\right)^{0.5} + \left(\gamma_{sv}{}^{P}\,\gamma_{lv}{}^{P}\right)^{0.5}$$

Calculation for a polymer's *polar surface energy component.*

1) Substitute the givens into the OWRK equation
   Water Test Liquid

$$0.5 \cdot 72.8\ \text{mN m}^{-1}\,(\cos\,83° + 1) = \left(40.1\ \text{mN m}^{-1} \cdot 21.8\ \text{mN m}^{-1}\right)^{0.5}$$
$$+ \left(\gamma_{sv}{}^{P}\,\text{mN m}^{-1} \cdot 51.0\ \text{mN m}^{-1}\right)^{0.5}$$

2) Solve for $\gamma_{sv}{}^{P}$.

$$0.5 \cdot 72.8\,\text{mN m}^{-1}\,(0.1219 + 1) = \left[874.18\,\left(\text{mN m}^{-1}\right)^{2}\right]^{0.5} + \left(\gamma_{sv}{}^{P}\,\text{mN m}^{-1} \cdot 51.0\,\text{mN m}^{-1}\right)^{0.5}$$

$$40.84\ \text{mN m}^{-1} = 29.57\ \text{mN m}^{-1} + \left(\gamma_{sv}{}^{P}\,\text{mN m}^{-1} \cdot 51.0\ \text{mN m}^{-1}\right)^{0.5}$$

$$11.27\ \text{mN m}^{-1} = +\left(\gamma_{sv}{}^{P}\,\text{mN m}^{-1} \cdot 51.0\ \text{mN m}^{-1}\right)^{0.5}$$

$$127.01\,\left(\text{mN m}^{-1}\right)^{2} = \gamma_{sv}{}^{P}\,\text{mN m}^{-1} \cdot 51.0\ \text{mN m}^{-1}$$

$$\gamma_{sv}{}^{P}\ \text{mN m}^{-1} = 127.01\,\left(\text{mN m}^{-1}\right)^{2}/51.0\,\text{mN m}^{-1} = 2.49\ \text{mN m}^{-1}$$

The polar surface energy component is $\underline{2.49\ \text{mN m}^{-1}}$

7.14 a) Calculate the pressure difference across the bubble film of a bubble having a diameter of 10.0 mm. b) Calculate the pressure inside of the 10 mm bubble. The bubble film is comprised of soapy water, $\gamma = 25.0\ \text{mN m}^{-1}$.

a) Calculate the pressure difference across a 10.0 mm bubble.

$$P_{int} - P_{ext} = (2\gamma)/r = \Delta P$$
$$\Delta P = \left(2 \cdot 25.0\,\text{mN m}^{-1}\right)/5.00\ \text{mm} = \left(50.0\,\text{mN} \cdot \text{m}^{-1} \div (5.00\,\cancel{\text{mm}})^{\cancel{-1}}\right)$$
$$\cdot\left(1000\,\cancel{\text{mm}} \cdot \text{m}^{-1}\right) = \underline{1.00\text{E}4\,\text{mN} \cdot \text{m}^{-2}}$$
$$\Delta P = 1.00\text{E}4\,\cancel{\text{mN}} \cdot \text{m}^{-2}\,(\text{N}/1000\,\cancel{\text{mN}}) = 1.00\text{E}1\,\text{N m}^{-2} = \underline{10.0\,\text{Pa}}\ \left(\text{where } 1\,\text{N m}^{-2} = 1\,\text{Pa}\right)$$

b) Calculate the pressure inside the 10 mm diameter bubble.
   Atmospheric pressure is *101,325 Pa at STP and is* P_{ext}.
   Given:

$$P_{int} - P_{ext} = 10.0\ \text{Pa}$$
$$\underline{P_{int}} = 10.0\ \text{Pa} + 101,325\ \text{Pa} = 101,335\ \text{Pa} = \underline{1.01\text{E}5\ \text{Pa with 3 SFs}}$$

7.15 Calculate the relative solubility increase predicted for a 10 nm radius particle (20 nm diameter).
Given:

$$\gamma = 30.0\,mN\,m^{-1} = 0.0300\,N\,m^{-1}$$

$$M = 500\,\mathrm{g\,mol^{-1}} = 0.500\,\mathrm{kg\,mol^{-1}}$$

$$r = 10.0\,\mathrm{nm} = 1\mathrm{E}-8\,\mathrm{m}$$

$$\rho = 1.00\,\mathrm{g\,cm^{-3}} = 1000\,\mathrm{kg\,m^{-3}}$$

$$R = 8.314\,\mathrm{N\,m\,mol^{-1}\,K^{-1}}$$

$$T = 298.15\,\mathrm{K}$$

$$\ln S/S_{\infty} = (2\gamma M)/(r\rho RT)$$

(For refresher for dimensional analysis, see Section 1.2.5, Dimensional Quantity Algebra or Dimensional Algebra)

$$\ln S/S_{\infty} = (2)\,0.0300\,\cancel{\mathrm{N}}\,\cancel{\mathrm{m}}^{-1}\cdot 0.500\,\mathrm{kg\,mol^{-1}})/\big(1\mathrm{E}\text{-}8\,\cancel{\mathrm{m}}\cdot 1000\,\cancel{\mathrm{kgm}}^{-3}$$
$$\cdot 8.314\,\cancel{\mathrm{N}}\cdot\cancel{\mathrm{m}}\cdot\cancel{\mathrm{mol}^{-1}}\cdot\cancel{\mathrm{K}}\cdot 298.15\,\cancel{\mathrm{K}}) = 1.210$$

Anti-natural log $1.210 = S/S_{\infty} = 3.35$ is the relative solubility increase.

7.16 Calculate the specific surface area ($m^2\,g^{-1}$) of a 20.0 nm diameter particle.
Given:

Material density = 1.00 g/cm^3
Specific Surface Area = 6/(density · diameter)
1 nm = 1E-9 m
1 cm^3 = 1E-6 m^3
Specific Surface Area = 6/(1 g/cm^3 · 20 nm)

(For refresher for dimensional analysis, see Section 1.2.5, Dimensional Quantity Algebra or Dimensional Algebra)

$$\text{Specific surface area} = 6/\big\{\big[1\mathrm{g}\,\cancel{\mathrm{cm}}^{-\cancel{3}}\cdot\,(\cancel{\mathrm{cm}}^{\cancel{3}}/1\mathrm{E}\text{-}6\,\mathrm{m}^{\cancel{3}2})\big]\cdot[20\,\cancel{\mathrm{nm}}\cdot(1\mathrm{E}\text{-}9\,\cancel{\mathrm{m}}/1\,\cancel{\mathrm{nm}})]\big\}$$

$$\underline{\text{Specific surface area}} = 300\,\underline{\mathrm{m^2\,g^{-1}}}$$

Chapter 8 Solved Problem Set

8.1 Define adsorbate, adsorbent, and adsorptive.

Adsorbate is a material unit, like atoms, molecules, or ions that is adhered to the surface of a substrate.

Adsorbent is the substrate to which the adsorbate is adhered.

Adsorptive are atoms, molecules, and ions that undergo the adsorption process to become adsorbates.

8.2 Fill out the table to identify the differences between physisorption and chemisorption.

Properties	**Physisorption**	**Chemisorption**
Nature	Reversible	Nonreversible
Bonding interaction	London dispersion force, Debye force, keeson force, hydrogen bonding, and ion–ion interactions	Chemical interactions involving orbital overlap and charge transfer (that is, covalent bond formation)
Adsorption enthalpies	5–50 kJ mol^{-1}[*]	40–800 kJ mol^{-1}
Saturation uptake	Mono and multilayer	Monolayer
Surface specificity	Often, but not required in all cases	Required
Temperature range	Favored at low temperature. Near or below condensation of gas. Adsorption is exothermic and its extent decreases with increases in temperature[1]	Increasing temperature increases the extent of adsorption, and effects may be complex.[2]
Adsorption kinetics	Fast	Variable. Activated process.

8.3 Fill out the table to identify the differences between adsorption and absorption.

Properties	**Adsorption (physical)**	**Absorption**
Phenomenon	Surface	Bulk
Enthalpy	Exothermic	Endothermic
Temperature	Favored at low temperatures	Favored at higher temperatures
Reaction Rate	Fast	Diffusion controlled
Concentration	Surface concentration differs from bulk concentration	At equilibrium, uniform concentration in bulk

8.4

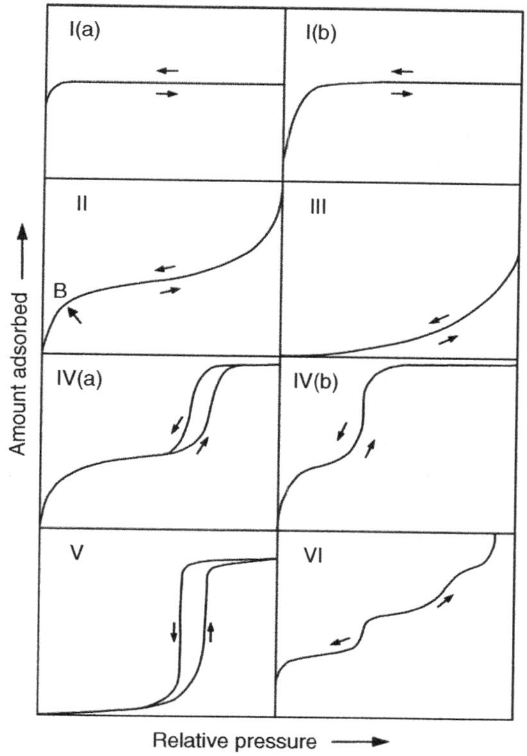

8.4a Describe the pertinent features of the Type I isotherm

Type I isotherms are reversible, and the curvature is concave to the x-axis. The shape is the result of microporous solids that have a small portion of nonmicroporous external surface. The limited amount of adsorption is governed by the external surface, not internal surface area or volume.

The rapid uptake at low relative pressures is due to strong adsorbent–adsorbate adhesive interactions and adsorptive molecular size. Slow uptake of adsorptive gas is demonstrated for adsorbents with a pore size over a broader range.

8.4b Describe the pertinent features of the Type II isotherm.

Nonporous or microporous solids are responsible for reversible Type II isotherms. A "sharp" concave knee at point B, represents the beginning of monolayer adsorbate formation that continues throughout the almost linear portion of the curve. The convex curvature to the P/P^o axis at higher relative pressures is indicative of adsorbate multilayers. At $P/P^o = 1$, the adsorbed layer increases without limit.

8.4c Describe the pertinent features of the Type III isotherm.

Nonporous or microporous solids are also responsible for the reversible Type II isotherms. The initial portion of the curve is convex to the P/P^o axis and exhibits a gradual increase in amount adsorbed. It is posited that slow increase is due to weak adsorbent–adsorbate adhesive forces and a lack of favorable surface energy interaction sites. At the saturation pressure, $P/P^o = 1$, the amount adsorbed reaches a limit.

8.4d Describe the pertinent features of the Type IV isotherm.

Reversible Type IV isotherm arises from 2 nm to 50 nm mesoporous substrate materials. The adsorption phase is similar in shape to the Type II isotherms, which suggest monolayer formation before multilayer form occurs. The overall adsorption–desorption profile is a function of solid substrate-adsorbate adhesive energy and the appearance of adsorbent-condensed adsorbate in the mesopores at pressures less than the saturation pressure, P^o. The plateau region beyond the convex curvature seen from the x-axis is characteristic of the Type IV profile.

8.4e Describe the pertinent features of the Type V isotherm.

Isotherm V's shape at low relative pressure is similar to isotherm III. As noted with Type III, the slow increase in amount adsorbed is due to weak adsorbent–adsorbate adhesive forces and a lack of favorable surface sites. At higher relative pressures, the slope of the adsorption curve increases sharply. The adsorption shape is most likely due to delayed condensation of microporous and mesoporous adsorbents. Limited adsorption is seen at higher relative vapor pressures. The sharp desorption is explained by the "narrow-neck ink bottle effect," where the neck diameter is uniform. Once evaporative conditions are met, the condensed phase undergoes rapid and uniform evaporation.

8.4f Describe the pertinent features of the Type VI isotherm.

Type VI isotherm shows superimposable adsorption–desorption as seen in Types I, II, II, and IVb. The arrows designate the start of the first, second, and third adsorption layers. At high P/P^o less than 1, the amount adsorbed does not reach a limit.

8.5a What is the Freundlich equation?

$$\log\ q_e = \log\ K_F + 1/n\ \log\ C_e$$

8.5b What is the significance of K_F *and* n?

K_F ($L/g_{adsorbent}$) is the adsorbent adsorption capacity and n (a dimensionless parameter) is the *adsorption intensity*, which is reflected in the shape of the adsorption curve. When $1/n$ is less than 1, and n is greater than 1, the adsorption curve is concave to the x-axis (C_e). This isotherm is similar to the Type I curves. This indicates *favorable adsorption*. When n equals 1, the Freundlich equation reduces to Henry's equation. When $1/n$ is greater than 1, and n is less than one, the adsorption curve is convex as observed from the x-axis. This curve is similar to isotherm Type III curves and implies *unfavorable adsorption*.

8.5c Using the Freundlich equation and the data table below, determine K_F and n for the adsorption of tetracaine onto activated charcoal at 25 °C.

Adsorption of Tetracaine onto Activated Charcoal (from Florence and Attwood, 2011)

Equilibrium concentration, C_e, mg L^{-1}	0.155	0.468	1.259	2.510	5.370
Amount adsorbed/adsorbent, q_e, mg g^{-1}	202.8	217.7	232.0	243.2	254.7

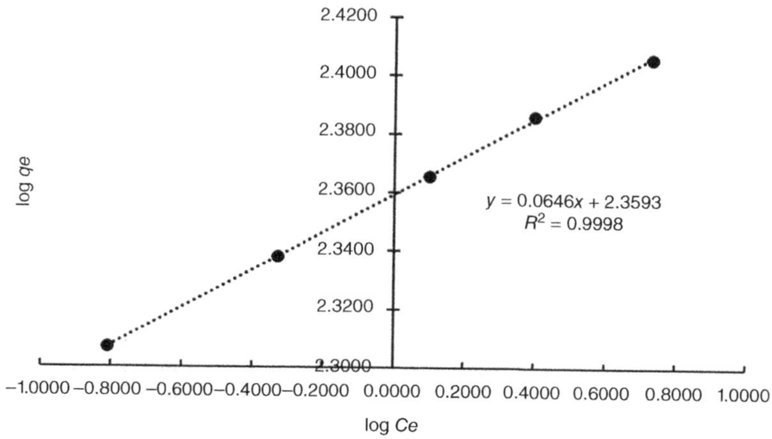

Adsorption of Tetracaine onto Activated Charcoal

Freundlich equation: $\log q_e = \log K_F + 1/n \log C_e$

K_F

The intercept of line is 2.359.

$$\log K_F = 2.359$$

Taking the antilog of both sides gives:

$K_F = 228.55$ or 229 L adsorptive solute solution/g adsorbent with 3 significant figures.

n

The slope of the line is $1/n$.

$$\text{Slope} = 0.0646 = 1/n$$

Solving for n:

$$n = 1/0.0646 = 15.5 \text{ with 3 significant figures.}$$

n is greater than 1. The adsorption curve will be similar to a Type I isotherm, showing a strong adsorption profile and a concave curve relative to the x-axis.

8.7 Using Langmuir isotherm shown below and the Langmuir equation:

$$\theta = A_{adsorbate\ bound}/A_{total} = K_L \cdot P/P^o{}_{adsorptive-vapor}/\left(1 + K_L \cdot P/P^o{}_{adsorptive-vapor}\right)$$

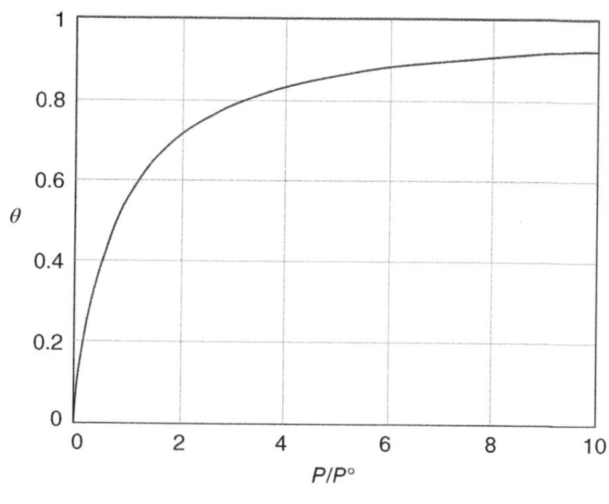

8.7a Where is the Henry's law portion of the curve?

At low $P/P^o_{adsorptive\text{-}vapor}$ values, the denominator approaches 1. The Langmuir equation is linear in this portion of the curve and is often referred to as Henry's law region.

8.7b What does it mean when the line asymptotically approaches 1

At high $P/P^o_{adsorptive\text{-}vapor}$ values concentrations, the denominator approaches $K_L \cdot P/P^o_{adsorptive\text{-}vapor}$ and the fraction of the total available adsorption sites that are bound increases asymptotically to 1. This indicates that adsorption becomes increasingly more difficult.

8.8 Provide the linear form of the Langmuir equation and state the meaning of each term.

$$V/\ (P/P^o) = V_m \cdot K_L - V \cdot K_L$$

plot of $V/\ (P/P^o)$ versus $V \cdot K_L$ results in a slope of K_L and intercept of $V_m \cdot K_L$. Once the slope is determined, V_m can be calculated from the intercept of $V_m \cdot K_L$. As a reminder K_L, is equal to k_a/k_d, the adsorption and desorption rate constants and is related to adsorption capacity of the adsorbent.

8.9 What are the four assumptions related to the Langmuir equation?

1) The surface consists of some number of adsorption sites.
2) All of the sites are energetically equal.
3) The adsorbed phase is confined to a monolayer.
4) Lateral interactions between adsorbate molecules do not occur.

8.10 The following adsorption of amitriptyline onto carbon black data is taken from Florence and Attwood (2011), which was extracted from Nambu et al. (1975).

8.10a Using the modified Langmuir-like equation, $C_e/q_e = 1/(q_{max} \cdot K_L) + C_e/q_{max}$ and the data given in the table later, determine K_L and q_{max}.

8.10b What physical meaning does q_{max} and K_L?

K_L, is equal to k_a/k_d or the ratio of the adsorption rate constant and the desorption rate constant. K_L units are typically L solution/g adsorptive solute. q_{max} is the maximum amount of adsorbed adsorptive and total number of adsorption sites.

Adsorption of Amitriptyline onto Carbon Black

Amount adsorbed/ adsorbent, q_e, mol g^{-1}	0.75	0.95	1.10	1.25	1.40	1.55	1.65
Equilibrium Concentration, C_e, mol L^{-1}	0.25	0.40	0.60	0.70	1.10	1.35	1.95
C_e/q_e, g L^{-1}	3.33	4.21	5.45	5.60	7.86	8.71	11.82

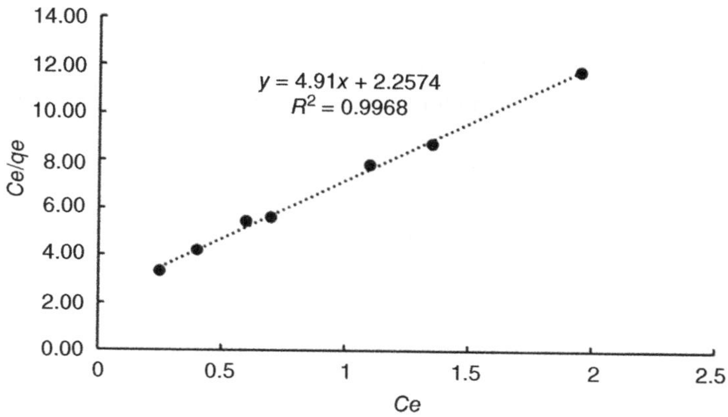

q_{max}
The slope is = $1/q_{max} = 4.91$ g mol^{-1}
Solving for q_{max} gives:
$q_{max} = 1/4.91$ g mol^{-1} = 2.0E-1 mol adsorbate/g adsorbent with 2 significant figures
K_L
The intercept is $1/(q_{max} \cdot K_L)$ is 2.257 g L^{-1}
Substituting for q_{max} and solving for K_L gives:

$$\left(1/2.0\text{E}-1\ \text{mol g}^{-1}\right) \cdot (1/K_L) = 2.257\ \text{g L}^{-1}$$

$$(1/K_L) = 2.257\ \text{g L}^{-1} \cdot 2.0\text{E}-1\ \text{mol g}^{-1} = 4.51\text{E}-1\ \text{mol L}^{-1}$$

Solving for K_L gives:

$$K_L = 1/4.51\text{E}-1\ \text{mol L}^{-1} = 2.2\ \text{L mol}^{-1}\ \text{with 2 significant figures}$$

8.11 True (T) or False (F)

T The Elovich isotherm is a good model for chemisorption.
T At higher adsorbate concentration, the Fritz and Schlunder five-parameter equation gives a similar curve to the Freundlich equation.
T The Hill isotherm is used to assess the influence of different adsorptives upon adsorbent binding sites.

8.12 List seven factors that can affect the rate and extent of adsorption

Adsorbent nature, adsorptive nature, pressure, concentration, solubility, temperature, and pH.

8.13 True or False Capillary condensation occurs below the saturation gas vapor pressure. This unique phenomenon is described the Kelvin-Laplace equation.

T

8.14 Describe the pertinent features of an H1 isotherm.

The H1 adsorption arm exhibits a Type IV(a) isotherm. It demonstrates delayed pore condensation leading to a sharp increase in amount adsorbed at higher relative pressures. During the post adsorption phase, as the relative pressure is decreased, the gas–liquid equilibrium is disturbed. Le Châtelier's principle predicts that the gas will leave the system in response to the stress caused by the decreased pressure. This desorption process is referred to as *equilibrium pore evaporation or pore evaporation*

8.15 Define deliquescence.

A *deliquescent* material has the ability to adsorb water vapor and dissolve in it.

8.16 Calculate the surface area of a drug given the following information.

Given:

- The cross-section area of N_2 is 0.162 nm^2

  $- 1\,nm^2 = 1E - 18\,m^2$

- Avogadro's number is 6.023E23 molecules per mole
- Nitrogen's molecular weight is 28.02 g mol^{-1}

  $- q_m = 140.124\,cm^3\,N_2/1g\text{ drug}$

- Nitogen's molar volume at standard temperature and pressure is 2.24E4 $cm^3\,mol^{-1}$
  Using dimensional analysis:

  $$\left(140.124\,\cancel{cm^3}\,N_2\,g^{-1}\,\text{drug}\right)\cdot\left(\cancel{1\,mol}/2.24E4\,\cancel{cm^3}\right)\cdot\left(6.023E23\,\cancel{molecule/mol}\right)\cdot$$
  $$\left(0.162\,\cancel{nm^2}/\cancel{molecule}\right)\cdot\left(1E\text{-}18\,m^2\cancel{nm^{-2}}\right) = 610.36\,m^2\,g^{-1}\,\text{drug}$$
  $$= \underline{610\,m^2\,g^{-1}\,\text{drug}}\text{ with 3 significant figures.}$$

Index

Note: *Italicized* and **bold** page numbers refer to figures and tables, respectively.

b

c

d

Printed and bound by CPI Group (UK) Ltd, Croydon, CR0 4YY
17/09/2026
14964259-0003